Sandra Tisdell–Clifford

DEVELOPMENTAL MATHEMATICS
BOOK 3

Founding authors
Allan Thompson · Effie Wrightson

Series editor
Robert Yen

NELSON
A Cengage Company

Australia · Brazil · Japan · Korea · Mexico · Singapore · Spain · United Kingdom · United States

Developmental Mathematics Book 3
5th Edition
Sandra Tisdell-Clifford

Publishing editor: Robert Yen
Project editor: Alan Stewart
Editor: Anna Pang
Art direction: Luana Keays
Text design: Sarah Hazell
Cover design: Sarah Hazell
Cover image: iStockphoto/shuoshu; shutterstock/javarman
Permissions researcher: Helen Mammides
Production controller: Erin Dowling
Typeset by: Cenveo Publisher Services

Any URLs contained in this publication were checked for currency
during the production process. Note, however, that the publisher
cannot vouch for the ongoing currency of URLs.

For product information and technology assistance,
in Australia call **1300 790 853**;
in New Zealand call **0800 449 725**

For permission to use material from this text or product, please email
aust.permissions@cengage.com

National Library of Australia Cataloguing-in-Publication Data
Tisdell-Clifford, Sandra, author.
Developmental maths. Book 3 / Sandra Tisdell-Clifford.

5th edition.
9780170351027 (paperback)
For secondary school age.

Mathematics--Australia--Textbooks.

510.76

Cengage Learning Australia
Level 7, 80 Dorcas Street
South Melbourne, Victoria Australia 3205

Cengage Learning New Zealand
Unit 4B Rosedale Office Park
331 Rosedale Road, Albany, North Shore 0632, NZ

For learning solutions, visit **cengage.com.au**

Printed in China by 1010 Printing International Limited.
6 7 8 23

CONTENTS

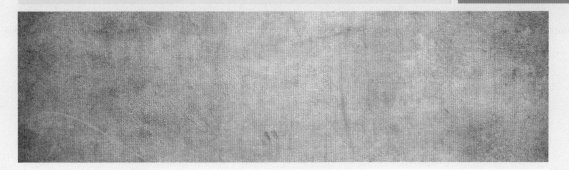

ISBN 9780170351027

= NSW additional content

* = NSW STAGE 5.2, # = NSW additional content

PREFACE

In schools for over four decades, *Developmental Mathematics* has been a unique, well-known and trusted Years 7–10 mathematics series aimed at developing key numeracy and literacy skills. This 5th edition of the series has been revised for the new Australian curriculum as well as the NSW syllabus Stages 4 and 5.1. The four books of the series contain short chapters with worked examples, definitions of key words, graded exercises, a language activity and a practice test. Each chapter covers a topic that should require about two weeks of teaching time.

Developmental Mathematics supports students with mathematics learning, encouraging them to experience more confidence and success in the subject. This series presents examples and exercises in clear and concise language to help students master the basics and improve their understanding. We have endeavoured to equip students with the essential knowledge required for success in junior high school mathematics, with a focus on basic skills and numeracy.

Developmental Mathematics Book 3 is written for students in Years 9–10, covering the Australian curriculum (mostly Year 9 content) and NSW syllabus (see the curriculum grids on the following pages and the teaching program on the NelsonNet teacher website). This book presents concise and highly-structured examples and exercises, with each new concept or skill on a double-page spread for convenient reading and referencing.

Students learning mathematics need to be taught by dynamic teachers who use a variety of resources. Our intention is that teachers and students use this book as their primary source or handbook, and supplement it with additional worksheets and resources, including those found on the *NelsonNet* teacher website (access conditions apply). We hope that teachers can use this book effectively to help students achieve success in secondary mathematics. Good luck!

ABOUT THE AUTHOR

Sandra Tisdell-Clifford teaches at Newcastle Grammar School and was the Mathematics coordinator at Our Lady of Mercy College (OLMC) in Parramatta for 10 years. Sandra is best known for updating *Developmental Mathematics* for the 21st century (4th edition, 2003) and writing its blackline masters books. She also co-wrote *Nelson Senior Maths 11 General for the Australian curriculum*, teaching resources for the NSW senior series *Maths in Focus* and the Years 7–8 homework sheets for *New Century Maths/NelsonNet*.

Sandra expresses her thanks and appreciation to the Headmaster and staff of Newcastle Grammar School and dedicates this book to her husband, Ray Clifford, for his support and encouragement. She also thanks series editor **Robert Yen** and editor Anna Pang at Cengage Learning for their leadership on this project.

Original authors **Allan Thompson** and **Effie Wrightson** wrote the first three editions of *Developmental Mathematics* (published 1974, 1981 and 1988) and taught at Smith's Hill High School in Wollongong. Sandra thanks them for their innovative pioneering work, which has paved the way for this new edition for the Australian curriculum.

ISBN 9780170351027

FEATURES OF THIS BOOK

- Each chapter begins with a table of contents and list of chapter outcomes
- Each teaching section of a chapter is presented clearly on a double-page spread

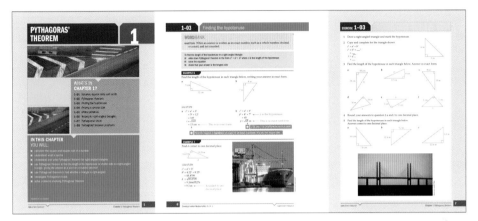

- The left page contains explanations, worked examples, and if appropriate, a Wordbank of mathematical terminology and a fact box
- The right page contains an exercise set, including multiple-choice questions, scaffolded solutions and realistic applications of mathematics
- Each chapter concludes with a **Language activity** (puzzle) that reinforces mathematical terminology in a fun way, and a Practice **test** containing non-calculator questions on general topics and specific topic questions grouped by chapter subheading

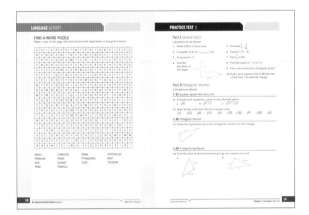

- Additional teaching resources can be downloaded from the NelsonNet teacher website at **www.nelsonnet.com.au**: worksheets, puzzle sheets, skillsheets, video tutorials, technology worksheets, teaching program, curriculum grids, chapter PDFs of this book
- Note: Complimentary access to NelsonNet is only available to teachers who use Developmental Mathematics as a core educational resource in their classroom. Contact your sales representative for information about access codes and conditions.

CURRICULUM GRID
AUSTRALIAN CURRICULUM

STRAND AND SUBSTRAND	DEVELOPMENTAL MATHEMATICS BOOK 1 CHAPTER		DEVELOPMENTAL MATHEMATICS BOOK 2 CHAPTER	
NUMBER AND ALGEBRA				
Number and place value	1	Integers and the number plane	1	Working with numbers
	3	Working with numbers	2	Primes and powers
	4	Factors and primes	4	Integers
	5	Powers and decimals		
	6	Multiplying and dividing decimals		
	9	Algebra and equations		
Real numbers	5	Powers and decimals	5	Decimals
	6	Multiplying and dividing decimals	11	Fractions
	7	Fractions	12	Percentages
	8	Multiplying and dividing fractions	16	Ratios and rates
	17	Percentages and ratios		
Money and financial mathematics	6	Multiplying and dividing decimals	12	Percentages
Patterns and algebra	9	Algebra and equations	6	Algebra
			15	Further algebra
Linear and non-linear relationships	1	Integers and the number plane	15	Further algebra
	9	Algebra and equations	17	Graphing lines
MEASUREMENT AND GEOMETRY				
Using units of measurement	12	Length and time	9	Length and time
	13	Area and volume	10	Area and volume
Shape	10	Shapes and symmetry		
Location and transformation	10	Shapes and symmetry	7	Angles and symmetry
Geometric reasoning	2	Angles	7	Angles and symmetry
	11	Geometry	8	Triangles and quadrilaterals
Pythagoras and trigonometry			3	Pythagoras' theorem
STATISTICS AND PROBABILITY				
Chance	16	Probability	14	Probability
Data representation and interpretation	14	Statistical graphs	13	Investigating data
	15	Analysing data		

Developmental Mathematics Book 3

ISBN 9780170351027

CURRICULUM GRID
AUSTRALIAN CURRICULUM

STRAND AND SUBSTRAND	DEVELOPMENTAL MATHEMATICS BOOK 3 CHAPTER		DEVELOPMENTAL MATHEMATICS BOOK 4 CHAPTER	
NUMBER AND ALGEBRA				
Real numbers	2	Whole numbers and decimals	1	Working with numbers
	3	Integers and fractions	2	Percentages
	6	Percentages	7	Ratios and rates
	7	Indices		
	16	Ratios and rates		
Money and financial mathematics	6	Percentages	3	Earning and saving money
Patterns and algebra	4	Algebra	4	Algebra
	7	Indices	10	Indices
Linear and non-linear relationships	9	Equations	13	Equations and inequalities
	14	Graphing lines	15	Coordinate geometry
			16	Graphing lines and curves
MEASUREMENT AND GEOMETRY				
Using units of measurement	12	Length and time	9	Length and time
	13	Area and volume	11	Area and volume
Geometric reasoning	8	Geometry	8	Congruent and similar figures
Pythagoras and trigonometry	1	Pythagoras' theorem	5	Pythagoras' theorem
	5	Trigonometry	6	Trigonometry
STATISTICS AND PROBABILITY				
Chance	15	Probability	14	Probability
Data representation and interpretation	11	Investigating data	12	Investigating data 

ISBN 9780170351027

SERIES OVERVIEW

PYTHAGORAS' THEOREM

1

IN THIS CHAPTER YOU WILL:

- calculate the square and square root of a number
- understand what a surd is
- understand and write Pythagoras' theorem for right-angled triangles
- use Pythagoras' theorem to find the length of the hypotenuse or shorter side in a right-angled triangle, giving the answer as a surd or a rounded decimal
- use Pythagoras' theorem to test whether a triangle is right-angled
- investigate Pythagorean triads
- solve problems involving Pythagoras' theorem

Shutterstock.com/isaravut

WORDBANK

squared A number multiplied by itself, for example, 5^2 is read '5 squared' and means $5 \times 5 = 25$.

square root The positive value which, if squared, will give that number; for example, $\sqrt{25}$ is read 'the square root of 25' and $\sqrt{25} = 5$ because $5^2 = 25$.

surd A square root whose answer is not an exact number. For example, $\sqrt{8} = 2.8284...$ is a surd because there isn't an exact number squared that is equal to 8. As a decimal, the digits of $\sqrt{8}$ run endlessly without any repeating pattern.

EXAMPLE 1

Evaluate each expression.

a 6^2 **b** 3.4^2 **c** $(-13)^2$

SOLUTION

a $6^2 = 36$ ◄─────── Enter 6 x^2 = or 6 × 6 = on the calculator

b $3.4^2 = 11.56$

c $(-13)^2 = 169$ ◄─────── Enter ((−) 13) x^2 = on the calculator

EXAMPLE 2

Evaluate each expression, correct to 2 decimal places if necessary.

a $\sqrt{49}$ **b** $\sqrt{5}$ **c** $\sqrt{2.56}$ **d** $\sqrt{82}$

SOLUTION

a $\sqrt{49} = 7$ ◄─────── Enter √ 49 = on the calculator

b $\sqrt{5} = 2.2360...$ ◄─────── Enter √ 5 = on the calculator
≈ 2.24

c $\sqrt{2.56} = 1.6$

d $\sqrt{82} = 9.0553...$
≈ 9.06

EXAMPLE 3

Which numbers in Example **2** are surds?

SOLUTION

The surds are $\sqrt{5}$ and $\sqrt{82}$ because they do not simplify to exact decimals.

ISBN 9780170351027

1 What is the meaning of 8^2? Select the correct answer **A**, **B**, **C** or **D**.

   A 8×2 **B** $8 + 8$ **C** 8×8 **D** 2×8

2 Evaluate each expression.

   a 3^2 **b** 7^2 **c** 11^2 **d** 14^2

   e 8.6^2 **f** 9^2 **g** 12.1^2 **h** 1.9^2

   i $(-4)^2$ **j** 16.2^2 **k** $(-8)^2$ **l** 9.6^2

3 Evaluate each expression.

   a $\sqrt{16}$ **b** $\sqrt{36}$ **c** $\sqrt{64}$ **d** $\sqrt{121}$

   e $\sqrt{196}$

4 What is the most accurate answer for the value of $\sqrt{58}$? Select **A**, **B**, **C** or **D**.

   A 7 **B** 7.616 **C** 8 **D** 7.6

5 Evaluate each expression correct to 2 decimal places.

   a $\sqrt{18}$ **b** $\sqrt{32}$ **c** $\sqrt{82}$ **d** $\sqrt{150}$

   e $\sqrt{220}$

6 Which one of the numbers below is a surd? Select **A**, **B**, **C** or **D**.

   A $\sqrt{9}$ **B** 7^2 **C** $\dfrac{2}{3}$ **D** $\sqrt{13}$

7 Is each statement true or false?

   a $10^2 = 100$ **b** $\sqrt{64} = 4$ **c** $1.1^2 = 11$ **d** $\sqrt{169} = 13$

   e $(-9)^2 = -81$

8 Find the area of this square.

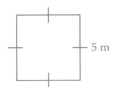

      5 m

9 If the area of a square is 49 m², what is the length of a side of the square?

10 Sketch a square with an area of 64 cm², showing the side length of the square.

WORDBANK

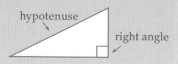

right-angled triangle A triangle with one angle exactly 90°. This angle is called the right angle.

hypotenuse The longest side of a right-angled triangle, the side opposite the right angle.

Pythagoras' theorem The rule or formula $c^2 = a^2 + b^2$ that relates the lengths of the sides of a right-angled triangle. Pythagoras was the ancient Greek mathematician who discovered this rule ('theorem' means rule).

PYTHAGORAS' THEOREM

In any right-angled triangle, the square of the hypotenuse is equal to the sum of the squares of the other two sides.
In the diagram, c is the length of the hypotenuse (longest side), and Pythagoras' theorem is $c^2 = a^2 + b^2$

EXAMPLE 4

State Pythagoras' theorem for each right-angled triangle below.

a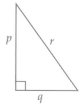

b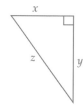

SOLUTION

a $r^2 = p^2 + q^2$ **b** $z^2 = x^2 + y^2$

 * Remember, Pythagoras' theorem begins with (hypotenuse)^2, and r is the hypotenuse here.

EXAMPLE 5

Test Pythagoras' theorem on each triangle below.

a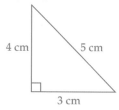

b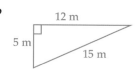

SOLUTION

a For Pythagoras' theorem to be true, hypotenuse^2 = sum of the squares of the other two sides.
Does $5^2 = 3^2 + 4^2$?
$25 = 9 + 16$ True, so Pythagoras' theorem is true.
This means that the triangle is right-angled.

b Does $15^2 = 5^2 + 12^2$?
$225 = 25 + 144$ False, so Pythagoras' theorem is not true.
This means that the triangle is not right-angled.

Developmental Mathematics Book 3

ISBN 9780170351027

EXERCISE 1–02

1 Which side of this right-angled triangle is the hypotenuse?
   Select the correct answer **A**, **B** or **C**.

   A a

   B b

   C c

2 What is Pythagoras' theorem for the triangle in question **1**?
   Select **A**, **B** or **C**.

   A $b^2 = a^2 + c^2$

   B $c^2 = a^2 + b^2$

   C $a^2 = b^2 + c^2$

3 Write Pythagoras' theorem for each right-angled triangle.

   a 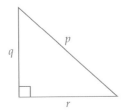   **b**    **c**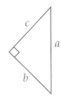

   d 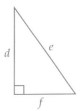   **e** 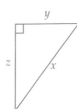   **f**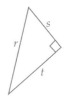

4 Test Pythagoras' theorem on each triangle below.

   a 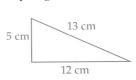   **b** 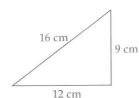   **c**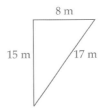

5 Which triangles in question **4** are right-angled?

ISBN 9780170351027

WORDBANK

exact form When an answer is written as an exact number, such as a whole number, decimal or a surd, and not rounded.

To find the length of the hypotenuse in a right-angled triangle:
- ◼ write down Pythagoras' theorem in the form $c^2 = a^2 + b^2$ where c is the length of the hypotenuse
- ◼ solve the equation
- ◼ check that your answer is the longest side

EXAMPLE 6

Find the length of the hypotenuse in each triangle below, writing your answer in exact form.

a
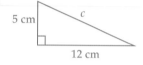

b
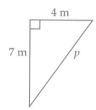

SOLUTION

a
$$c^2 = a^2 + b^2$$
$$= 5^2 + 12^2$$
$$= 169$$
$$c = \sqrt{169}$$
$$= 13 \text{ cm} \longleftarrow \text{This is in exact form.}$$

✳ From the diagram, a hypotenuse of length 13 cm looks reasonable. It is also the longest side.

b
$$c^2 = a^2 + b^2$$
$$p^2 = 4^2 + 7^2 \longleftarrow p \text{ is the hypotenuse.}$$
$$= 65$$
$$p = \sqrt{65} \text{ m} \longleftarrow \text{This is in exact surd form.}$$

✳ $\sqrt{65}$ does not simplify to an exact decimal

EXAMPLE 7

Find d correct to one decimal place.

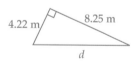

SOLUTION

$$c^2 = a^2 + b^2$$
$$d^2 = 4.22^2 + 8.25^2$$
$$= 85.8709$$
$$d = \sqrt{85.8709}$$
$$= 9.266655276$$
$$\approx 9.3 \text{ m} \longleftarrow \text{Rounded to one decimal place.}$$

Shutterstock.com/hichie81

1 Draw a right-angled triangle and mark the hypotenuse.

2 Copy and complete for the triangle shown

$c^2 = a^2 + b^2$

$c^2 = 6^2 + \underline{}^2$

$= \underline{}$

$c = \sqrt{\underline{}}$

$= \underline{}$

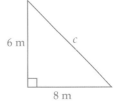

3 Find the length of the hypotenuse in each triangle below. Answer in exact form.

a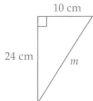

b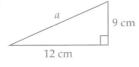

c

d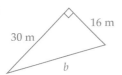

e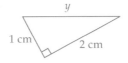

f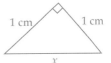

4 Round your answers to question **3 e** and **f** to one decimal place.

5 Find the length of the hypotenuse in each triangle below.
Answer correct to one decimal place.

a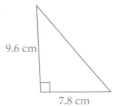

b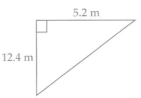

c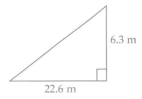

Shutterstock.com/JK Photo

To find the length of a shorter side in a right-angled triangle:
- ▨ write down Pythagoras' theorem in the form $c^2 = a^2 + b^2$ where c is the length of the hypotenuse
- ▨ rearrange the equation so that the shorter side is on the LHS (left-hand side)
- ▨ solve the equation
- ▨ check that your answer is shorter than the hypotenuse

EXAMPLE 8

Find the length of the unknown side in each triangle below. Answer in exact form.

a

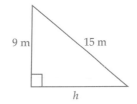

b

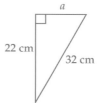

SOLUTION

a $15^2 = h^2 + 9^2$ ◄── 15 is the hypotenuse.
$h^2 + 9^2 = 15^2$ ◄── Rearranging the equation
$h^2 = 15^2 - 9^2$ so that h is on the LHS.
$= 144$
$h = \sqrt{144}$
$= 12$ m

b $32^2 = a^2 + 22^2$
$a^2 + 22^2 = 32^2$
$a^2 = 32^2 - 22^2$
$= 540$
$a = \sqrt{540}$ cm

✱ | From the diagram, a length of 12 m looks reasonable. It is also shorter than the hypotenuse, 15 m.

EXAMPLE 9

Find the length of the unknown side in each triangle below. Answer correct to one decimal place.

a

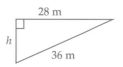

b

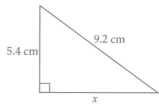

SOLUTION

a $36^2 = h^2 + 28^2$
$h^2 + 28^2 = 36^2$
$h^2 = 36^2 - 28^2$
$= 512$
$h = \sqrt{512}$
$= 22.627\ 417...$
≈ 22.6 m ◄── Rounded to 1 decimal place.

b $9.2^2 = x^2 + 5.4^2$
$x^2 + 5.4^2 = 9.2^2$
$x^2 = 9.2^2 - 5.4^2$
$= 55.48$
$x = \sqrt{55.48}$
$= 7.448\ 489...$
≈ 7.4 cm

1 To find the length of the shorter side, a, in the triangle below, which rule is easier to use? Select the correct answer **A**, **B** or **C**.

A $25^2 = a^2 + 15^2$ **B** $25^2 = a^2 - 15^2$ **C** $a^2 = 15^2 - 25^2$

2 What is the length of the hypotenuse in the triangle above?

    A 15   **B** 25   **C** None of these

3 Copy and complete to find b.

$$20^2 = b^2 + 12^2$$
$$b^2 + \underline{\quad}^2 = 20^2$$
$$b^2 = 20^2 - \underline{\quad}^2$$
$$= 400 - \underline{\quad}$$
$$b = \sqrt{\underline{\quad}}$$
$$= \underline{\quad}$$

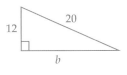

4 Find the value of each pronumeral. Leave your answers in exact form.

a

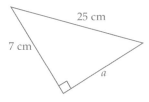

b

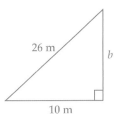

c

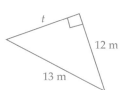

d

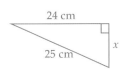

5 Find, correct to one decimal place, the value of each pronumeral.

a

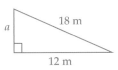

b

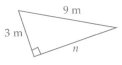

c

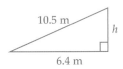

d

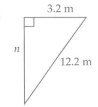

For this right-angled triangle:
- to find the length of the hypotenuse, use $c^2 = a^2 + b^2$ where c is the hypotenuse
- to find the length of one of the shorter sides, use the shortcut $b^2 = c^2 - a^2$ to find side b or $a^2 = c^2 - b^2$ to find side a.

EXAMPLE 10

Find the length of the unknown side in each triangle below. Leave your answer in exact form.

a
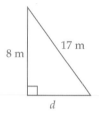
8 m   17 m d

b
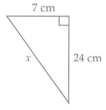
7 cm x 24 cm

c
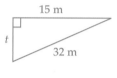
15 m t 32 m

SOLUTION

a d is a shorter side
Square and subtract

$d^2 = 17^2 - 8^2$

❋ the largest side (hypotenuse) goes first

$= 289 - 64$
$= 225$
$d = \sqrt{225}$
$= 15$ m

b x is the hypotenuse
Square and add

$x^2 = 7^2 + 24^2$
$= 625$
$x = \sqrt{625}$
$= 25$ cm

c t is a shorter side
Square and subtract

$t^2 = 32^2 - 15^2$
$= 799$
$t = \sqrt{799}$ m

ISBN 9780170351027

1 To find the length of the hypotenuse, h, in this triangle, which rule is easier to use? Select the correct answer **A**, **B**, **C** or **D**.

   A $h^2 = i^2 + j^2$

   B $h^2 = i^2 - j^2$

   C $j^2 = h^2 - i^2$

   D $i^2 = h^2 + j^2$

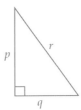

2 To find the length of the shorter side, p, in this triangle, which rule is correct?

   A $p^2 = q^2 + r^2$

   B $p^2 = r^2 - q^2$

   C $q^2 = r^2 - p^2$

   D $r^2 = p^2 + q^2$

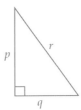

3 Find the length of the unknown side in each triangle. Leave your answers in exact form.

   a
   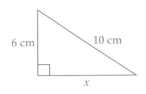

   b
   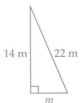

   c
   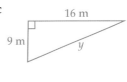

   d
   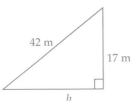

   e
   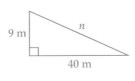

   f
   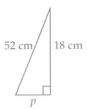

4 Find, correct to one decimal place, the value of each pronumeral.

   a

   b
   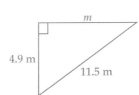

   c

5 A ladder is placed against a building to reach a window on the second floor.

   What is the name for the side of the triangle where the ladder is positioned?

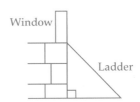

If the sides of a triangle follow the rule $c^2 = a^2 + b^2$, then the triangle must be right-angled. This is called the **converse** of Pythagoras' theorem, the theorem used in reverse.

To prove that a triangle is right-angled:
- ■ substitute the lengths of its sides into the rule $c^2 = a^2 + b^2$
- ■ if it is true, then the triangle is right-angled
- ■ if it is false, then the triangle is not right-angled

EXAMPLE 11

Test whether each triangle is right-angled.

a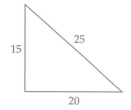

b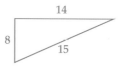

SOLUTION

Substitute the lengths of the sides into the rule $c^2 = a^2 + b^2$.

a Does $25^2 = 15^2 + 20^2$?
  $625 = 225 + 400$
  $625 = 625$ ◄——— Yes

  This triangle is right-angled.

  ✴ | The right angle is opposite the hypotenuse, 25, between the sides marked 15 and 20.

b Does $15^2 = 8^2 + 14^2$?
  $225 = 64 + 196$
  $225 \neq 260$ ◄——— No

  This triangle is not right-angled.

Alamy/Mike Pawley

1 Explain in words how you can prove that a triangle is right-angled.

2 In the triangle below, which angle is the right angle? Select the correct answer **A**, **B** or **C**.

   A $\angle A$ **B** $\angle B$ **C** $\angle C$

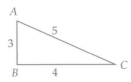

3 What is Pythagoras' theorem for the triangle above?

   A $4^2 = 3^2 + 5^2$ **B** $3^2 = 4^2 + 5^2$ **C** $5^2 = 3^2 + 4^2$

4 Copy and complete to test if this triangle is right-angled.

Does $17^2 = 15^2 + \underline{\quad}^2$?

      $289 = \underline{\quad} + 64$

      $289 = \underline{\qquad}$

So the triangle $\underline{\qquad}$ right-angled.

5 Test whether each triangle is right-angled.

a
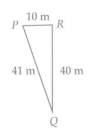

b
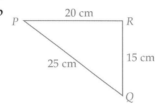

c
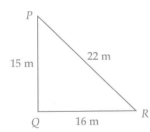

d
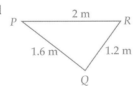

e
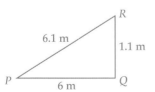

f
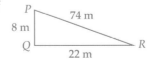

6 For each right-angled triangle in question **5**, state which angle is the right angle.

WORDBANK

Pythagorean triad A set of 3 numbers that follows Pythagoras' theorem, $c^2 = a^2 + b^2$.

To prove that a group of 3 numbers is a Pythagorean triad:
- ▨ substitute the numbers into the rule $c^2 = a^2 + b^2$ where c is the largest number
- ▨ if it is true, then the numbers form a Pythagorean triad
- ▨ if it is false, then the numbers do not form a Pythagorean triad

EXAMPLE 12

Test whether each set of numbers form a Pythagorean triad.

a {15, 20, 25}   **b** {8, 12, 13}

SOLUTION

a Does $25^2 = 15^2 + 20^2$?   **b** Does $13^2 = 8^2 + 12^2$?

> ✱ Always substitute the largest number for c.

$625 = 225 + 400$ $169 = 64 + 144$
$625 = 625$ ◀—— true $169 \neq 208$ ◀—— false

So {15, 20, 25} is a Pythagorean triad   So {8, 12, 13} is not a Pythagorean triad.

If the three numbers in a Pythagorean triad are multiplied by the same value, then the new numbers also form a Pythagorean triad.

EXAMPLE 13

By multiplying each number of the Pythagorean triad {7, 24, 25} by a number of your choice, find two more triads.

SOLUTION

{7 × 2, 24 × 2, 25 × 2} = {14, 48, 50} ◀—— Multiplying {7, 24, 25} by 2.

> ✱ Check that $50^2 = 14^2 + 48^2$

{7 × 5, 24 × 5, 25 × 5} = {35, 120, 125} ◀—— Multiplying {7, 24, 25} by 5.

Shutterstock.com/gui jun peng

1   Copy and complete:

A Pythagorean triad is a set of _____ numbers which follow _____ theorem.

2   To prove that a set of numbers form a Pythagorean triad, which variable in the formula $c^2 = a^2 + b^2$ should be substituted by the largest number? Select the correct answer **A**, **B** or **C**.

A a　　　　　**B** b　　　　　**C** c

3   Which set of numbers is a Pythagorean triad? Select **A**, **B**, **C** or **D**.

A {3, 4, 4}　　　**B** {3, 4, 5}　　　**C** {3, 4, 6}　　　**D** {3, 4, 7}

4   Copy and complete:

$13^2 = 5^2 + 12^2$

$169 = \text{\_\_\_} + 144$

$169 = \text{\_\_\_}$

So {5, 12, 13} _____ a Pythagorean triad.

5   Test whether each set of numbers form a Pythagorean triad.

a　{9, 40, 41}　　**b**　{7, 20, 25}　　**c**　{12, 16, 20}

d　{11, 50, 52}　　**e**　{5, 6, 7}　　**f**　{7, 24, 25}

g　{ 8, 15, 17}　　**h**　{4, 8, 12}

6   If {5, 12, 13} is a Pythagorean triad, find two more Pythagorean triads using multiples of {5, 12, 13}.

7   **a**　Prove that {16, 30, 34} is a Pythagorean triad.

b　Which Pythagorean triad is this a multiple of?

8   Write down 2 more Pythagorean triads using {7, 24, 25}.

9   Is each statement true or false?

a　{30, 40, 50} is a Pythagorean triad.

b　{9, 12, 15} is a multiple of {3, 4, 5}.

c　There are only 3 Pythagorean triads.

10 Draw a right-angled triangle with sides 6 cm, 8 cm and 10 cm.

Is {6, 8, 10} a Pythagorean triad?

Alamy/Dinodia Photos

ISBN 9780170351027

1-08 Pythagoras' theorem problems

To solve a problem using Pythagoras' theorem:
- draw a diagram if it is not given and draw a right-angled triangle
- identify the unknown value
- use $c^2 = a^2 + b^2$ to solve an equation
- answer the problem in words.

EXAMPLE 14

Toby is flying a kite at a height of 32 m above the ground. He is standing on the ground 24 m away from the ground level of the kite. How long is the piece of string that he is using to fly the kite?

SOLUTION

$c^2 = a^2 + b^2$ ⟵——————— c is the length of the string and is the hypotenuse of the triangle.
$c^2 = 24^2 + 32^2$ ⟵——————— Substitute in the numbers given in the question.
$ = 1600$
$c = \sqrt{1600}$
$ = 40$

The length of the string is 40 m.

EXAMPLE 15

A window in a building is 6 m above the ground. A ladder is placed 9 m from the base of the building so that it reaches the window. How long is the ladder (correct to 1 decimal place)?

SOLUTION

Let the length of the ladder be x m.

$x^2 = 6^2 + 9^2$
$ = 117$
$x = \sqrt{117}$
$ = 10.8166 \ldots$
$ \approx 10.8$ m

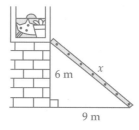

The ladder is 10.8 m long.

EXERCISE 1-08

1   Gemma leans a ladder against a 6 m high wall so that it reaches the top of it. She places the ladder 2.5 m from the base of the wall. Which is the correct diagram for the problem below? Select **A, B** or **C**.

A

6 m
2.5 m

B

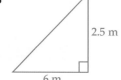

2.5 m
6 m

C

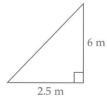

6 m
2.5 m

ISBN 9780170351027

2 Find the length of the ladder for the problem described in question **1**.

3 Find the length of the diagonal in this rectangle.

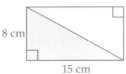

8 cm

15 cm

4 In this rectangle, the diagonal is 13 cm long and one side measures 12 cm. What is the length of the other side?

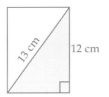

13 cm

12 cm

5 What length of wood is needed to make the cross-bar of this garden gate?

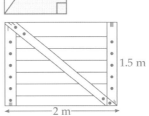

1.5 m

2 m

6 A rectangular playground is 95 m from one corner, across the playground, to the opposite corner. Its longest side measures 73 m.

   a Draw a sketch to represent the playground.

   b Find the length of the shorter side (correct to the nearest metre).

7 A 3-metre ladder leans against a wall. The foot of the ladder is 1.5 metres from the wall. How high up the wall does the ladder reach? Answer correct to one decimal place.

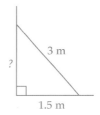

3 m

?

1.5 m

8 An empty block of land measures 12 m by 20 m. What is the shortest distance from *A* to *C*, to the nearest metre?

9 For the block of land in question **8**, how much farther is it to walk from *A* to *B* then *B* to *C*, rather than from *A* to *C*?

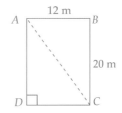

12 m

A   *B*

20 m

D   *C*

10 Tim was stranded on an island 3.5 km east of a lighthouse. He knew there was a town on the mainland 12.8 km south of the lighthouse. How far (correct to one decimal place) would he have to swim to reach the town if he swam in a straight line from where he was?

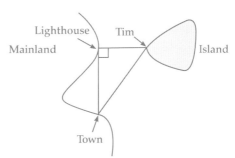

Lighthouse   Tim

Mainland   Island

Town

FIND-A-WORD PUZZLE

Make a copy of this page, then find all the words listed below in this grid of letters.

A	O	J	S	R	O	O	T	W	Y	P	R	T	G	B	K	J	E	Z	G
W	A	B	B	A	W	J	M	X	E	B	X	M	D	A	I	H	S	S	T
F	P	H	V	Q	R	Q	M	T	X	D	J	A	F	Y	U	N	U	V	A
C	B	D	R	U	E	O	C	O	N	V	E	R	S	E	H	X	N	D	G
A	F	Z	U	N	M	X	G	B	W	G	Q	Y	H	D	S	V	E	G	L
K	F	F	I	Q	C	B	U	A	S	V	L	G	R	E	U	F	T	I	V
A	E	L	R	D	Y	Y	D	X	H	M	A	S	Y	Q	G	L	O	U	I
W	T	P	V	Q	Q	P	L	A	K	T	Y	A	D	E	Q	O	P	O	S
N	G	W	B	B	E	K	E	N	I	D	Y	R	Q	V	N	V	Y	R	T
W	C	M	U	S	Z	L	Q	J	E	R	G	P	H	O	I	X	H	P	Y
K	Q	Q	L	O	G	S	B	A	D	A	T	Z	J	R	Z	U	T	L	F
C	C	Q	E	N	Y	Q	A	E	S	W	F	Y	T	P	U	B	Y	P	S
O	R	S	A	U	V	Q	M	T	G	O	K	D	X	H	E	K	S	R	Z
F	G	I	D	Y	U	O	M	S	M	R	L	J	N	D	E	W	K	O	J
B	R	V	D	K	F	Q	Q	V	F	Q	D	G	I	F	V	O	I	B	U
T	Q	Z	C	Z	S	U	E	L	G	N	A	S	A	G	W	L	R	L	Y
O	I	H	B	J	A	Q	I	C	E	P	L	Z	U	T	J	D	G	E	C
A	E	L	D	R	G	P	Y	H	C	K	Z	L	E	R	Q	W	E	M	M
D	Q	O	E	F	C	X	J	B	U	L	C	A	N	K	D	D	N	P	V
B	G	A	W	L	E	I	W	U	I	U	G	U	A	R	C	A	I	B	I

ANGLE	CONVERSE	DRAW	HYPOTENUSE
PROBLEM	PROVE	PYTHAGORAS	ROOT
SIDE	SQUARE	SURD	THEOREM
TRIAD	TRIANGLE		

Part A General topics

Calculators are not allowed.

1 Write 2108 in 12-hour time.

2 Complete: 60.8 cm = _____ mm

3 Evaluate 44 × 5.

4 Find the perimeter of this shape.

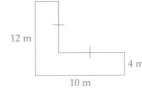

5 Evaluate $\dfrac{2}{7} - \dfrac{3}{14}$.

6 Expand $-2(x - 4)$.

7 Find $\dfrac{2}{3}$ of $36.

8 Find the mean of 1, 2, 8, 3, 6.

9 How many faces has a triangular prism?

10 Kathy pays a grocery bill of $83.45 with a $100 note. Calculate the change.

Part B Pythagoras' theorem

Calculators are allowed.

1–01 Squares, square roots and surds

11 Evaluate each expression, correct to two decimal places.

   a $\sqrt{34}$ b $\sqrt{6^2 + 3^2}$ c $\sqrt{7.6^2 - 2.4^2}$

12 Select all the surds from this list of square roots.

   $\sqrt{16}$ $\sqrt{324}$ $\sqrt{82}$ $\sqrt{169}$ $\sqrt{28}$ $\sqrt{225}$ $\sqrt{144}$ $\sqrt{75}$ $\sqrt{289}$ $\sqrt{68}$

1–02 Pythagoras' theorem

13 Name the hypotenuse and write Pythagoras' theorem for this triangle.

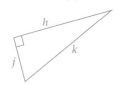

1–03 Finding the hypotenuse

14 Find the value of each pronumeral, giving your answer as a surd.

   a

   b

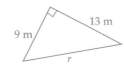

1-04 Finding a shorter side

15 Find the value of each pronumeral, giving your answer correct to two decimal places.

a
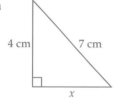

b
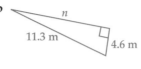

1-05 Mixed problems

16 Find the value of each pronumeral, correct to one decimal place.

a

b

c
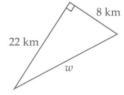

1-06 Testing for right-angled triangles

17 Test whether each triangle is right-angled.

a
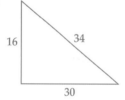

b
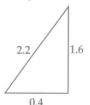

1-07 Pythagorean triads

18 Test whether each set of numbers is a Pythagorean triad.

   a {1.8, 2.4, 3.0}   **b** {7, 24, 26}

1-08 Pythagoras' theorem problems

19 Find, correct to one decimal place, the length of the longest umbrella that can fit inside a suitcase measuring 60 cm long and 46 cm wide.

20 Find the perimeter of this trapezium.

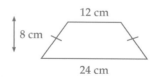

WHOLE NUMBERS AND DECIMALS

2

WHAT'S IN CHAPTER 2?

IN THIS CHAPTER YOU WILL:

- add, subtract, multiply and divide mentally with whole numbers
- round decimals and money amounts
- add, subtract, multiply and divide decimals
- understand and use terminating and recurring decimals
- convert fractions to recurring decimals

Shutterstock.com/Tatiana53

WORDBANK

sum The answer to an addition (+) of two or more numbers.

mental Using the mind, not a calculator.

evaluate To find the value or amount.

EXAMPLE 1

Find each sum using mental calculation.

a $12 + 36$ **b** $12 + 36 + 45$

SOLUTION

a $12 + 36 = 10 + 2 + 30 + 6$ or $12+$
$= 10 + 30 + 2 + 6$ $\underline{36}$
$\underline{48}$

 adding the tens and units separately

$= 40 + 8$
$= 48$

b $12 + 36 + 45 = 48 + 45$ ⟵—— using the answer 48 from part **a**
$= 40 + 40 + 8 + 5$ or $4^{1}8+$
$= 80 + 13$ $\underline{4\,5}$
$= 93$ $\underline{9\,3}$

When adding numbers mentally:
- remember that numbers can be added in any order
- if one of the numbers is close to 10, 20, 30, … , split it up
- look for unit digits that add up to 10, such as 6 and 4

EXAMPLE 2

Evaluate each sum mentally by splitting up the second number.

a $24 + 12$ **b** $56 + 99$ **c** $147 + 298$

SOLUTION

a $24 + 12 = 24 + 10 + 2$ **b** $56 + 99 = 56 + 100 - 1$ **c** $147 + 298 = 147 + 300 - 2$
$= 34 + 2$ $= 156 - 1$ $= 447 - 2$
$= 36$ $= 155$ $= 445$

EXAMPLE 3

Evaluate each sum mentally by pairing up numbers that have units digits adding to 10.

a $22 + 46 + 18 + 4$ **b** $221 + 386 + 479$

SOLUTION

a $22 + 46 + 18 + 4$ **b** $221 + 386 + 479$
$= (22 + 18) + (46 + 4)$ $= (221 + 479) + 386$
$= 40 + 50$ $= 700 + 386$
$= 90$ $= 1086$

ISBN 9780170351027

1   Which expression gives the same answer as 52 + 45? Select the correct answer **A**, **B**, **C** or **D**.

   A 52 + 54   **B** 50 + 2 + 45   **C** 45 + 25   **D** 52 + 40 + 3

2   Evaluate each sum using mental calculation.

   a 12 + 7   **b** 24 + 15   **c** 32 + 5   **d** 23 + 125

   e 15 + 24   **f** 16 + 28   **g** 46 + 13   **h** 58 + 19

   i 72 + 36   **j** 13 + 46   **k** 35 + 87   **l** 36 + 72

3   Write down which sums in question **2** are the same.

4   Evaluate each sum mentally.

   a 28 + 54   **b** 54 + 28   **c** 28 + 54 + 66   **d** 54 + 28 + 66 + 41

5   **a** How could you use your answer to **4a** to find **4b**?

   b How could you use your answer to **4c** to find **4d**?

6   Is each statement true or false?

   a 28 + 27 = 27 + 28   **b** 14 + 23 = 23 − 14

   c 112 + 57 = 57 + 122   **d** 134 + 78 = 78 + 134

7   Copy and complete each expression.

   a 502 = 500 + ___   **b** 903 = 900 + ___

   c 498 = 500 − ___   **d** 299 = ___ − 1

8   Which expression gives the same answer as 275 + 699? Select **A**, **B**, **C** or **D**.

   A 257 + 699   **B** 275 + 969

   C 275 + 700 − 1   **D** 275 + 680 + 9

9   Evaluate each sum mentally.

   a 354 + 502   **b** 243 + 903   **c** 627 + 498   **d** 1246 + 299

   e 257 + 401   **f** 783 + 199   **g** 1256 + 603   **h** 3243 + 398

10 Evaluate each sum mentally by pairing up numbers that have units digits adding to 10.

   a 12 + 83 + 8 + 17   **b** 34 + 45 + 5 + 6

   c 28 + 52 + 32 + 8   **d** 16 + 35 + 24 + 5

   e 243 + 56 + 17   **f** 421 + 512 + 9

   g 381 + 274 + 19   **h** 508 + 63 + 112

11 A magic square has all rows, columns and diagonals adding to the same number. Copy and complete each magic square.

a

3		6
	5	
		7

b

4		7
		3
		8

WORDBANK

difference The result of subtracting two numbers.

EXAMPLE 4

Find each difference using mental calculation.

a 85 − 24 **b** 248 − 63

SOLUTION

a 85 − 24 = 80 + 5 − 20 − 4 or 8 5 −
2 4
✱ subtracting the tens and units separately 6 1

= 80 − 20 + 5 − 4
= 60 + 1
= 61

b 248 − 63 = 200 + 40 + 8 − 60 − 3 or $^1\!\cancel{2}$ ^{1}4 8 −
= 200 + 40 − 60 + 8 − 3 6 3
= 100 + 140 − 60 + 8 − 3 1 8 5

✱ 40 − 60 gives a negative answer, so change 40 to 140 by taking 100 from 200

= 100 + 80 + 5
= 185

EXAMPLE 5

Evaluate each difference mentally by splitting up the second number.

a 36 − 9 **b** 44 − 21 **c** 846 − 28

SOLUTION

a 36 − 9 = 36 − 10 + 1 **b** 44 − 21 = 44 − 20 − 1 **c** 846 − 28 = 846 − 30 + 2

✱ To subtract 9, subtract 10 and add 1 ✱ To subtract 21, subtract 20 and subtract 1 ✱ To subtract 28, subtract 30 and add 2

= 26 + 1 = 24 − 1 = 816 + 2
= 27 = 23 = 818

EXAMPLE 6

Use a number line to evaluate 365 − 284.

SOLUTION

Draw a number line and jump along it from 284 to 365 using bridges. Write the size of each bridge and add the sizes.

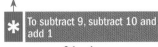

6 + 10 + 60 + 5 = 81

So 365 − 284 = 81

1 Which expression is the same as 57 − 12? Select the correct answer **A, B, C** or **D**.

    A 57 − 10 − 2   **B** 57 − 10 + 2

    C 50 + 7 − 10 + 2   **D** 57 + 10 − 2

2 Evaluate each difference using mental calculation.

    a 15 − 4   **b** 28 − 12   **c** 38 − 24   **d** 54 − 26

    e 42 − 36   **f** 86 − 34   **g** 74 − 58   **h** 96 − 37

    i 258 − 34   **j** 451 − 15   **k** 728 − 65   **l** 924 − 86

3 Copy and complete each line of working.

    a 84 − 9   **b** 358 − 41

       = 84 − _____ + 1   = 358 − _____ − 1

       = 74 + _____   = 318 − _____

       = _____   = _____

4 Which expression is the same as 158 − 21? Select **A, B, C** or **D**.

    A 158 − 20 + 1   **B** 185 − 20 − 1

    C 158 + 20 − 1   **D** 158 − 20 − 1

5 Evaluate each difference mentally by splitting up the second number.

    a 72 − 9   **b** 54 − 11   **c** 68 − 21   **d** 97 − 8

    e 82 − 31   **f** 96 − 9   **g** 121 − 19   **h** 132 − 9

    i 426 − 18   **j** 512 − 11   **k** 489 − 21   **l** 524 − 32

6 Use the number line below to jump from 164 to 203 and complete the statement below.

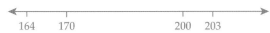

    203 − 164 = 6 + _____ + 3 = _____.

7 Use a number line to evaluate each difference.

    a 63 − 38   **b** 96 − 25   **c** 124 − 86   **d** 154 − 116

    e 236 − 189   **f** 275 − 238   **g** 321 − 286   **h** 455 − 398

8 Is each statement true or false?

    a 63 − 29 = 35   **b** 85 − 41 = 44

    c 125 − 78 = 47   **d** 236 − 89 = 145

9 Evaluate mentally and write down your answer: 5678 − 401 − 129 − 78

WORDBANK

rounding decimals To write a number with approximately the same value using fewer digits.

The place value for decimals is:

7	2	3	5	.	4	1	8
Thousands (1000s)	Hundreds (100s)	Tens (10s)	Units (1s)	Decimal point	Tenths $\left(\frac{1}{10}s\right)$	Hundredths $\left(\frac{1}{100}s\right)$	Thousandths $\left(\frac{1}{1000}s\right)$

1 decimal place is 0. __ $= \dfrac{\square}{10}$ **1** number after the point, rounding to the nearest **tenth**.

2 decimal places is 0. __ __ $= \dfrac{\square}{100}$ **2** numbers after the point, rounding to the nearest **hundredth**.

3 decimal places is 0. __ __ __ $= \dfrac{\square}{1000}$ **3** numbers after the point, rounding to the nearest **thousandth**.

To **round decimals** to a certain number of decimal places, look at the next digit.
If it is 5 or more, round the decimal **up.** If it is less than 5, round the decimal **down.**

EXAMPLE 7

Round each decimal to the number of decimal places shown in square brackets.

a 0.412 [1]   **b** 2.386 [2]   **c** 1.6525 [3]

SOLUTION

a For 1 decimal place, look at the next digit 1. It is less than 5, so leave the digit 4 as it is.
0.412 ≈ 0.4

b For 2 decimal places, look at the 6. It is more than 5, so round the digit 8 up to 9.
2.386 ≈ 2.39

c For 3 decimal places, look at the 5. It is 5 or more, so round the digit 2 up to 3.
1.6525 ≈ 1.653

EXAMPLE 8

Round each money amount as required.

a $30.75 (nearest dollar)   **b** $2.483 (nearest cent)   **c** $0.568 (nearest cent)

SOLUTION

a $30.75 ≈ $31 ⟵——— 75c is more than 50c, half a dollar, so round up.

b $2.483 ≈ $2.48 ⟵——— 3 is less than 5, so leave the 8 alone.

c $0.568 ≈ 57c ⟵——— 8 is more than 5, so take the digit 6 up to 7.

1 Round 3.426 18 to 2 decimal places. Select the correct answer **A**, **B**, **C** or **D**.

   A 3.42   **B** 3.4   **C** 3.43   **D** 3.426

2 Write the decimal 0.63 as a fraction. Select **A**, **B**, **C** or **D**.

   A $\dfrac{63}{10}$ **B** $\dfrac{63}{100}$ **C** $\dfrac{63}{1000}$ **D** $\dfrac{6.3}{100}$

3 Is each statement true or false?

   a $4.6 < 4$ **b** $2.8 > 2$ **c** $7.4 > 7$

   d $3 < 3.1$ **e** $9.4 < 9$

4 Round each decimal to the nearest whole number.

   a 3.2   **b** 6.8   **c** 12.7   **d** 21.3

   e 25.4   **f** 7.28   **g** 15.52   **h** 120.6

   i 263.9   **j** 18.44

5 Round each decimal to 1 decimal place.

   a 0.42   **b** 1.68   **c** 0.927   **d** 1.73

   e 3.22   **f** 0.652   **g** 2.17   **h** 3.428

   i 12.472   **j** 1.96

6 Round each decimal to 2 decimal places.

   a 0.216   **b** 1.423   **c** 5.518   **d** 11.319

   e 126.224   **f** 0.4281   **g** 3.268   **h** 0.9163

   i 7.234   **j** 1.896

7 Round each decimal to 3 decimal places.

   a 0.2638   **b** 1.7312   **c** 5.8436

   d 11.3214   **e** 126.2323

8 Round each amount of money to the nearest dollar.

   a $4.72   **b** $52.48   **c** $26.92   **d** $45.23

   e $28.99   **f** $10.42   **g** $231.68   **h** $500.91

   i $227.29   **j** $401.89

9 Round each amount of money to the nearest cent.

   a $5.712   **b** $12.428   **c** $32.568   **d** $4.231

   e $218.929   **f** $110.425   **g** $1.684   **h** $502.917

   i $22.296   **j** $43.897

10 Round $486.846 to the nearest:

   a ten dollars   **b** dollar   **c** ten cents   **d** cent

Adding and subtracting decimals

TO ADD OR SUBTRACT DECIMALS:

- ▨ write the decimals underneath each other in columns
- ▨ make sure the decimal points line up underneath each other: 'points under points' (PUP)
- ▨ fill in the gaps with 0s
- ▨ add or subtract the digits in columns
- ▨ place the decimal point directly underneath
- ▨ check your answer by estimating

EXAMPLE 9

Evaluate each expression.

a 5.6 + 12.432

b 265.84 − 18.66

SOLUTION

a 5.600 + ◄——— Gaps filled in with 0s
    12.432
    ——————
    18.032

b 2⁵6¹5.⁷8¹4 ◄——— Use trading to subtract
     1 8. 6 6
     ——————
    2 4 7. 1 8

points under points or PUP

Check by estimating:

5.6 + 12.432 ≈ 6 + 12 = 18
(18.032 is close to 18)

265.84 − 18.66 ≈ 266 − 19 = 247
(247.18 is close to 247)

EXAMPLE 10

Find the change given from a $50 note if I buy food costing $42.75.

SOLUTION

To calculate $50 − $42.75, use a number line to jump from $42.75 to $50.

So the change from $50 is $0.25 + $2 + $5 = $7.25.

Shutterstock.com/Dmitry Kalinovsky

Developmental Mathematics Book 3

ISBN 9780170351027

1 Find the sum of 5.2 and 13.86. Select **A**, **B**, **C** or **D**.

   A 18.06   **B** 18.16   **C** 19.16   **D** 19.06

2 Find the difference between 18.6 and 9.23. Select **A**, **B**, **C** or **D**.

   A 9.37   **B** 9.47   **C** 8.37   **D** 8.47

3 What must be added to:

   a 0.01 to make 0.04?   **b** 0.62 to make 0.88?

   c 1.25 to make 1.78?   **d** 4.64 to make 8?

4 What must be subtracted from:

   a 9.5 to make 9.3?   **b** 12.6 to make 10.4?

   c 18.6 to make 15.7?   **d** 24.8 to make 22?

5 Evaluate each expression.

   a $12.5 + 3.48$ **b** $38.65 + 12.4$ **c** $128.8 + 4.568$

   d $5.68 + 126.9$ **e** $28.94 - 6.8$ **f** $362.8 - 23.9$

   g $48.64 - 19.826$ **h** $263.5 - 115.87$ **i** $18.62 + 9.68$

   j $27.9 - 18.42$ **k** $216.8 + 90.237$ **l** $489.2 - 68.9$

6 Find each missing number.

   a $4.7 + \_\_ = 5$ **b** $3.8 + \_\_ = 7.9$ **c** $\_\_ + 5.1 = 6$

   d $\_\_ + 2.4 = 3.6$ **e** $1.23 + \_\_ = 1.29$ **f** $5.64 + \_\_ = 7.86$

   g $\_\_ + 2.8 = 3.1$ **h** $\_\_ + 8.76 = 10.62$ **i** $\_\_ - 1.2 = 2.6$

   j $\_\_ - 5.5 = 6.1$ **k** $12.34 - \_\_ = 10.24$ **l** $21.45 - \_\_ = 9.66$

7 Use a number line to evaluate each difference.

   a $150 - 126.50$ **b** $80 - 62.70$ **c** $15 - 7.25$ **d** $100 - 42.60$

   e $180 - 124.78$ **f** $58 - 26.99$ **g** $186 - 165.35$ **h** $200 - 148.82$

8 Emma was shopping in the supermarket and bought the following items.

   • 1 box of chocolates for $8.65

   • 2 litres of milk for $3.48

   • 2.5 kg apples for $5.26

   a Estimate the cost of the groceries.

   b If Emma's mother gave her a $20 note, would she have enough money to pay for the items?

   c Find the total cost of the groceries, correct to the nearest 5 cents.

   d Calculate Emma's change from $20.

9 A 5 kg tin of biscuits costing $22.95 was reduced by $8.56. What was the new price?

10 Find the cost of a European holiday correct to the nearest dollar if the flights are $2565.85, the Eurail pass is $428.24 and the accommodation and meals are $1890.50.

WORDBANK

product The answer to a multiplication (×) of two or more numbers.

Multiplication table

×	1	2	3	4	5	6	7	8	9	10
1	1	2	3	4	5	6	7	8	9	10
2	2	4	6	8	10	12	14	16	18	20
3	3	6	9	12	15	18	21	24	27	30
4	4	8	12	16	20	24	28	32	36	40
5	5	10	15	20	25	30	35	40	45	50
6	6	12	18	24	30	36	42	48	54	60
7	7	14	21	28	35	42	49	56	63	70
8	8	16	24	32	40	48	56	64	72	80
9	9	18	27	36	45	54	63	72	81	90
10	10	20	30	40	50	60	70	80	90	100

Mental multiplication strategies

Multiplying by	Strategy
2	Double
4	Double twice
5	Multiply by 10, then halve
8	Multiply by (10 − 2)
9	Multiply by (10 − 1)
10	Add a 0 to the end
11	Multiply by (10 + 1)
12	Multiply by (10 + 2)

EXAMPLE 11

Evaluate each product mentally.

a 32×5 **b** 23×9 **c** $4 \times 18 \times 25$

d 64×20 **e** 58×4 **f** 43×11

SOLUTION

a $32 \times 5 = 32 \times 10 \div 2$

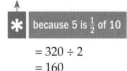 because 5 is $\frac{1}{2}$ of 10

$= 320 \div 2$
$= 160$

b $23 \times 9 = 23 \times (10 - 1)$
$= 23 \times 10 - 23 \times 1$
$= 230 - 23$
$= 207$

c $4 \times 18 \times 25 = 18 \times 4 \times 25$

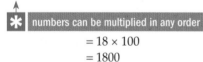 numbers can be multiplied in any order

$= 18 \times 100$
$= 1800$

d $64 \times 20 = 64 \times 2 \times 10$
$= 128 \times 10$
$= 1280$

e $58 \times 4 = 58 \times 2 \times 2$

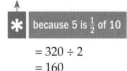 double twice

$= 116 \times 2$
$= 232$

f $43 \times 11 = 43 \times (10 + 1)$
$= 43 \times 10 + 43 \times 1$
$= 430 + 43$
$= 473$

1 Which expression can be used to evaluate 47×8 mentally? Select the correct answer **A**, **B**, **C** or **D**.

    A $47 \times (10 - 1)$ **B** $47 \times (10 - 2)$

    C $47 \times (10 + 1)$ **D** $47 \times (10 + 2)$

2 Which expression can be used to evaluate 19×11 mentally? Select **A**, **B**, **C** or **D**.

    A $19 \times (10 - 1)$ **B** $19 \times (10 - 2)$

    C $19 \times (10 + 1)$ **D** $19 \times (10 + 2)$

3 Write the answer to each product.

    a 5×6 **b** 8×3 **c** 7×4 **d** 9×2

    e 5×8 **f** 4×6 **g** 9×7 **h** 4×9

    i 6×7 **j** 10×3 **k** 3×7 **l** 8×9

    m 8×6 **n** 9×3 **o** 7×5 **p** 8×2

    q 5×9 **r** 3×6

4 Is each statement true or false?

    a $27 \times 4 = 27 \times 2 \times 2$ **b** $54 \times 9 = 54 \times (10 + 1)$

    c $42 \times 11 = 42 \times (10 + 1)$ **d** $36 \times 12 = 36 \times (10 + 2)$

    e $73 \times 5 = 73 \times 10 \div 2$ **f** $93 \times 8 = 93 \times (10 - 1)$

5 Evaluate each product using a mental strategy.

    a 28×5 **b** 46×11

    c 63×9 **d** 82×4

    e 35×10 **f** 52×8

    g 75×2 **h** 97×12

    i 17×20 **j** 57×10

    k 74×40 **l** 68×90

    m 27×300 **n** 48×600

    o 82×500 **p** 91×4000

6 Copy and complete each line of working.

    a $2 \times 33 \times 5 = 33 \times 2 \times \underline{\quad}$ **b** $4 \times 42 \times 25 = 42 \times 4 \times \underline{\quad}$

                    $= 33 \times \underline{\quad}$ $= 42 \times \underline{\quad}$

                    $= \underline{\quad}$ $= \underline{\quad}$

7 Evaluate each product by changing the order.

    a $2 \times 27 \times 5$ **b** $25 \times 15 \times 4$

    c $2 \times 26 \times 50$ **d** $5 \times 14 \times 20$

    e $2 \times 54 \times 5$ **f** $50 \times 58 \times 2$

    g $4 \times 36 \times 25$ **h** $5 \times 82 \times 200$

Multiplying decimals

To multiply a decimal by 10, 100 or 1000, move the decimal point to the **right**, since the number is getting **larger**.

- To multiply by 10, move the decimal point 1 place to the right
- To multiply by 100, move the decimal point 2 places to the right
- To multiply by 1000, move the decimal point 3 places to the right

EXAMPLE 12

Evaluate each product.

a 6.42×10

b 0.2834×100

c 28.6×1000

SOLUTION

a $6.42 \times 10 = 64.2$

 point moves 1 right

b $0.2834 \times 100 = 28.34$

 point moves 2 right

c $28.6 \times 1000 = 28.600 \times 1000 = 28\ 600$ ⟵——— Adding zeros to the decimal to allow the point to move

 point moves 3 right

TO MULTIPLY DECIMALS:

- multiply the numbers without the decimal points
- count the total number of decimal places in the numbers
- write the answer using this number of decimal places
- check your answer by estimating

EXAMPLE 13

Evaluate each product.

a 0.6×4

b 2.41×5

c 3.86×0.7

SOLUTION

a $6 \times 4 = 24$

b $\begin{array}{r} ^2241 \times \\ 5 \\ \hline 1205 \end{array}$

c $\begin{array}{r} ^63^486 \times \\ 7 \\ \hline 2702 \end{array}$

0.6 has 1 decimal place
4 has no decimal places
Total decimal places = 1
So $0.6 \times 4 = 2.4$

2.41 has 2 decimal places
5 has no decimal places
Total decimal places = 2
So $2.41 \times 5 = 12.05$

3.86 has 2 decimal places
0.7 has 1 decimal place
Total decimal places = 3
So $3.86 \times 0.7 = 2.702$

Check by estimating:

$0.6 \times 4 \approx \frac{1}{2} \times 4 = 2$
(2.4 is close to 2)

$2.41 \times 5 \approx 2 \times 5 = 10$
(12.05 is close to 10)

$3.86 \times 0.7 \approx 4 \times 1 = 4$
(2.702 is close to 4)

ISBN 9780170351027

1. When multiplying a decimal by 100, where does the decimal point move? Select the correct answer **A, B, C** or **D**.

   A 1 place right   **B** 2 places right   **C** 1 place left   **D** 2 places left

2. Evaluate the product 2.35 × 1000. Select **A, B, C** or **D**.

   A 235   **B** 23 500   **C** 23.5   **D** 2350

3. Evaluate each product.

   a 2.6 × 100   **b** 5.4 × 10   **c** 6.8 × 1000   **d** 0.45 × 10

   e 22.65 × 1000   **f** 15.42 × 10   **g** 0.38 × 100   **h** 20.415 × 10

4. Is each statement true or false?

   a 3.27 has 3 decimal places   **b** 0.47 has 2 decimal places

   c 9.04 has 1 decimal place

5. Count the total number of decimal places in each expression.

   a 0.5 × 9   **b** 2.1 × 8   **c** 0.7 × 0.2   **d** 0.4 × 0.03

6. Evaluate each product in question **5**.

7. Evaluate each product.

   a 5.6 × 4   **b** 2.8 × 3   **c** 28.5 × 6   **d** 32.4 × 8

   e 28.4 × 0.3   **f** 37.65 × 0.5   **g** 18.4 × 0.02   **h** 12.46 × 0.07

   i 120.2 × 0.9   **j** 0.48 × 0.02   **k** 203.1 × 0.05   **l** 128.03 × 0.4

8. If 128 × 13 = 1664, evaluate each product, then check your answer by estimating.

   a 1.28 × 13   **b** 12.8 × 0.13   **c** 1.28 × 1.3   **d** 0.128 × 1.3

9. If Ruby worked in a flower shop and earned $21.80 per hour, how much would she earn in 6 hours? Estimate your answer first, then calculate the exact amount.

10. Jonah planted a vegetable garden that was 8.6 m long and 0.9 m wide. Estimate the area of his garden before calculating the exact area.

Shutterstock.com/Gcictoria

WORDBANK

quotient The result of dividing ($\div$) a number by another number. For example, for $8 \div 4 = 2$, the quotient is 2.

Mental division strategies

Dividing by	Strategy
2	Halve
4	Halve twice
5	Divide by 10, then double
8	Halve 3 times
10	Move the decimal point 1 place left, or for a whole number ending in 0, drop a 0 from the end of the number
20	Divide by 10, then halve
100	Move the decimal point 2 places left, or for a whole number ending in 0s, drop two 0s from the end of the number

EXAMPLE 14

Evaluate each quotient mentally.

a $624 \div 4$ b $840 \div 5$ c $1280 \div 20$

SOLUTION

a $624 \div 4 = 624 \div 2 \div 2$

 halving 624 twice

$= 312 \div 2$
$= 156$

b $840 \div 5 = 840 \div 10 \times 2$

 divide by 10 and double

$= 84 \times 2$
$= 168$

c $1280 \div 20 = 1280 \div 10 \div 2$

✱ divide by 10, then halve

$= 128 \div 2$
$= 64$

EXAMPLE 15

Evaluate $14\,670 \div 6$ by short division.

SOLUTION

The steps of short division are shown below.

$$6 \overline{)14^{2}670} \quad \overset{2}{}$$

$14 \div 6 = 2$ remainder 2

$$6 \overline{)14^{2}6^{2}70} \quad \overset{2\ 4}{}$$

$26 \div 6 = 4$ remainder 2

$$6 \overline{)14^{2}6^{2}7^{3}0} \quad \overset{2\ 4\ 4}{}$$

$27 \div 6 = 4$ remainder 3

$$6 \overline{)14^{2}6^{2}7^{3}0} \quad \overset{2\ 4\ 4\ 5}{}$$

$30 \div 6 = 5$ no remainder

So $14\,670 \div 6 = \textbf{2445}$.

1   What is the answer to a division called? Select **A**, **B**, **C** or **D**.

   A difference   **B** quotient   **C** product   **D** sum

2   What is a quick way to divide a number by 20? Select **A**, **B**, **C** or **D**.

   A halve it   **B** double it

   C divide by 10 twice   **D** divide by 10 then halve

3   Write the answer to each quotient.

   a 15 ÷ 3   **b** 24 ÷ 6   **c** 32 ÷ 8   **d** 25 ÷ 5

   e 42 ÷ 7   **f** 56 ÷ 8   **g** 45 ÷ 9   **h** 20 ÷ 4

   i 22 ÷ 2   **j** 90 ÷ 10   **k** 63 ÷ 9   **l** 64 ÷ 8

   m 30 ÷ 6   **n** 27 ÷ 3   **o** 35 ÷ 7   **p** 72 ÷ 8

   q 28 ÷ 4   **r** 36 ÷ 9

4   Is each statement true or false?

   a 128 ÷ 2 = 64 ÷ 8   **b** 96 ÷ 4 = 48 ÷ 2   **c** 96 ÷ 8 = 24 ÷ 2

   d 150 ÷ 10 = 15   **e** 2800 ÷ 100 = 2.8   **f** 54 ÷ 10 = 0.54

   g 236 ÷ 100 = 2.36

5   Evaluate each quotient using a mental strategy.

   a 86 ÷ 2   **b** 156 ÷ 4   **c** 256 ÷ 8   **d** 428 ÷ 4

   e 184 ÷ 8   **f** 306 ÷ 2   **g** 80 ÷ 5   **h** 280 ÷ 20

   i 400 ÷ 5   **j** 620 ÷ 20   **k** 260 ÷ 5   **l** 340 ÷ 20

6   Evaluate each quotient using a mental strategy.

   a 1200 ÷ 100   **b** 540 ÷ 10   **c** 330 ÷ 10   **d** 5100 ÷ 100

   e 245 ÷ 10   **f** 218 ÷ 100   **g** 29 ÷ 10   **h** 321 ÷ 100

   i 43 ÷ 100   **j** 6.2 ÷ 10   **k** 0.48 ÷ 10   **l** 3.6 ÷100

7   Use short division to find each quotient, then check your answer using a calculator.

   a 1782 ÷ 3   **b** 4550 ÷ 7   **c** 2844 ÷ 9   **d** 2868 ÷ 6

8   At a Christmas sale, all decorations were reduced by half.
   On Boxing Day, these same decorations were reduced by
   half again.

   a What number should you divide by to find the final
        sale price?

   b Find the final sale price of each item below.

   Tinsel $4.20   Baubles $2.60

   Gift wrap $6.80   Cards $3.60

To **divide a decimal by 10, 100 or 1000,** move the decimal point to the **left** since the number is getting **smaller.**
- ■ To divide by 10, move the decimal point 1 place to the left
- ■ To divide by 100, move the decimal point 2 places to the left
- ■ To divide by 1000, move the decimal point 3 places to the left

EXAMPLE 16

Evaluate each quotient.

a $32.5 \div 10$ **b** $14.68 \div 100$ **c** $27.023 \div 1000$

SOLUTION

a $3\,2.5 \div 10 = 3.25$

✱ point moves 1 left

b $14.68 \div 100 = 0.1468$

✱ point moves 2 left

c $27.023 \div 1000 = 0\,0\,2\,7.023 \div 1000 = 0.027\,023$ ◄——— Adding zeros to the left of the decimal to allow the point to move

✱ point moves 3 left

- ■ To **divide a decimal by a whole number,** use short division, then write the answer with the decimal point in the same column as the original decimal
- ■ To **divide a decimal by another decimal,** move the points in both decimals the same number of places to the right so that you are dividing by a **whole number**
- ■ Check your answer by estimating

EXAMPLE 17

Evaluate each quotient.

a $42.8 \div 4$ **b** $128.65 \div 0.5$

SOLUTION

a
$\dfrac{10.7}{4\overline{)42.^28}}$

✱ decimal points in the same column

$42.8 \div 4 = \mathbf{10.7}$

b  $128.\,6\,5 \div 0.\,5 = 1286.5 \div 5$

✱ move both points 1 right so 0.5 is a whole number, 5

$\dfrac{2\ 5\ 7.3}{5\overline{)12^28^36.^15}}$

$128.65 \div 0.5 = \mathbf{257.3}$

Check by estimating:

$42.8 \div 4 \approx 40 \div 4 = 10$
(10.7 is close to 10)

$1286.5 \div 5 \approx 1300 \div 5 = 260$
(257.3 is close to 260)

1 When dividing a decimal by 100, where does the decimal point move? Select the correct answer **A**, **B**, **C** or **D**.

    A 1 place right   **B** 2 places right   **C** 1 place left   **D** 2 places left

2 Evaluate the quotient $36.4 \div 1000$. Select **A**, **B**, **C** or **D**.

    A 3.64   **B** 36 400   **C** 0.0364   **D** 0.364

3 Evaluate each quotient.

    a $3.8 \div 100$ **b** $25.6 \div 10$

    c $72.4 \div 1000$ **d** $780 \div 10$

    e $128.3 \div 1000$ **f** $12.78 \div 100$

    g $0.95 \div 10$ **h** $25 \div 1000$

4 Copy and complete this division.

$$5\overline{)64.5}$$

5 Evaluate each quotient.

    a $45.6 \div 6$ **b** $86.48 \div 8$ **c** $92.7 \div 3$ **d** $112.5 \div 5$

    e $126.4 \div 4$ **f** $721.14 \div 7$ **g** $129.6 \div 9$ **h** $328.5 \div 2$

6 Copy and complete this division.

$$68.46 \div 0.3 = \boxed{} \div 3 = 3\overline{)\boxed{}}$$

7 Evaluate each quotient by making the second decimal a whole number.

    a $3.6 \div 0.9$ **b** $2.48 \div 0.4$

    c $3.618 \div 0.3$ **d** $25.56 \div 0.6$

    e $12.45 \div 0.05$ **f** $0.688 \div 0.08$

    g $47.384 \div 0.04$ **h** $128.1 \div 0.07$

    i $0.92 \div 0.8$ **j** $12.45 \div 0.03$

    k $234 \div 0.9$ **l** $38.94 \div 0.06$

8 Sam made a mistake when calculating this quotient and put the decimal point in the wrong place: $37.128 \div 8 = 0.4641$. What is the correct answer?

9 A length of timber 4.76 m long is cut into 4 equal pieces. What is the length of each piece?

10 Corrine spent $31.40 at the school canteen last week. What was her average spending per day if she spent at the canteen every school day?

WORDBANK

recurring decimal A decimal with digits that repeat or recur, such as $0.\dot{3} = 0.3333...$ and $10.\dot{2}\dot{6} = 10.262626....$

terminating decimal A decimal that ends or terminates, such as 0.512 or 2.96.

Recurring decimals are written with **dots** placed over the repeating digits. If a **series** of digits are repeated, then a dot is placed over the **first** and **last** digits in the series.

EXAMPLE 18

Write each recurring decimal using dot notation.

a 0.424242...   **b** 2.366666...   **c** 0.415415...

SOLUTION

Where a group of digits recur, just put a dot above the first and last digits.

a $0.424242... = 0.\dot{4}\dot{2}$ **b** $2.366666... = 2.3\dot{6}$ **c** $0.415415... = 0.\dot{4}1\dot{5}$

When using **short division,** if there are still remainders after dividing into the last digit, you need to add 0s to the end of the decimal to continue dividing, until there are no remainders.

EXAMPLE 19

Evaluate each quotient and state the type of decimal in each answer.

a 28.3 ÷ 4   **b** 356.8 ÷ 3

SOLUTION

a
$$\begin{array}{r} 7.0\,7\,5 \\ 4\overline{)28.3\,^30^20} \end{array}$$

✱ It was necessary to add 2 zeros.

28.3 ÷ 4 = **7.075**, a terminating decimal.

b
$$\begin{array}{r} 11\,8.\,9\,3\,\,3\,3... \\ 3\overline{)35\,^26.\,^28\,^10\,^10\,^10...} \end{array}$$

✱ It was necessary to add 3 or more zeros to see a repeating pattern.

356.8 ÷ 3 = $118.9\dot{3}$, a recurring decimal.

EXAMPLE 20

Convert each fraction to a decimal.

a $\dfrac{2}{5}$ **b** $\dfrac{5}{6}$

SOLUTION

a $\dfrac{2}{5}$ means 2 ÷ 5

$$\begin{array}{r} 0.4 \\ 5\overline{)2.0} \end{array}$$

$\dfrac{2}{5} = 0.4$

b $\dfrac{5}{6}$ means 5 ÷ 6

$$\begin{array}{r} 0.\,8\,3\,3\,3... \\ 6\overline{)5.\,^50^20^20^20...} \end{array}$$

$\dfrac{5}{6} = 0.8\dot{3}$

1   Write 3.544444... as a recurring decimal. Select the correct answer **A**, **B**, **C** or **D**.

   A 3.54   **B** 3.5̇4̇   **C** 3.544   **D** 3.54̇

2   What type of decimal is the answer to 28.416 ÷ 4? Select **A**, **B**, **C** or **D**.

   A non-terminating   **B** recurring

   C terminating   **D** none of these

3   Is each statement true or false?

   a 0.325555... = 0.325̇   **b** 0.454545... = 0.4̇5̇   **c** 2.135135... = 2.1̇35̇

4   Write each recurring decimal using dot notation.

   a 0.7777...   **b** 0.626262...

   c 2.11111...   **d** 3.46666....

   e 5.232323...   **f** 0.528528...

   g 7.23333...   **h** 12.425555...

5   Evaluate each quotient using short division, adding zeros to the decimal if necessary.

   a 12.75 ÷ 4   **b** 32.6 ÷ 5

   c 285.4 ÷ 3   **d** 22.6 ÷ 8

   e 182.6 ÷ 6   **f** 78.4 ÷ 9

   g 22.7 ÷ 2   **h** 65.84 ÷ 7

6   State the type of decimal for each quotient found in question **5**.

7   **a** How do you change $\frac{3}{8}$ into a decimal without the use of a calculator?

   b Convert $\frac{3}{8}$ to a decimal and check your answer using a calculator.

   c What type of decimal is it?

8   Convert each fraction below to a decimal, then check your answer using a calculator.

   a $\frac{1}{4}$ **b** $\frac{1}{3}$ **c** $\frac{5}{8}$ **d** $\frac{3}{5}$

   e $\frac{1}{6}$ **f** $\frac{2}{7}$ **g** $\frac{5}{12}$ **h** $\frac{7}{8}$

9   Jamiela had a $900 gift card and her sister had spent one-third of it.

   a What fraction of the money is left on the gift card?

   b Write this fraction as a decimal.

   c How much money does Jamiela have left to spend on her gift card?

iStockphoto/YinYang

WORD PUZZLE

List the letters in order, for the clues below to spell out a phrase relating to this topic.

The first 0.2 of DESCENDING

The first 0.1 of CELLOPHANE

The first 0.6 of IMAGE

The first 0.2 of LUNCH

The first 0.4 of DIGIT

The last 0.25 of DEAL

The last 0.5 of STEM

The first 0.2 of MANAGEMENT

123RF/konstantin32

PRACTICE TEST 2

Part A General topics

Calculators are not allowed.

1 Evaluate 80×9.

2 Complete: $12 \text{ km} = \underline{\hspace{1cm}}$ m.

3 List all the factors of 6.

4 Simplify $-x + y - y + 2x$

5 Find the range of the scores: 12, 5, 7, 9, 8, 7, 10.

6 Given that $15 \times 6 = 90$, evaluate 1.5×6.

7 Find 5% of $240.

8 Find the value of d if the perimeter of this rectangle is 46 m.

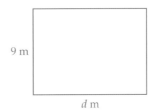

9 m

d m

9 Simplify $4 \times r \times 9 \times v$

10 Copy and complete $\dfrac{4}{7} = \dfrac{}{28}$.

Part B Whole numbers and decimals

Calculators are not allowed.

2-01 Mental addition

11 Use a mental strategy to evaluate each sum.

    a $27 + 11$

    b $72 + 9$

    c $23 + 421 + 27$

12 Which equation is correct? Select the correct answer **A**, **B**, **C** or **D**.

    A $42 + 8 = 42 + 10 - 1$

    B $42 + 9 = 42 + 10 - 1$

    C $42 + 11 = 42 + 10 - 1$

    D $42 + 12 = 42 + 10 - 1$

2-02 Mental subtraction

13 Use a mental strategy to evaluate each difference.

    a $58 - 9$

    b $128 - 31$

2-03 Rounding decimals and money

14 Round each decimal to the required number of decimal places in the square brackets.

    a   2.682 [2]

    b   0.9246 [3]

2-04 Adding and subtracting decimals

15 Evaluate each expression.

    a $2.45 + 13.8$

    b $28.63 - 8.92$

ISBN 9780170351027

2–05 Mental multiplication

16 Evaluate each product using a mental strategy.

    a 48×20

    b $25 \times 53 \times 4$

2–06 Multiplying decimals

17 Evaluate each product.

    a 47.6×100

    b 0.4×0.07

    c 15.8×0.06

18 Is this statement true or false? $0.08 \times 0.007 = 0.000\,56$

2–07 Mental division

19 Evaluate each quotient using a mental strategy.

    a $220 \div 5$

    b $6780 \div 20$

    c $4688 \div 8$

2–08 Dividing decimals

20 Evaluate each quotient.

    a $16.42 \div 10$

    b $68.6 \div 4$

    c $228.75 \div 0.5$

21 Is this statement true or false? $246.78 \div 0.8 = 24.678 \div 8$

2–09 Terminating and recurring decimals

22 Write as recurring decimals using dot notation.

    a $6.122222...$ **b** $0.484848...$ **c** $12.412412...$

23 Convert $\dfrac{2}{9}$ to a decimal.

INTEGERS AND FRACTIONS

3

WHAT'S IN CHAPTER 3?

IN THIS CHAPTER YOU WILL:

- add, subtract, multiply and divide integers
- understand and use the order of operations
- understand and convert between improper fractions and mixed numerals
- calculate equivalent fractions and simplify fractions
- find a fraction of a quantity
- compare and order fractions
- add, subtract, multiply and divide fractions

Shutterstock.com/Chones

WORDBANK

integer A positive or negative whole number, or zero, such as –5, 7 and 10.

- Integers can be **added** and **subtracted** using a number line
- Move **right** if **adding** a positive integer, move **left** if **subtracting** a positive integer
- Adding a **negative integer** is the same as **subtracting its opposite,** for example, 10 + (–1) = 10 – 1 = 9
- Subtracting a **negative integer** is the same as **adding its opposite,** for example, 3 – (–4) = 3 + 4 = 7

EXAMPLE 1

Evaluate each sum.

a –6 + 7   **b** 2 + (–5)   **c** –2 + (–3)

SOLUTION

a Start at –6 and move 7 units to the right.

–6 + 7 = 1.

OR using the calculator:

b Start at 2 and move 5 units to the left.

2 + (–5) = –3

OR using the calculator: 2 ➕ ⊖ 5 🟰

c Start at –2 and move 3 units to the left.

–2 + (–3) = –5

OR using the calculator: ⊖ 2 ➕ ⊖ 3 🟰

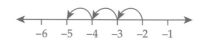

EXAMPLE 2

Evaluate each difference.

a 3 – 4   **b** 4 – (–3)   **c** –3 – 4

SOLUTION

a Start at 3 and move 4 units to the left.

3 – 4 = –1.

OR using the calculator: 3 ➖ 4 🟰

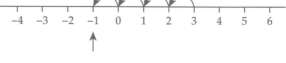

b 4 – (–3) = 4 + 3. Start at 4 and move 3 units to the right.

4 – (–3) = 7.

OR using the calculator: 4 ➖ ⊖ 3 🟰

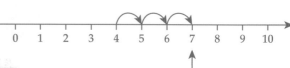

c Start at –3 and move 4 units to the left.

–3 – 4 = –7.

OR using the calculator:

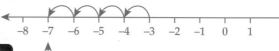

1   Evaluate –7 + 2. Select the correct answer **A, B, C** or **D**.

   A 5   **B** –5   **C** –9   **D** 9

2   Which expression gives the same answer as –3 – (–4)? Select **A, B, C** or **D**.

   A –3 + (–4)   **B** 3 + 4   **C** –3 + 4   **D** 3 + (–4)

3   Copy and complete each statement.

   a To evaluate –4 + 6, start at –4 on the number line and move ___ units right.

   b To evaluate 3 + (–7), start at 3 on the number line and move ___ units ____.

   c To evaluate –5 – 8, start at –5 on the number line and move ___ units ____.

   d To evaluate 4 + (–2), start at ___ on the number line and move ___ units ____.

4   Evaluate each sum and check your answer on a calculator.

   a –2 + 7   **b** 4 + (–5)   **c** –3 + 5   **d** 8 + (–6)

   e –6 + 6   **f** 7 + (–8)   **g** –4 + 3   **h** –8 + 11

   i –7 + 10   **j** –13 + 13   **k** –5 + 12   **l** 16 + (–7)

5   Copy and complete each statement.

   a The difference 5 – (–3) is the same as the sum 5 + __.

   b The difference 5 – 3 is the same as the sum ____.

   c The difference –5 – 3 is the same as the sum ____.

6   Evaluate each difference and check your answer on a calculator.

   a 9 – 6   **b** 3 – (–4)   **c** –2 – 7   **d** 4 – 8

   e 2 – (–6)   **f** –5 – 2   **g** 3 – 10   **h** 4 – (–5)

   i –1 – (–2)   **j** 12 – (–4)   **k** –3 – (–9)   **l** 8 – (–13)

7   Evaluate each expression.

   a –3 + 5 – (–4)   **b** 12 – (–5) + 6   **c** –5 – (–7) + 4   **d** 10 + (–6) – (–8)

   e 20 – (–12) + 6   **f** –18 + (–9) – (–3) **g** 15 + 7 – (–13)   **h** –10 – (–4) + (–5)

8   Toby had $28 in his money box. He added $15 to it on Monday, spent $8 from it on Wednesday and put his pocket money of $16 in it on Saturday. How much is in his money box now?

9   Elena entered a lift from the 3rd floor. She went up 6 floors, down 11 floors and back up 15 floors. She then exited the lift. What floor was she on?

10 When Sophia went to bed it was 18°C in her room. She woke up at 2 a.m. and the temperature had dropped by 6°. By the time she woke in the morning the temperature had risen 10° since 2 a.m. What was the temperature now?

When multiplying or dividing integers:
- ■ **two positive** integers give a **positive** answer (+ + = +)
- ■ a **positive** and a **negative** integer give a **negative** answer (+ − = −)
- ■ **two negative** integers give a **positive** answer (− − = +)

EXAMPLE 3

Evaluate each expression.

a -4×6 **b** $-3 \times (-6)$ **c** $5 \times (-8) \times (-2)$ **d** $(-8)^2$

e $-35 \div 7$ **f** $-18 \div (-3)$ **g** $-240 \div 80$ **h** $-48 \div (-2)^2$

SOLUTION

a $-4 \times 6 = -24$

b $-3 \times (-6) = 18$

 ✱ Check these answers are correct on your calculator.

c $5 \times (-8) \times (-2) = -40 \times (-2)$
$= 80$

d $(-8)^2 = -8 \times (-8)$
$= 64$

e $-35 \div 7 = -5$

f $-18 \div (-3) = 6$

g $-240 \div 80 = -3$

h $-48 \div (-2)^2 = -48 \div 4$
$= -12$

iStockphoto/aloha_17

1   If $18 \times (-4) = -72$, what is $-72 \div (-4)$? Select the correct answer **A**, **B**, **C** or **D**.

   A -18 **B** 4 **C** 18 **D** -4

2   If $-91 \div 13 = -7$, what is -7×13? Select **A**, **B**, **C** or **D**.

   A 13 **B** 91 **C** -13 **D** -91

3   Is each statement true or false?

   a $-5 \times 3 = 15$ **b** $-15 \div 3 = -5$

   c $15 \div (-5) = 3$ **d** $-3 \times (-5) = 15$

4   Copy and complete this table.

×	-2	5	-6	8	-4
3					
-7					
-9					
10					

5   Evaluate each product, then check your answers on a calculator.

   a -4×3 **b** $6 \times (-5)$ **c** $-7 \times (-3)$ **d** $8 \times (-2)$

   e $-9 \times (-4)$ **f** 3×9 **g** -6×7 **h** -5×5

   i $8 \times (-9)$ **j** $-3 \times (-10)$ **k** $(-4)^2$ **l** $-2 \times 7 \times (-3)$

   m $(-6)^2$ **n** $4 \times (-7) \times 6$ **o** $(-5)^2 \times (-4)$ **p** $-3 \times (-3)^2 \times 5$

6   Evaluate each quotient, then check your answers on a calculator.

   a $-24 \div 3$ **b** $16 \div (-8)$

   c $-27 \div (-3)$ **d** $48 \div (-6)$

   e $-49 \div (-7)$ **f** $36 \div (-9)$

   g $-64 \div 8$ **h** $-35 \div 5$

   i $81 \div (-9)$ **j** $-300 \div (-10)$

   k $280 \div (-7)$ **l** $-560 \div 80$

   m $72 \div (-3)^2$ **n** $-120 \div 60$

   o $64 \div (-2)^2$ **p** $-2400 \div (-120)$

7   Evaluate each expression.

   a $-72 \div 9 \times (-4)$ **b** $54 \div (-9) \times 3$

   c $-8 \times (-6) \div 4$ **d** $5 \times (-4) \div (-10)$

8   Sally opened a new business and was shocked to find she was losing $250 each week. If this trend occurred for the first 6 weeks, how much money had she lost?

WORDBANK

brackets Grouping symbols around expressions. Round brackets ( ) or square brackets [ ] can be used.

When evaluating **mixed expressions** with more than one operation, calculate using this **order of operations**:
- brackets ( ) first,
- then powers (x^y) and square roots ($\sqrt{\ }$)
- then multiplication (×) and division (÷) from left to right
- then addition (+) and subtraction (–) from left to right.

EXAMPLE 4

Evaluate each mixed expression using the order of operations.

a $20 - 9 \div (-3)$

b $36 + (-5) \times (-9)$

c $(-8 + 2) \times (-32 \div 8)$

d $6.8 \div 4 + (-3) \times 0.6$

e $(-8)^2 \div (-4) + \sqrt{16} \times (-2)$

f $\dfrac{-48 + 16}{36 - 20}$

SOLUTION

a No brackets

Division first:

$20 - 9 \div (-3) = 20 - (-3)$
$\qquad\qquad\qquad = 23$

b No brackets

Multiplication first:

$36 + (-5) \times (-9) = 36 + 45$
$\qquad\qquad\qquad\quad = 81$

c Brackets first:

$(-8 + 2) \times (-32 \div 8) = -6 \times (-4)$
$\qquad\qquad\qquad\qquad\quad = 24$

d Divide and multiply first

$6.8 \div 4 + (-3) \times 0.6$
$= 1.7 + (-1.8)$
$= -0.1$

e Powers and square root first

$(-8)^2 \div (-4) + \sqrt{16} \times (-2)$
$= 64 \div (-4) + 4 \times (-2)$
$= -16 + (-8)$
$= -24$

f Work out top and bottom first

$\dfrac{-48 + 16}{36 - 20} = \dfrac{-32}{16}$
$\qquad\qquad\quad = -2$

Shutterstock.com/CHEN WS

1 For $22 - 8 \times (-3) + 7$, which operation is performed first? Select the correct answer **A, B, C** or **D**.

    A $22 - 8$ **B** $8 \times (-3)$ **C** $(-3) + 7$ **D** $22 + 7$

2 Which one of these comes first in the order of operations? Select **A, B, C** or **D**.

    A multiplication   **B** division

    C brackets   **D** addition

3 Which operation is done first in each expression below?

    a $12 - 5 \times (-3)$ **b** $40 + (-8) \div (-2)$

    c $120 - (-63) \div 7$ **d** $43 + 48 \div (-8)$

    e $-18 + (-6) \times 4$ **f** $80 - 25 \div (-5)$

    g $19 \times (-2) + (-7)$ **h** $22 + (-8) \times (-7)$

    i $-12 \times (-6) - (-8)$

4 Evaluate each expression in question **3**.

5 Evaluate each expression.

    a $(-6 \times 2) + [18 \div (-3)]$ **b** $-45 + (-8 + 4) \times (-5)$

    c $140 - (-6 + 10) \div (-2)$ **d** $80 - (-9) \times (-3 - 4)$

    e $[(-8 \times (-9)] \div (-2 - 4)$ **f** $\dfrac{-56}{-7} \times (-4 + 8)$

6 Evaluate each expression using your calculator.

    a $\{400 - [-15 + (-35)]\} \div 90$ **b** $-252 \div [-10 + (-8 \times 3) + 6]$

    c $80 - [-24 \times (-3 + 5)] + (-6)$ **d** $[280 + (-15 + 45)] - 24$

7 Evaluate each expression.

    a $(-4)^2 \times (-3) + 12$ **b** $-54 \div (-9) \times 2.6$

    c $22.8 - (-6)^2 + 8.2$ **d** $28.7 - (-4) \times 16.6$

    e $\sqrt{25} \times (-8) + (-3)^2$ **f** $23.8 - 12.3 \div (-3)$

    g $\sqrt{49} + (-9)^2 \times (-3)$ **h** $\dfrac{42 - 16}{39 \div 3}$

    i $\dfrac{28.4 - 12.2}{48 \div (-12)}$

8 A tennis club charges \$125 per year membership, \$9.50 per visit on weekdays and \$22 per visit on weekends. How much will it cost Peter over a year if he plays twice during the week and once every second weekend?

EXAMPLE 5

a   Change $\dfrac{8}{5}$ to a mixed numeral.   b   Change $2\dfrac{3}{4}$ to an improper fraction.

SOLUTION

a $\dfrac{8}{5} = 8 \div 5$

 $= 1$ remainder 3

 $= 1\dfrac{3}{5}$

✱ Write the remainder in the numerator of the fraction

b $2\dfrac{3}{4} = \dfrac{2 \times 4}{4} + \dfrac{3}{4}$

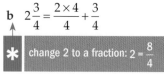 ✱ change 2 to a fraction: $2 = \dfrac{8}{4}$

 $= \dfrac{8}{4} + \dfrac{3}{4}$

 $= \dfrac{11}{4}$

OR on a calculator, enter: 8 [a^b/c] 5 [=]

OR on a calculator, enter: 2 [a^b/c] 3 [a^b/c] 4 [=] [d/c] *

*The improper fraction key [d/c] or [▪ ◱] may require the [SHIFT] or [2ndF] key.

To find equivalent fractions, multiply the numerator and denominator by the same number.

EXAMPLE 6

Find 3 equivalent fractions for $\dfrac{2}{3}$.

SOLUTION

$\dfrac{2}{3} = \dfrac{2 \times 2}{3 \times 2} = \dfrac{4}{6}$ $\dfrac{2}{3} = \dfrac{2 \times 3}{3 \times 3} = \dfrac{6}{9}$ $\dfrac{2}{3} = \dfrac{2 \times 4}{3 \times 4} = \dfrac{8}{12}$

To simplify a fraction, divide both the numerator and the denominator by the same number, preferably the HCF (highest common factor).
Continue dividing until there are no common factors left except for 1 in the numerator and the denominator.

EXAMPLE 7

Simplify each fraction.

a $\dfrac{14}{16}$ b $\dfrac{18}{24}$

SOLUTION

a $\dfrac{14}{16} = \dfrac{14 \div 2}{16 \div 2} = \dfrac{7}{8}$

b $\dfrac{18}{24} = \dfrac{18 \div 6}{24 \div 6} = \dfrac{3}{4}$

OR on a calculator, enter: 14 [a^b/c] 16 [=]

OR enter: 18 [a^b/c] 24 [=]

1 Change $\dfrac{7}{4}$ to a mixed numeral. Select the correct answer **A, B, C** or **D**.

   A $1\dfrac{3}{4}$ **B** $1\dfrac{1}{4}$ **C** $2\dfrac{1}{4}$ **D** $1\dfrac{3}{7}$

2 Change $3\dfrac{2}{3}$ to an improper fraction. Select **A, B, C** or **D**.

   A $\dfrac{18}{3}$ **B** $\dfrac{8}{3}$ **C** $\dfrac{5}{3}$ **D** $\dfrac{11}{3}$

3 Which of these fractions are improper fractions?

   $\dfrac{5}{4}$ $2\dfrac{1}{3}$ $\dfrac{3}{5}$ $\dfrac{12}{5}$ $1\dfrac{2}{3}$ $\dfrac{9}{8}$ $\dfrac{8}{5}$ $3\dfrac{3}{4}$

4 Change each improper fraction to a mixed numeral in simplified form.

   a $\dfrac{5}{4}$ **b** $\dfrac{7}{2}$ **c** $\dfrac{8}{5}$ **d** $\dfrac{13}{8}$

   e $\dfrac{7}{3}$ **f** $\dfrac{15}{7}$ **g** $\dfrac{15}{6}$ **h** $\dfrac{11}{9}$

5 Change each mixed numeral to an improper fraction.

   a $1\dfrac{1}{2}$ **b** $3\dfrac{1}{3}$ **c** $4\dfrac{2}{5}$ **d** $1\dfrac{7}{8}$

   e $2\dfrac{3}{5}$ **f** $3\dfrac{1}{6}$ **g** $1\dfrac{8}{9}$ **h** $2\dfrac{4}{5}$

6 Bobbie bought 5 cakes and cut them into 6 slices each. Her friends ate $3\dfrac{5}{6}$ of the cakes. How many pieces did they eat?

7 Is each statement true or false?

   a $\dfrac{8}{10} = \dfrac{4}{5}$ **b** $\dfrac{12}{15} = \dfrac{4}{3}$ **c** $\dfrac{12}{20} = \dfrac{3}{5}$ **d** $\dfrac{20}{30} = \dfrac{2}{3}$ **e** $\dfrac{18}{42} = \dfrac{6}{7}$

8 Simplify each fraction and check your answers on a calculator.

   a $\dfrac{8}{20}$ **b** $\dfrac{10}{25}$ **c** $\dfrac{16}{20}$ **d** $\dfrac{12}{16}$ **e** $\dfrac{10}{24}$

   f $\dfrac{24}{60}$ **g** $\dfrac{42}{64}$ **h** $\dfrac{36}{50}$ **i** $\dfrac{28}{56}$ **j** $\dfrac{72}{96}$

9 On his birthday, Zac bought a huge slab of pizza cut into 18 pieces. Zac ate 6 pieces and Eliza ate 4 pieces.

   a How many pieces were left for Zac's family?

   b What fraction of the pizza did Zac eat?

   c What fraction did Eliza eat?

EXAMPLE 8

Find:

a $\dfrac{1}{4}$ of 36

b $\dfrac{2}{3}$ of 120

c $\dfrac{4}{5}$ of 180

SOLUTION

a $\dfrac{1}{4}$ of $36 = \dfrac{1}{4} \times 36$

$\quad = 36 \div 4$
$\quad = 9$

b $\dfrac{2}{3}$ of $120 = \dfrac{2}{3} \times 120$

$\quad = 120 \div 3 \times 2$
$\quad = 40 \times 2$
$\quad = 80$

c $\dfrac{4}{5}$ of $180 = \dfrac{4}{5} \times 180$

$\quad = 180 \div 5 \times 4$
$\quad = 36 \times 4$
$\quad = 144$

EXAMPLE 9

Find:

a $\dfrac{1}{3}$ of 2 days

b $\dfrac{3}{5}$ of 1 m

c $\dfrac{3}{4}$ of 0.8 L

SOLUTION

a $\dfrac{1}{3}$ of 2 days $= \dfrac{1}{3} \times 48$ hours

✱ Convert to smaller metric units if needed

$\quad = 16$ hours

b $\dfrac{3}{5}$ of 1 m $= \dfrac{3}{5} \times 100$ cm

$\quad = 60$ cm

c $\dfrac{3}{4}$ of 0.8 L $= \dfrac{3}{4} \times 800$ mL

$\quad = 600$ mL

Shutterstock.com/Elena Mayne

1 Which expression gives $\dfrac{1}{12}$ of 1 day in hours? Select the correct answer **A**, **B**, **C** or **D**.

    A 1 ÷ 12   **B** 60 ÷ 12   **C** 24 ÷ 12   **D** 24 × 12

2 Find $\dfrac{3}{8}$ of 48. Select **A**, **B**, **C** or **D**.

    A 12   **B** 18   **C** 20   **D** 32

3 Is each statement true or false?

    a $\dfrac{1}{3}$ of 21 = 7   **b** $\dfrac{1}{5}$ of 100 = 25   **c** $\dfrac{3}{4}$ of 120 = 90   **d** $\dfrac{5}{8}$ of 24 = 16

4 Find:

    a $\dfrac{1}{5}$ of 200   **b** $\dfrac{1}{7}$ of 56

    c $\dfrac{3}{4}$ of 80   **d** $\dfrac{5}{8}$ of 640

    e $\dfrac{1}{6}$ of 1 hour   **f** $\dfrac{2}{3}$ of 15 m

    g $\dfrac{4}{5}$ of \$100   **h** $\dfrac{3}{8}$ of 32 L

    i $\dfrac{2}{5}$ of 2 m   **j** $\dfrac{3}{5}$ of 1 kg

    k $\dfrac{5}{8}$ of 1 day   **l** $\dfrac{5}{12}$ of \$1080

5 **a** Liam earns \$960 per week and saves $\dfrac{2}{3}$ of it. How much does he save?

    b What fraction of his wage does Liam spend?

6 **a** Mel earns \$840 per week and spends $\dfrac{5}{12}$ of it. How much does she spend?

    b If Mel saves the rest of her wage, what fraction does she save?

7 When Cameron and Olivia made a business profit of \$36 000, Cameron received $\dfrac{1}{4}$ of it, Olivia took $\dfrac{5}{12}$ and the remainder went to shareholders.

    a How much money did Cameron receive?

    b What was Olivia's share of the money?

    c What fraction went to the shareholders?

8 Kane divided his 12-hectare farm into the following areas: $\dfrac{1}{3}$ for cow paddock, $\dfrac{1}{4}$ for vegetable garden and the remainder for his house.

    a What area of land is devoted to the vegetable garden?

    b What area is for the cows?

    c How much area remains for his house?

WORDBANK

ascending order From smallest to largest, to go **up** in size.

descending order From largest to smallest, to go **down** in size.

> The symbol for 'is greater than'

< The symbol for 'is less than'

lowest common multiple (LCM) The smallest multiple of two or more numbers, for example, the LCM of 4 and 6 is 12.

To **order or compare fractions,** we must first write them with the **same denominator,** then compare their **numerators.** A common denominator can be found by multiplying the denominators of both fractions together or by using the **lowest common multiple (LCM)** of the denominators.

EXAMPLE 10

Copy and complete each statement using the sign < or >.

a $\dfrac{5}{8}\underline{}\dfrac{2}{3}$ b $\dfrac{5}{7}\underline{}\dfrac{3}{5}$ c $\dfrac{2}{3}\underline{}\dfrac{8}{9}$

SOLUTION

a For denominators 8 and 3, a common denominator is $8 \times 3 = 24$.

$\dfrac{5}{8}=\dfrac{15}{24}$ and $\dfrac{2}{3}=\dfrac{16}{24}$

So $\dfrac{5}{8}<\dfrac{2}{3}$ ⟵ because 15 < 16 ('15 is less than 16')

b For denominators 7 and 5, a common denominator is $7 \times 5 = 35$.

$\dfrac{5}{7}=\dfrac{25}{35}$ and $\dfrac{3}{5}=\dfrac{21}{35}$

So $\dfrac{5}{7}>\dfrac{3}{5}$ ⟵ because 25 > 21 ('25 is more than 21')

c For denominators 3 and 9, the lowest common multiple (LCM) is 9.

$\dfrac{2}{3}=\dfrac{6}{9}$ and $\dfrac{8}{9}=\dfrac{8}{9}$

So $\dfrac{2}{3}<\dfrac{8}{9}$

EXAMPLE 11

List these fractions in ascending order. $\dfrac{3}{4},\dfrac{7}{12},\dfrac{1}{2},\dfrac{2}{3}$.

SOLUTION

Make all fractions have the same denominator, 12.

$\dfrac{3}{4}=\dfrac{9}{12},\dfrac{7}{12}=\dfrac{7}{12},\dfrac{1}{2}=\dfrac{6}{12},\dfrac{2}{3}=\dfrac{8}{12}$

In ascending order: $\dfrac{6}{12},\dfrac{7}{12},\dfrac{8}{12},\dfrac{9}{12}$ which is: $\dfrac{1}{2},\dfrac{7}{12},\dfrac{2}{3},\dfrac{3}{4}$

ISBN 9780170351027

1   Which fraction is the smallest? Select the correct answer **A**, **B**, **C** or **D**.

   A $\dfrac{1}{3}$ **B** $\dfrac{1}{2}$ **C** $\dfrac{1}{5}$ **D** $\dfrac{1}{4}$

2   Which fraction in question **1** is the largest? Select **A**, **B**, **C** or **D**.

3   Is each statement true or false?

   a $\dfrac{1}{3} < \dfrac{2}{5}$ **b** $\dfrac{3}{4} > \dfrac{2}{3}$ **c** $\dfrac{9}{12} < \dfrac{3}{5}$ **d** $\dfrac{2}{7} < \dfrac{3}{14}$

   e $\dfrac{4}{5} > \dfrac{3}{8}$ **f** $\dfrac{5}{8} = \dfrac{18}{24}$ **g** $\dfrac{5}{6} = \dfrac{10}{12}$ **h** $\dfrac{5}{8} < \dfrac{5}{6}$

4   List these fractions in ascending order: $\dfrac{6}{7}, \dfrac{1}{5}, \dfrac{7}{10}, \dfrac{11}{70}$.

5   List these fractions in descending order: $\dfrac{3}{8}, \dfrac{1}{12}, \dfrac{5}{6}, \dfrac{3}{4}$.

6   Copy and complete each statement using the sign <, > or =.

   a $\dfrac{1}{2} \underline{} \dfrac{3}{4}$ **b** $\dfrac{2}{3} \underline{} \dfrac{5}{8}$ **c** $\dfrac{5}{7} \underline{} \dfrac{15}{21}$ **d** $\dfrac{3}{8} \underline{} \dfrac{5}{4}$

   e $\dfrac{1}{5} \underline{} \dfrac{1}{6}$ **f** $\dfrac{3}{5} \underline{} \dfrac{7}{8}$ **g** $\dfrac{2}{3} \underline{} \dfrac{3}{2}$ **h** $\dfrac{3}{4} \underline{} \dfrac{12}{16}$

7   Place each fraction in the correct position on the number line.

   $\dfrac{1}{2}$ $\dfrac{3}{4}$ $\dfrac{1}{3}$ $\dfrac{5}{6}$ $\dfrac{7}{12}$

   ⟵────┬──┬──┬──┬──┬──┬──┬──┬──┬──┬──┬──⟶
       0   1

8   Rena had a box of chocolates and gave her daughter $\dfrac{1}{4}$ of them and her son $\dfrac{2}{9}$ of them.
   Did her son or daughter have the greater share?

iStockphoto/Iisegagne

WORDBANK

lowest common denominator (LCD) The lowest common multiple (LCM) of the denominators of two or more fractions, for example, the LCD of $\frac{1}{4}$ and $\frac{5}{6}$ is 12.

TO ADD OR SUBTRACT FRACTIONS:

■ convert them (if needed) so that they have the same denominator, preferably the lowest common denominator (LCD)

■ add or subtract the numerators and keep the denominators the same.

EXAMPLE 12

Evaluate each expression.

a $\frac{1}{4} + \frac{1}{3}$

b $\frac{2}{5} + \frac{3}{15}$

c $4 + 1\frac{2}{3}$

d $\frac{3}{4} - \frac{3}{8}$

e $\frac{11}{12} - \frac{2}{3}$

f $6 - \frac{3}{5}$

SOLUTION

a $\frac{1}{4} + \frac{1}{3} = \frac{3}{12} + \frac{4}{12}$

$*$ LCD of 4 and 3 is 12

$= \frac{7}{12}$

OR on a calculator, enter:

1 $\boxed{a^{b}/_{c}}$ 4 $\boxed{+}$ 1 $\boxed{a^{b}/_{c}}$ 3 $\boxed{=}$

b $\frac{2}{5} + \frac{3}{15} = \frac{6}{15} + \frac{3}{15}$

$*$ LCD of 5 and 15 is 15

$= \frac{9}{15}$

$= \frac{3}{5}$

c $4 + 1\frac{2}{3} = 4 + 1 + \frac{2}{3}$

$= 5\frac{2}{3}$

d $\frac{3}{4} - \frac{3}{8} = \frac{6}{8} - \frac{3}{8}$

$= \frac{3}{8}$

e $\frac{11}{12} - \frac{2}{3} = \frac{11}{12} - \frac{8}{12}$

$= \frac{3}{12}$

$= \frac{1}{4}$

f $6 - \frac{3}{5} = 5 + 1 - \frac{3}{5}$

$*$ split 6 into 5 + 1

$= 5 + \frac{2}{5}$

$= 5\frac{2}{5}$

1   What is the lowest common denominator of $\frac{2}{5}$ and $\frac{1}{4}$? Select the correct answer **A**, **B**, **C** or **D**.

   A 5   **B** 4   **C** 40   **D** 20

2   What is the LCD of $\frac{1}{3}$ and $\frac{5}{12}$? Select **A**, **B**, **C** or **D**.

   A 3   **B** 12   **C** 4   **D** 36

3   Copy and complete each line of working.

   a $\frac{1}{4} = \frac{}{12}$ b $\frac{2}{3} = \frac{}{12}$ c $\frac{1}{4} + \frac{2}{3} = \frac{}{12} + \frac{}{12}$

$$= \frac{}{12}$$

   d $\frac{3}{4} = \frac{}{12}$ e $\frac{1}{3} = \frac{}{12}$ f $\frac{3}{4} - \frac{1}{3} = \frac{}{12} - \frac{}{12}$

$$= \frac{}{12}$$

4   Evaluate each sum, then use your calculator to check your answers.

   a $\frac{1}{2} + \frac{1}{3}$ b $\frac{1}{5} + \frac{1}{4}$ c $\frac{1}{3} + \frac{2}{5}$ d $\frac{3}{8} + \frac{1}{4}$

   e $\frac{3}{5} + \frac{5}{8}$ f $\frac{2}{7} + \frac{3}{4}$ g $3 + \frac{4}{5}$ h $\frac{7}{8} + \frac{3}{4}$

   i $2\frac{2}{3} + 3$ j $1\frac{1}{4} + \frac{1}{3}$

5   Evaluate each difference, then use your calculator to check your answers.

   a $\frac{3}{4} - \frac{1}{2}$ b $\frac{4}{5} - \frac{1}{3}$ c $\frac{5}{8} - \frac{1}{4}$ d $\frac{7}{9} - \frac{1}{3}$

   e $\frac{5}{12} - \frac{1}{4}$ f $\frac{7}{8} - \frac{2}{3}$ g $4 - \frac{1}{3}$ h $\frac{11}{12} - \frac{5}{6}$

   i $2\frac{1}{2} - \frac{1}{4}$ j $3\frac{3}{4} - \frac{2}{3}$

6   Use your calculator to evaluate each expression.

   a $2\frac{1}{4} + 3\frac{2}{5}$ b $4\frac{3}{8} - 2\frac{1}{3}$ c $5\frac{5}{8} + 2\frac{3}{5}$ d $6\frac{1}{8} - 2\frac{3}{4}$

   e $1\frac{7}{12} + 4\frac{1}{2} - 3\frac{1}{3}$

7   The ingredients of a Christmas cake are $\frac{1}{4}$ mixed fruit, $\frac{1}{3}$ flour, $\frac{1}{12}$ eggs and the rest is made up of milk and sugar.

   a   What fraction of the ingredients are mixed fruit, flour and eggs?

   b   Find the fraction left for milk and sugar.

   c   If I cut the cake into 12 pieces and Amy eats $\frac{1}{6}$ of it, how many pieces does she eat?

WORDBANK

reciprocal The reciprocal of a fraction is the fraction turned upside down. For example, the reciprocal of $\frac{2}{3}$ is $\frac{3}{2}$.

TO MULTIPLY FRACTIONS:

■ simplify numerators with denominators (if possible) by dividing by a common factor
■ then **multiply numerators** and **multiply denominators**

EXAMPLE 13

Evaluate each product.

a $\quad \frac{3}{4} \times \frac{3}{5}$

b $\quad \frac{4}{5} \times \frac{10}{11}$

c $\quad \frac{3}{8} \times \frac{6}{15}$

SOLUTION

a $\quad \dfrac{3}{4} \times \dfrac{3}{5} = \dfrac{3 \times 3}{4 \times 5}$

$\qquad = \dfrac{9}{20}$

b $\quad \dfrac{4}{5} \times \dfrac{10}{11} = \dfrac{4 \times \cancel{10}^{2}}{\cancel{5} \times 11}$

$\qquad = \dfrac{8}{11}$

c $\quad \dfrac{3}{8} \times \dfrac{6}{15} = \dfrac{\cancel{3}^{1} \times \cancel{6}^{3}}{\cancel{8}^{4} \times \cancel{15}^{5}}$

$\qquad = \dfrac{3}{20}$

OR on a calculator, enter:

3 $\boxed{a^{b}/_{c}}$ 4 $\boxed{\times}$ 3 $\boxed{a^{b}/_{c}}$ 5 $\boxed{=}$

To divide by a fraction $\dfrac{a}{b}$, multiply by its reciprocal $\dfrac{b}{a}$.

EXAMPLE 14

Evaluate each quotient.

a $\quad \frac{3}{4} \div \frac{7}{8}$

b $\quad \frac{5}{6} \div \frac{20}{3}$

c $\quad \frac{4}{5} \div 8$

SOLUTION

a $\quad \dfrac{3}{4} \div \dfrac{7}{8} = \dfrac{3}{4} \times \dfrac{8}{7}$

$\qquad = \dfrac{3}{\cancel{4}^{1}} \times \dfrac{\cancel{8}^{2}}{7}$

$\qquad = \dfrac{6}{7}$

b $\quad \dfrac{5}{6} \div \dfrac{20}{3} = \dfrac{5}{6} \times \dfrac{3}{20}$

$\qquad = \dfrac{\cancel{5}^{1}}{\cancel{6}^{2}} \times \dfrac{\cancel{3}^{1}}{\cancel{20}^{4}}$

$\qquad = \dfrac{1}{8}$

c $\quad \dfrac{4}{5} \div 8 = \dfrac{4}{5} \times \dfrac{1}{8}$

✱ The reciprocal of 8 is $\dfrac{1}{8}$

$\qquad = \dfrac{\cancel{4}^{1}}{5} \times \dfrac{1}{\cancel{8}^{2}}$

$\qquad = \dfrac{1}{10}$

OR on a calculator, enter:

3 $\boxed{a^{b}/_{c}}$ 4 $\boxed{\div}$ 7 $\boxed{a^{b}/_{c}}$ 8 $\boxed{=}$

ISBN 9780170351027

1 Evaluate the product $\frac{1}{4} \times \frac{2}{3}$. Select the correct answer **A**, **B**, **C** or **D**.

    A $\frac{3}{7}$ **B** $\frac{3}{12}$ **C** $\frac{1}{6}$ **D** $\frac{2}{7}$

2 Evaluate the quotient $\frac{2}{3} \div \frac{4}{3}$. Select **A**, **B**, **C** or **D**.

    A 2   **B** $\frac{8}{9}$ **C** 1   **D** $\frac{1}{2}$

3 Copy and complete this statement.

$$\frac{2}{3} \times \frac{1}{5} = \frac{\underline{\quad} \times 1}{3 \times \underline{\quad}} = \underline{\quad}$$

4 Evaluate each product, then check your answers on a calculator.

    a $\frac{1}{4} \times \frac{1}{3}$ **b** $\frac{2}{3} \times \frac{3}{4}$ **c** $\frac{3}{8} \times \frac{1}{5}$ **d** $\frac{7}{12} \times \frac{3}{8}$

    e $\frac{4}{5} \times \frac{5}{8}$ **f** $\frac{5}{6} \times \frac{4}{9}$ **g** $\frac{4}{9} \times 18$ **h** $\frac{7}{8} \times \frac{4}{21}$

    i $\frac{3}{5} \times 12$ **j** $\frac{7}{15} \times \frac{5}{14}$

5 Copy and complete this statement.

$$\frac{2}{3} \div \frac{1}{5} = \frac{\underline{\quad}}{3} \times \frac{\underline{\quad}}{1} = \frac{\underline{\quad}}{\underline{\quad}} = 3\frac{\underline{\quad}}{\underline{\quad}}$$

6 Evaluate each quotient, then check your answers on a calculator.

    a $\frac{1}{4} \div \frac{1}{2}$ **b** $\frac{2}{3} \div \frac{4}{3}$ **c** $\frac{3}{8} \div \frac{1}{5}$ **d** $\frac{7}{12} \div \frac{14}{8}$

    e $\frac{4}{5} \div \frac{16}{5}$ **f** $\frac{5}{6} \div \frac{4}{9}$ **g** $\frac{4}{9} \div 8$ **h** $\frac{7}{8} \div \frac{14}{24}$

    i $\frac{3}{5} \div 12$ **j** $\frac{6}{15} \div \frac{12}{14}$

7 Use your calculator to evaluate each expression.

    a $2 \times 3\frac{2}{5}$ **b** $3\frac{3}{8} \times 2\frac{1}{2}$ **c** $5\frac{3}{8} \div 2$ **d** $5\frac{1}{8} \div 2\frac{3}{4}$

    e $2\frac{1}{4} \times 3 \times 1\frac{3}{8}$

8 Bianca was running a business and worked out that the wages were $\frac{1}{4}$ of the takings for a week and the profit was $\frac{1}{2}$ of the wages.

    a What fraction of the takings was the profit?

    b If the takings for one week were \$8400, how much were the wages?

    c What was the profit for this week?

WORD PUZZLE

List the letters in order for the clues below to spell out a phrase relating to this topic.

The first $\dfrac{2}{7}$ of FREEWAY

The first $\dfrac{2}{5}$ of ACTOR

The middle $\dfrac{1}{7}$ of DAYTIME

The middle $\dfrac{1}{5}$ of POINT

The last $\dfrac{2}{11}$ of EXCLAMATION

The first $\dfrac{2}{9}$ of FRIGHTENS

The last $\dfrac{1}{3}$ of SPOKEN

The last $\dfrac{1}{4}$ of JAZZ

The first $\dfrac{1}{6}$ of YELLOW

Part A General topics

Calculators are not allowed.

1 Round 2.189 to two decimal places.

2 Find the reciprocal of $1\frac{1}{4}$.

3 Simplify $8x + 4y + 7x - 2y$

4 Find the area of the triangle.

6 cm

8 cm

5 Convert $\frac{36}{8}$ to a mixed numeral.

6 Evaluate $8.4 - 4.25$

7 Find 15% of $40.

8 Simplify $\frac{9x^2y}{3y}$

9 Write these integers in descending order:

$$-5 \quad 7 \quad 0 \quad 6 \quad 3 \quad -2$$

10 Find the mean of these scores correct to 1 decimal place:

$$12 \quad 10 \quad 9 \quad 15 \quad 17 \quad 13$$

Part B Integers and fractions

Calculators are allowed.

3-01 Adding and subtracting integers

11 Evaluate each sum.

    a $-5 + 3$ **b** $-7 + (-4)$ **c** $-12 + 6 + (-8)$

12 Evaluate $-7 - (-3)$. Select the correct answer **A**, **B**, **C** or **D**.

    A -4 **B** -10 **C** 10 **D** 4

3-02 Multiplying and dividing integers

13 Evaluate each expression.

    a $-4 \times (-7)$ **b** $56 \div (-8)$ **c** $9 \times (-3) \times (-2)^2$

3-03 Order of operations

14 For $28 - 6 \times (-4) + (-12)$, which operation would you perform first? Select **A**, **B**, **C** or **D**.

    A $28 - 6$ **B** $6 \times (-4)$ **C** $-4 + (-12)$ **D** $28 + (-12)$

15 Evaluate each expression.

    a $24 + (-3) \times 6$ **b** $56 - (-8) \div 2$ **c** $9 \times (-3) + (-20) \div 5$

3-04 Fractions

16 Change each fraction to a mixed numeral.

    a $\frac{5}{4}$ **b** $\frac{13}{8}$ **c** $\frac{54}{20}$

17 Simplify each fraction.

    a $\frac{8}{12}$ **b** $\frac{20}{25}$ **c** $\frac{24}{72}$

3-05 Fraction of a quantity

18 Find:

a $\frac{2}{5}$ of 75 b $\frac{3}{4}$ of \$120 c $\frac{3}{8}$ of 4 m

3-06 Ordering fractions

19 Is each statement true or false?

a $\frac{1}{3} < \frac{2}{5}$ b $\frac{3}{4} > \frac{7}{8}$ c $\frac{3}{5} = \frac{15}{25}$

3-07 Adding and subtracting fractions

20 Evaluate each expression.

a $\frac{3}{5} + \frac{2}{3}$ b $\frac{5}{6} - \frac{3}{8}$ c $2\frac{1}{2} + \frac{1}{4}$ d $5 - \frac{7}{8}$

3-08 Multiplying and dividing fractions

21 Evaluate each expression.

a $\frac{3}{5} \times \frac{2}{9}$ b $\frac{5}{6} \div \frac{7}{18}$ c $2\frac{1}{2} \times \frac{15}{4}$ d $5 \div \frac{15}{8}$

22 Georgia made a cake and cut it in half. She then cut each piece into quarters.

a How many pieces did she cut the cake into?

b Amber ate $3\frac{1}{2}$ pieces. What fraction of the cake did Amber eat?

ALGEBRA

4

IN THIS CHAPTER YOU WILL:

- convert worded descriptions into algebraic expressions
- substitute into algebraic expressions
- add, subtract, multiply and divide algebraic terms
- expand algebraic expressions
- factorise algebraic terms and expressions

Shutterstock.com/Sergey Nivens

WORDBANK

pronumeral or variable A letter of the alphabet such as a, b, c, x or y that represents a number.

algebraic expression A relationship of pronumerals, numbers and operations written in algebraic form. For example, $2x - 3, 5a + 2b, x^2 - 3x$

Mathematical word	Meaning
sum, total, increase or plus	add (+)
difference or decrease	subtract (−)
product	multiply (×)
quotient	divide (÷)
twice or double	multiply by 2 (× 2)
triple	multiply by three (× 3)
square	multiply by itself (x^2)

EXAMPLE 1

Write each statement as an algebraic expression.

a   Twice x b   The product of a and b c   Increase b by 5

d   The quotient of m and n e   Triple the sum of a and b f m squared

SOLUTION

a $2x$ b ab c $b + 5$

d $\dfrac{m}{n}$ e $3(a + b)$ f m^2

EXAMPLE 2

a   Write an algebraic expression for the area of the rectangle.

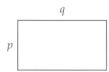

b   Farmer Bob has x pigs. Write an expression for the total number of legs on the pigs.

SOLUTION

a   Area of the rectangle $= p \times q$
                          $= pq$

b   Number of legs $= 4 \times x$
                    $= 4x$

1   To find twice the sum of two numbers, which is the correct order? Select the correct answer **A**, **B**, **C** or **D**.

  A   multiply by 2 and then add
  B   divide by 2 and then add
  C   add and then multiply by 2
  D   add and then divide by 2

2   How many mm in x cm? Select **A**, **B**, **C** or **D**.

  A $10x$
  B $100x$
  C $\dfrac{x}{10}$
  D $\dfrac{x}{100}$

3   Write the algebraic expression for:

  a   the sum of a and b
  b   the difference between a and b
  c   the product of a and b
  d   the quotient of a and b

4   Is each statement true or false?

  a   The product of 3, a, b and d is $3abd$.
  b   The sum of m, 6, n and 12 is $m + n - 18$.
  c   The difference between $7w$ and $5v$ is $35w - v$.
  d   The quotient of $8a$ and $5b$ is $\dfrac{8a}{5b}$.

5   Write each statement as an algebraic expression.

  a   triple a less 8
  b   12 decreased by twice b
  c   double the sum of m and n
  d   triple the difference of x and y
  e   twice n plus 7
  f w squared less 5
  g   triple y plus w
  h   double the sum of x and y
  i   the square of the product of m and n
  j   24 less g squared
  k   120 plus triple d
  l m squared plus double n
  m   12 less triple b
  n   increase h by twice j

6   Write an algebraic expression for the cost of:

  a   5 apples at a cents each
  b   6 pizzas at $q each
  c d hours work at $15 per hour
  d   1 share when 3 winners win $r
  e k pencils at $w each
  f   1 bag if 12 bags cost $d

7   Write an algebraic expression for the perimeter of each shape.

  a   **b**   **c**

8   Write down an algebraic expression for the area of each figure in question **7**.

9   If Eva is m years old and Jake is twice her age, find in terms of m:

  a   Jake's age
  b   Eva's age 4 years ago
  c   Jake's age in 10 years time
  d   Eva's age in 8 years time

10 Old McDonald had a farm with x cows and y chickens. Find the total number of:

  a   cows' legs
  b   chicken legs
  c   legs
  d   heads

WORDBANK

substitution Replacing a variable with a given value in an algebraic expression to find the value of it.

formula An algebraic rule using variables and an equal sign, such as $A = \frac{1}{2}bh$ for the area of a triangle.

EXAMPLE 3

If $x = 4$ and $y = -3$, evaluate each algebraic expression.

a $3x + 2y$ **b** $5xy$ **c** $3xy^2$

SOLUTION

a $3x + 2y = 3 \times 4 + 2 \times (-3)$
$= 6$

b $5xy = 5 \times 4 \times (-3)$
$= -60$

c $3xy^2 = 3 \times 4 \times (-3)^2$
$= 108$

EXAMPLE 4

The formula $A = \frac{1}{2}bh$ gives the area of a triangle with a base b and a height h.

Find the area of the triangle using this formula.

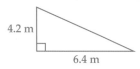

4.2 m

6.4 m

SOLUTION

$A = \frac{1}{2}bh$

$= \frac{1}{2} \times 6.4 \times 4.2$ ⟵ base $= b = 6.4$ and height $= h = 4.2$

$= 13.44 \text{ m}^2$

EXAMPLE 5

Complete this table of values using the formula $y = 2x + 1$.

x	-1	0	1	2
y				

SOLUTION

x	-1	0	1	2
y	-1	1	3	5

$2 \times (-1) + 1$ $\quad 2 \times 0 + 1$ $\quad 2 \times 1 + 1$ $\quad 2 \times 2 + 1$

1 Evaluate $6bc$ if $b = 3$ and $c = -4$. Select the correct answer **A**, **B**, **C** or **D**.

    A -72 **B** 72 **C** -24 **D** 84

2 Evaluate $5m^2$ if $m = -3$. Select **A**, **B**, **C** or **D**.

    A -45 **B** 15 **C** -15 **D** 45

3 If $x = 3$ and $y = -2$, are the following true or false?

    a $x + y = -1$ **b** $2x - y = 10$ **c** $3xy = 18$ **d** $\dfrac{6x}{y} = -9$

4 Find the value of each algebraic expression if $a = 2$ and $b = -5$.

    a $2ab$ **b** $2a - 3b$ **c** ab^2 **d** $4b - 2a$

    e $3(2a + b)$ **f** $6a^2$ **g** $12b - a$ **h** $3ab^2$

5 Copy and complete this table.

	$3m - n$	$2m^2$	$\dfrac{4m}{n}$	$3(2m + n)$	$6mn^2$
$m = 2, n = 4$					
$m = -1, n = 3$					
$m = 5, n = -2$					
$m = -1, n = -4$					
$m = 3, n = -2$					

6 Evaluate each formula.

    a $P = 2l + 2w$ where $l = 3.6$ and $w = 2.7$ **b** $A = \dfrac{1}{2}bh$ where $b = 12$ and $h = 8$

    c $A = lw$ where $l = 8.2$ and $w = 5.6$ **d** $V = lwh$ where $l = 8.6$, $w = 5$ and $h = 2.1$

    e $c = \sqrt{a^2 + b^2}$ where $a = 5$ and $b = 12$ **f** $f = \dfrac{uv}{u + v}$ where $u = 4$ and $v = 6$

7 Copy and complete each table of values.

    a $y = 3x - 1$ **b** $y = 12 - x$

x	-1	0	1	2
y				

x	-1	0	1	2
y				

    c $y = 2x^2$ **d** $y = \dfrac{2x}{3} + 1$

x	-2	0	2	4
y				

x	-3	0	3	6
y				

8 The formula for the volume of a cylinder is $V = \pi r^2 h$, where r is the radius of its base and h is its perpendicular height. Find, correct to one decimal place, the volume of a cylinder with radius 4 cm and height 6.2 cm.

Adding and subtracting terms

WORDBANK

like terms Terms with exactly the same variables, for example, $2a$ and $3a$, $7y$ and $-3y$, $4ab$ and $-2ba$.

- ■ Only like terms can be added and subtracted.
- ▓ The sign in front of a term belongs to it.
- ■ x means $1x$, the '1' does not need to be written.

EXAMPLE 6

Simplify:

a $6x - 4x$ **b** $-3y - y$ **c** $3x^2 - 4x + 2x^2$ **d** $5a \ 2b + a - 6b$

SOLUTION

a $6x - 4x = 2x$ **b** $-3y - y = -4y$

Group together like terms, including the signs in front of them.

c $\underline{3x^2} - 4x \underline{+ 2x^2} = 3x^2 + 2x^2 - 4x$ **d** $\underline{5a} - 2b \underline{+ a} - 6b = 5a + a - 2b - 6b$
$\qquad\qquad\qquad = 5x^2 - 4x$ $= 6a - 8b$

EXAMPLE 7

Write an algebraic expression for the perimeter of this rectangle.

$2x + 6$

$3x$

SOLUTION

Perimeter $= 2x + 6 + 3x + 2x + 6 + 3x$ ◄——— Perimeter is the sum of the four sides
$\qquad\qquad = 2x + 3x + 2x + 3x + 6 + 6$ ◄——— Group like terms
$\qquad\qquad = 10x + 12$

iStockphoto/Chiyacat

Developmental Mathematics Book 3

ISBN 9780170351027

1 Which are the like terms in the expression $2x - 3y - x - 3$? Select the correct answer **A, B, C** or **D**.

   A $2x - 3y$ **B** $2x - x$ **C** $3y - x$ **D** $-3y - 3$

2 Simplify $2x - 3y - x - 3$. Select **A, B, C** or **D**.

   A $x + 3y - 3$ **B** $3x - 6y$ **C** $x - 6y$ **D** $x - 3y - 3$

3 In each list, write down the like terms.

   a $3x, 3y, 2x, 2$ **b** $4a, a, 4b, 4$

   c $5w, 5v, -2w, 5$ **d** $2ab, 3ba, -2a, -2b$

   e $2a^2, 2ab, 2b^2, 2ba$ **f** $-6m, 5n, 12mn, 18m$

4 Is each statement true or false?

   a $3x - 5x = 2x$ **b** $7y + y = 8y$ **c** $8a + 8 - 5a = 8 - 3a$

5 Simplify each algebraic expression.

   a $5a - 2a$ **b** $4x + 2x$

   c $8w - 2w - 3$ **d** $12m - 5m$

   e $-6y + 8y$ **f** $12n - 14n + 7$

   g $7a - 9a$ **h** $12m + 4m - m$

   i $16t - 20t + t$

6 Copy and complete:

   a $4m - 2n - 6m + n = 4m - \underline{\quad} - 2n + \underline{\quad}$
                                $= -2m - \underline{\quad}$

   b $8x + 5y - 2y - x = 8x - \underline{\quad} + 5y - \underline{\quad}$
                            $= 7x + \underline{\quad}$

7 Simplify each algebraic expression.

   a $2x - y + 5x$ **b** $-6y - 2x + 8x$

   c $4m - 2n + 7m - n$ **d** $6m - n + 4m$

   e $4x - 2y - x + 6y$ **f** $12a - 4b - b - 3a$

   g $14r - 3s + r - 5s$ **h** $2ab - ba + 6 - 3ab$

   i $2xy - 7y - 3yx + 7$

8 Simplify an algebraic expression for the perimeter of each shape.

   a   **b**   **c**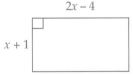

9 Explain in words why $5ab - 2b + 4ab^2$ cannot be simplified.

TO MULTIPLY ALGEBRAIC TERMS:

✳ They do not have to be like terms!

- ▪ multiply the numbers first
- ▪ then multiply the variables
- ▪ write the variables in alphabetical order

EXAMPLE 8

Simplify each expression.

a $4m \times 2n$ **b** $-6a \times 3b$ **c** $2mn \times (-8n)$

d $2x^2 \times 3x$ **e** $5a \times (-2ab)$ **f** $-x^2y \times (-5yx)^2$

SOLUTION

a $4m \times 2n = 4 \times 2 \times m \times n$
$$= 8mn$$

b $-6a \times 3b = -6 \times 3 \times a \times b$
$$= -18ab$$

c $2mn \times (-8n) = 2 \times (-8) \times mn \times n$
$$= -16mnn$$
$$= -16mn^2$$

d $2x^2 \times 3x = 6x^2 \times x$
$$= 6x^3$$

e $5a \times (-2ab) = -10a \times ab$
$$= -10a^2b$$

f $-x^2y \times (-5yx)^2 = -x^2y \times 25y^2x^2$
$$= -25x^4y^3$$

EXAMPLE 9

Find an algebraic expression for the area of each shape.

a
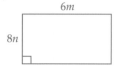
6m
8n

b
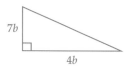
7b
4b

SOLUTION

a Area of a rectangle $= 6m \times 8n$
$$= 48mn$$

b Area of triangle $= \dfrac{1}{2} \times 4b \times 7b$
$$= 14b^2$$

Alamy/Candy/AppleRed Images

1   Simplify $5x \times 3yz$. Select the correct answer **A**, **B**, **C** or **D**.

   A $53xyz$ **B** $15yz$ **C** $15xz$ **D** $15xyz$

2   Simplify $-8a \times (-3ab)$. Select **A**, **B**, **C** or **D**.

   A $-24ab$ **B** $24a^2b$ **C** $-24a^2b$ **D** $24ab^2$

3   When multiplying algebraic terms, is each statement true or false?

   a Only like terms can be multiplied.

   b A positive multiplied by a negative is negative.

   c Multiply all the numbers together.

   d Multiply all the variables together.

4   Copy and complete each line of working.

   a $5a \times 3b = 5 \times \_\_ \times a \times \_\_$
   $= 15 \_\_$

   b $-6x \times 8y = -6 \times \_\_ \times x \times \_\_$
   $= -48 \_\_$

   c $5a^2 \times 2a = 5 \times \_\_ \times a^2 \times \_\_$
   $= 10 \_\_$

   d $-4mn \times (-7m^2) = -4 \times \_\_ \times mn \times \_\_$
   $= 28 \_\_$

5   Simplify each algebraic expression.

   a $2 \times 8m$

   b $3x \times 4y$

   c $-6m \times 3n$

   d $5a \times (-4b)$

   e $4 \times (-3cd)$

   f $-2w \times (-5v)$

   g $-9ab \times (-3)$

   h $-2xy \times (-3x)$

   i $-4m \times 3n \times (-5)$

   j $-6ab \times (-4ba)$

   k $4a^2 \times 3a$

   l $5m \times 3m^3$

   m $-2w^2 \times (-3w)$

   n $5xy \times (-4y^2)$

   o $8ab^2 \times (-3b)$

   p $-6xy^2 \times (-3x^2)$

   q $-4r^3 \times 8rs^2$

   r $6 \times 4w^2 \times (-3w)$

   s $-7m \times (-3) \times m^3$

   t $4ab \times (-3b) \times (-2a)$

6   Find a simplified algebraic expression for the area of each shape.

   a

   $3x$

   b
   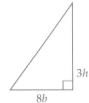
   $3h$
   $8b$

   c

   $3w$
   $9v$

ISBN 9780170351027

TO DIVIDE ALGEBRAIC TERMS:

✳ They do not have to be like terms!

- divide the numbers first
- then divide the variables
- write the answer in fraction form

EXAMPLE 10

Simplify each quotient.

a $25xy \div 5x$ **b** $24mn^2 \div (-8n)$ **c** $\dfrac{-12ab}{10ac}$

SOLUTION

Write each quotient as a fraction and simplify.

a $25xy \div 5x = \dfrac{25xy}{5x}$
$= 5y$

b $24mn^2 \div -8n = \dfrac{24mn^2}{-8n}$
$= -3mn$

c $\dfrac{-12ab}{10ac} = -\dfrac{6b}{5c}$

TO SIMPLIFY ALGEBRAIC EXPRESSIONS INVOLVING MIXED OPERATIONS:

- first **multiply** and **divide** from left to right
- then **add** and **subtract** from left to right

EXAMPLE 11

Simplify each expression using the order of operations.

a $12m - 6m \div 3$ **b** $2a \times 4 - a \times 5$ **c** $24xy \div (-8x) \times 4y$

SOLUTION

a $12m - 6m \div 3 = 12m - 2m$
$= 10m$

b $2a \times 4 - a \times 5 = 8a - 5a$
$= 3a$

 ✳ Do × and ÷ first, from left to right

c $24xy \div (-8x) \times 4y = -3y \times 4y$
$= -12y^2$

Alamy/Cultura Creative

1 Simplify $15ab \div (-3b)$. Select the correct answer **A, B, C** or **D**.

   A $-5a$ **B** $5ab$ **C** $5a$ **D** $-5ab$

2 Simplify $-4a + (-12ab) \div 2b$. Select **A, B, C** or **D**.

   A $10ab$ **B** $2a$ **C** $-10ab$ **D** $-10a$

3 Is each statement true or false?

   a $\dfrac{-15x}{-5} = -3x$ b $\dfrac{12ab}{-4b} = -3a$

   c $\dfrac{-18xy}{6yz} = \dfrac{-3x}{z}$ d $\dfrac{8uv}{-12vw} = \dfrac{2u}{3w}$

4 Simplify each quotient.

   a $8a \div 2$ b $9b \div (-3)$

   c $-12d \div (-6)$ d $\dfrac{15ab}{-5}$

   e $24bc \div (-8)$ f $50xy \div (-25)$

   g $\dfrac{-36mn}{9n}$ h $-27ab \div (-3bc)$

   i $\dfrac{24vw}{-6wx}$ j $18r^2 \div (-9r)$

   k $150xy \div (-25y^2)$ l $\dfrac{-84w^2}{7wy}$

5 Copy and complete each statement.

   a $\dfrac{32x^3y}{8xy^2} = \dfrac{4x^\square}{y}$ b $\dfrac{15a^4b}{3ab^3} = \dfrac{5a^\square}{b^\square}$ c $\dfrac{28m^4n^2}{35mn^4} = \dfrac{4m^\square}{5n^\square}$

6 Simplify each expression.

   a $\dfrac{18ab^5}{9b}$ b $\dfrac{24x^3y^2}{6xy}$ c $\dfrac{-25m^3n}{5mn^4}$

7 Which operation would you perform first for each expression below?

   a $12 - 3x \times 4$ b $6 + 14a \div 7$ c $-24ab \div (-6b) \times 2a$

8 Simplify each expression in question 7.

9 Simplify each expression using the order of operations.

   a $18 + 2x \times 6$ b $18m - 15m \div 3$

   c $3a \times 4 - a \times 6$ d $48xy \div (-8x) \times 3y$

   e $18mn \times 2m \div 6n$ f $24 - (-8xy) \div 2x$

   g $4m \times 3 - m \times (-6)$ h $64y \div 4 - 8 \times 2y$

WORDBANK

expand To remove the brackets or grouping symbols in an algebraic expression.

When mentally multiplying 14 by 11, we know that

$$14 \times 11 = 14 \times (10 + 1)$$
$$= 14 \times 10 + 14 \times 1$$
$$= 140 + 14$$
$$= 154$$

This idea can be used to expand algebraic expressions.

> **To expand an algebraic expression with brackets,** the term outside the brackets must be multiplied by every term inside the brackets.
>
> $a (b + c) = ab + ac$

EXAMPLE 12

Expand each algebraic expression.

a $2(x + 4)$ **b** $5(2a - 1)$ **c** $-3 (5x - 2)$ **d** $a(4a + 3)$

SOLUTION

a $2(x + 4) = 2 \times x + 2 \times 4$
$= 2x + 8$

b $5(2a - 1) = 5 \times 2a + 5 \times (-1)$
$= 10a - 5$

c $-3 (5x - 2) = -3 \times 5x + (-3) \times (-2)$
$= -15x + 6$

d $a(4a + 3) = a \times 4a + a \times 3$
$= 4a^2 + 3a$

EXAMPLE 13

Expand and simplify each expression.

a $4(2x - 3) - 2(x + 5)$ **b** $a(3a + 4) - (2a - 5)$

SOLUTION

a $4(2x - 3) - 2(x + 5) = 4 \times 2x + 4 \times (-3) + (-2) \times x + (-2) \times 5$ ⟵ Expand brackets first
$= 8x - 12 - 2x - 10$
$= 6x - 22$

b $a(3a + 4) - (2a - 5) = a(3a + 4) - 1(2a - 5)$ ⟵ $-(2a - 5)$ means $-1(2a - 5)$
$= a \times 3a + a \times 4 + (-1) \times 2a + (-1) \times (-5)$
$= 3a^2 + 4a - 2a + 5$
$= 3a^2 + 2a + 5$

ISBN 9780170351027

1 Expand $3(2a - 5)$. Select the correct answer **A, B, C** or **D**.

    A $6a - 5$ **B** $6a - 15$ **C** $2a - 15$ **D** $32a - 5$

2 Expand $2m(3m - 4)$. Select **A, B, C** or **D**.

    A $6m - 4$ **B** $6m^2 - 4$ **C** $6m^2 - 8m$ **D** $2m^2 - 8m$

3 Copy and complete each expansion.

    a $4(3x - 2) = 4 \times 3x + 4 \times \underline{\quad}$
                      $= 12x + \underline{\quad}$

    b $-6(2a - 1) = -6 \times 2a + (-6) \times \underline{\quad}$
                      $= -12a + \underline{\quad}$

    c $a(3a + 2) = a \times 3a + a \times \underline{\quad}$
                      $= 3a^2 + \underline{\quad}$

    d $-2x(3x - 1) = -2x \times 3x + (-2x) \times \underline{\quad}$
                      $= -6x^2 + \underline{\quad}$

4 Expand each expression.

    a $5(x + 2)$ **b** $3(a - 4)$ **c** $4(2a - 1)$ **d** $8(3x + 2)$

    e $6(4 - 2a)$ **f** $a(2a - 3)$ **g** $x(3x + 4)$ **h** $2m(m - 6)$

    i $3w(2w + 5)$ **j** $4a(3a - 6)$ **k** $-2(4a + 3)$ **l** $-5(8 - 3m)$

    m $-(2x - 8)$ **n** $-(3a + 9)$ **o** $-6(2w - 8)$

5 **a** Check that $5(x + 2) = 5x + 10$ by substituting $x = 7$ into both sides of the equation and testing whether the values are equal.

    b Substitute another value of x into both sides and check whether the values are still equal.

6 Copy and complete each expansion.

    a $3(2y - 1) + 5y = 6y - \underline{\quad} + \underline{\quad}$
                   $= \underline{\quad} - \underline{\quad}$

    b $12 + 4(2m - 1) = 12 + \underline{\quad} - \underline{\quad}$
                     $= \underline{\quad} + \underline{\quad}$

    c $2(3x - 4) - 3(x + 4) = 6x - \underline{\quad} - 3x - \underline{\quad}$
                        $= \underline{\quad} - \underline{\quad}$

    d $5(2a - 2) + 7(3a + 1) = 10a - \underline{\quad} + \underline{\quad} + 7$
                        $= \underline{\quad} - \underline{\quad}$

7 Expand and simplify each expression.

    a $2(x - 4) + 3(x - 5)$ **b** $5(x + 3) - 2(x + 3)$

    c $3(a + 6) - 2(a - 8)$ **d** $2(3x - 4) - 3(x - 6)$

    e $4(2x + 3) - 3(x + 7)$ **f** $5(2a + 6) - 3(2a - 8)$

    g $4(2x + 3) - 5(2x - 4)$ **h** $3(2x + 1) - 5(x + 8)$

    i $3(3a + 4) - 5(2a - 6)$

8 Expand and simplify $7(2w - 4v) - 3(2v + 5w)$.

WORDBANK

highest common factor (HCF) The largest number or algebraic term that divides into two or more numbers or algebraic terms evenly.

The HCF of 10 and 15 is 5 because 5 is the largest factor of both 10 and 15.
The HCF of $4ab$ and $6b$ is $2b$ because $2b$ is the largest factor of both $4ab$ and $6b$.

EXAMPLE 14

Write 6 factors of $12mn^2$.

SOLUTION

The factors of 12 are 1, 2, 3, 4, 6 and 12.

Some factors of mn^2 are m, n, mn and n^2.

All of these factors are also factors of $12mn^2$.

So choose any 6 of the above or products of them, such as 3, 12, $3m$, $2mn$, $3n^2$ and $12n$.

EXAMPLE 15

Find the highest common factor of 24 and 32.

SOLUTION

The factors of 24 are <u>1</u>, <u>2</u>, 3, <u>4</u>, 6, <u>8</u>, 12 and 24. ⟵ Listing the factors of 24 and 32 separately.

The factors of 32 are <u>1, 2, 4, 8</u>, 16 and 32. ⟵ The common factors are underlined.

The highest common factor of 24 and 32 is 8. ⟵ The largest underlined factor.

TO FIND THE HIGHEST COMMON FACTOR (HCF) OF ALGEBRAIC TERMS:

▨ find the HCF of the numbers
▨ find the HCF of the variables
▨ multiply the HCFs together

EXAMPLE 16

Find the highest common factor of:

a $8a$ and $12abc$ **b** $15st$ and $25st^2$

SOLUTION

a The HCF of 8 and 12 is 4.   **b** The HCF of 15 and 25 is 5.

✱ | Find the HCFs of the numbers and variables separately

The HCF of a and ab is a.   The HCF of st and st^2 is st.
The HCF of $8a$ and $12abc$ is $4 \times a = 4a$ The HCF of $15st$ and $25st^2$ is $5 \times st = 5st$

ISBN 9780170351027

1   What are the factors of 6? Select the correct answer **A, B, C** or **D**.

    A 1, 2, 6   **B** 2, 3, 6   **C** 1, 2, 3, 6   **D** 2, 3, 4, 6

2   What is the highest common factor of 6 and 15? Select **A, B, C** or **D**.

    A 5   **B** 3   **C** 6   **D** 15

3   Copy and complete each statement.

    a The factors of 8 are: 1, 2, __, __

       The factors of 20 are: 1, 2, __, __, __, __

       The HCF of 8 and 20 is ___

    b The factors of 16 are: 1, 2, 4, __, __

       The factors of 40 are: 1, 2, __, __, __, __, __, __

       The HCF of 16 and 40 is ___.

4   Find the highest common factor of each pair of numbers.

    a 6 and 10   **b** 15 and 18

    c 12 and 20   **d** 15 and 45

    e 16 and 30   **f** 18 and 24

    g 16 and 60   **h** 36 and 45

5   Write 4 factors of each algebraic term.

    a $9ab$ **b** $14mn$ **c** $8x^2$ **d** $12bc$

    e $4m^2$ **f** $16b^2$ **g** $20a^3$ **h** $6m^4$

6   Copy and complete the solution to find the highest common factor of $16xy^2$ and $32xy$.

    The HCF of 16 and 32 is: ___

    The HCF of xy^2 and xy is: ___

    The HCF of $16xy^2$ and $32xy$ is: ___ × ___ = ___

7   Find the highest common factor for each pair of terms.

    a $8a$ and $12a$ **b** $6m$ and $8n$

    c $10b$ and 6 **d** $9u$ and $15v$

    e $16m$ and $8mn$ **f** $12a$ and $4ab$

    g $8w$ and $6w^2$ **h** $14m^2$ and $18mn$

    i abc and bc^2 **j** mno and mn^2

    k $15rs$ and $12s^2$ **l** $18uv$ and $24uvw$

8   Find the highest common factor of $12abc$, $16bc^2$ and $20a^2bc$.

WORDBANK

factorise To insert brackets or grouping symbols in an algebraic expression by taking out the highest common factor (HCF); factorising is the opposite of expanding.

$$5(2x - 1) = 10x - 5$$

expanding →

factorising ←

TO FACTORISE AN ALGEBRAIC EXPRESSION:
- find the HCF of all the terms and write it in front of the brackets
- divide each term by the HCF and write the answers inside the brackets

$$ab + ac = a(b + c)$$

To check the answer is correct, expand it.

EXAMPLE 17

Factorise each expression.

a $3x - 9$ **b** $12ab + 16bc$ **c** $24m^2 - 8mn$ **d** $9xy^2 - 15x^2y$

SOLUTION

a $3x - 9 = 3(x - 3)$ ← HCF = 3   **b** $12ab + 16bc = 4b(3a + 4c)$ ← HCF = $4b$

Check each answer by expanding: Check each answer by expanding:

$3(x - 3) = 3x - 9$ $4b(3a + 4c) = 12ab + 16bc$

c $24m^2 - 8mn = 8m(3m - n)$ ← HCF = $8m$ **d** $9xy^2 - 15x^2y = 3xy(3y - 5x)$ ← HCF = $3xy$

Check each answer by expanding: Check each answer by expanding:

$8m(3m - n) = 24m^2 - 8mn$ $3xy(3y - 5x) = 9xy^2 - 15x^2y$

Alamy/ONOKY-Photononstop

Shutterstock.com/Air Images

1 Factorise $15x - 25$. Select the correct answer **A**, **B**, **C** or **D**.

  A $5(x - 3)$ **B** $3(5x - 5)$

  C $5(3x - 3)$ **D** $5(3x - 5)$

2 Factorise $20mn + 28m^2$. Select **A**, **B**, **C** or **D**.

  A $4(5mn + 7m^2)$ **B** $4m(5n + 7m)$

  C $4n(5m + 7n)$ **D** $7m(4n + 4m)$

3 Copy and complete each factorisation.

  a $3a + 3b = 3(\underline{\quad})$ **b** $5x - 10y = 5(\underline{\quad})$

  c $6m + 18n = 6(\underline{\quad})$ **d** $4b^2 - 4 = 4(\underline{\quad})$

  e $9a + 3b = \underline{\quad}(3a + b)$ **f** $15x - 12y = \underline{\quad}(5x - \underline{\quad})$

  g $7u + 14v = \underline{\quad}(u + \underline{\quad})$ **h** $8n^2 - 2 = \underline{\quad}(\underline{\quad} - 1)$

4 Factorise each expression, and check your answer by expanding.

  a $2a + 12$ **b** $6p - 18$

  c $5n + 20$ **d** $8y - 64$

  e $12a + 60$ **f** $16x - 48$

  g $18a + 54$ **h** $14c - 28$

  i $15a + 75$ **j** $15x - 5y$

  k $a^2 + ab$ **l** $xy + yz$

  m $mn + n^2$ **n** $9p - 27q$

  o $xyz - 3xy$

5 Copy and complete each factorisation.

  a $3a^2 + 6ab = 3a(\underline{\quad})$ **b** $15xy - 10y^2 = 5y(\underline{\quad})$

  c $16mno + 24noq = 8no(\underline{\quad})$ **d** $48b^2 - 16bc = \underline{\quad}(3b - \underline{\quad})$

  e $27ab + 18bc = \underline{\quad}(3a + \underline{\quad})$ **f** $25x^2 - 15xy = \underline{\quad}(5x - \underline{\quad})$

6 Factorise each expression, and check your answer by expanding.

  a $4ap - 12pq$ **b** $16rs + 12st$

  c $28xy - 42y^2$ **d** $64abc - 32bcd$

  e $8m^2 + 48mn$ **f** $22b^2 - 11ab$

  g $36fg + 18g^2$ **h** $xyz - 7uvx$

  i $14a^2 b - 21b^2 c$ **j** $54w^2 + 9uvw$

  k $24r^2t - 16tv^2$ **l** $75mn^2 + 50nm^2$

7 Factorise $42a^2bc - 14b^2c + 28abc$, then expand your answer to check that it is correct.

TO FACTORISE AN ALGEBRAIC EXPRESSION WITH A NEGATIVE FIRST TERM:
- ▨ include the negative sign when finding the HCF and write it in front of the brackets
- ▨ divide each term by the HCF and write the answers inside the brackets

To check that the answer is correct, expand it.

EXAMPLE 18

Factorise each expression.

a $-12x - 36$ b $-8mn - 16m$ c $-27ab + 18bc^2$

SOLUTION

a $-12x - 36 = -12(x + 3)$ ←—— HCF $= -12$

Check by expanding:

$-12(x + 3) = -12x - 36$

b $-8mn - 16m = -8m(n + 2)$ ←—— HCF $= -8m$

Check by expanding:

$-8m(n + 2) = -8mn - 16m$

c $-27ab + 18bc^2 = -9b(3a - 2c^2)$ ←—— HCF $= -9b$

Check by expanding:

$-9b(3a - 2c^2) = -27ab + 18bc^2$

✱ Note that the signs in each bracket are different from the sign in the question as you are dividing each term by a negative number.

Shutterstock.com/bikeriderlondon

1 Factorise $-12x - 24$. Select the correct answer **A**, **B**, **C** or **D**.

   A $12(x - 2)$ **B** $-12(x - 2)$

   C $-12(x + 2)$ **D** $6(-2x + 4)$

2 Factorise $-15mn + 30m^2$. Select **A**, **B**, **C** or **D**.

   A $-5(3mn + 6m^2)$ **B** $-15m(n - 2m)$

   C $15m(n - 2m)$ **D** $-15m(n + 2m)$

3 Find the negative HCF of:

   a -5 and $-10a$ **b** -4 and $-12x$

   c $-6m$ and $-18n$ **d** $-5w$ and -30

   e $-15b$ and $-10a$ **f** $-4xy$ and $16x$

   g $-24mn$ and $18n$ **h** $-5uv$ and $35vw$

4 Copy and complete each factorisation.

   a $-4a - 4b = -4(\underline{})$ **b** $-5x - 10y = -5(\underline{})$

   c $-16m - 8n = -8(\underline{})$ **d** $-4b^2 - 4 = -4(\underline{})$

   e $-9a + 3b = \underline{}(3a - b)$ **f** $-35x - 7y = \underline{}(5x + \underline{})$

   g $-7u + 14v = \underline{}(u - \underline{})$ **h** $-8n^2 + 2 = \underline{}(\underline{} - 1)$

5 Factorise each expression, then check your answer by expanding.

   a $-2a - 18$ **b** $-6p - 24$ **c** $-4n - 20$

   d $-7y - 49$ **e** $-12a - 60$ **f** $-16x + 32$

   g $-18a + 36$ **h** $-4c - 28$ **i** $-15a + 75$

   j $-25x - 50y$ **k** $-a^2 - ac$ **l** $-xy + yz$

   m $-mn + n^2$ **n** $-3p - 27q$ **o** $-xyz + 5xy$

6 Copy and complete each factorisation.

   a $-3a^2 - 6ab = -3a(\underline{})$ **b** $-5xy - 20y^2 = -5y(\underline{})$

   c $-8mno + 24noq = -8no(\underline{})$ **d** $-32b^2 - 16bc = \underline{}(2b + \underline{})$

   e $-9ab + 18bc = \underline{}(a - \underline{})$ **f** $-50x^2 - 25xy = \underline{}(2x + \underline{})$

7 Factorise each expression, then check your answer by expanding.

   a $-3ap - 12pq$ **b** $-8rs - 12st$

   c $-7xy - 42y^2$ **d** $-16abc + 32bcd$

   e $-6m^2 + 48mn$ **f** $-33b^2 - 11ab$

   g $-6fg + 18g^2$ **h** $-xyz - 7uvx$

   i $-7a^2b - 21b^2c$ **j** $-6w^2 + 9uvw$

   k $-8r^2t - 16tv^2$ **l** $-15mn^2 + 45nm^2$

8 Factorise $-24a^2bc - 12b^2c + 36abc$, then check your answer by expanding.

MIX AND MATCH

Match each question on the left with an algebraic expression on the right.

1	Twice the sum of x and 5	**A**	$9a - 5b$
2	The product of $4a$ and $2b$	**B**	$6mn$
3	The value of $2x - y$ if $x = 2$ and $y = -1$	**C**	$2(x + 5)$
4	The value of $3ab$ if $a = -2$ and $b = 6$	**E**	$9a + 6a^2$
5	$4a - 2b + 5a - 3b$	**G**	5
6	$5x - 7y - 6x + 3y$	**I**	$10ab$
7	$2m \times 3n$	**L**	$-x - 4y$
8	$5a \times 2b$	**N**	$4a - 13b$
9	Expand $4(2x \ \ 3)$	**R**	-36
10	Expand $3a(3 + 2a)$	**S**	$8ab$
11	Expand $5(2a - 3b) - 2(3a - b)$	**T**	$8x - 12$

In your book, use the matched question numbers and answer letters to decode this phrase relating to this chapter:

5-7-2-9-4-5-1-9   5-6-3-10-7-4-5   5-11-9-8-1-2

Getty Images/SW Productions

Part A General topics

Calculators are not allowed.

1   Evaluate $15 \times 3 + 15 \times 7$

2   How many degrees are there in a
    straight angle?

3   Simplify $\dfrac{9ab}{3a}$.

4   Find the volume of the prism.

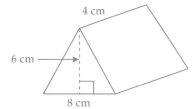

5   Find 25% of $120.

6   Evaluate $11.25 + $23.90

7   Find the median of
    6, 3, 2, 6, 5, 4, 6.

8   How many axes of symmetry has a
    rectangle?

9   Arrange these decimals in ascending
    order:
    9.95, 9.909, 9.91, 9.9

10 Complete the number pattern:
    1, 3, 9, ___, 81, ___

Part B Algebra

Calculators are allowed.

4-01 From words to algebraic expressions

11 Convert each description into an algebraic expression.
    a   Twice a, minus b.
    b   The product of $3x$ and $4y$.
    c   Triple the sum of m and n.
    d   The quotient of $7a$ and $3b$.

12 What is the perimeter of a square of side $3b$? Select the correct answer **A, B, C** or **D**.
    A $3b^2$ **B** $6b$ **C** $9b^2$ **D** $12b$

4-02 Substitution

13 If $a = -3$ and $b = 0.2$, evaluate each expression.
    a $4ab$ b $20 - 3a + b$ c $6a^2 - b$

4-03 Adding and subtracting terms

14 Simplify each expression.
    a $-3w + 2v - 8w$
    b $12ab - 3b - b + ba$
    c $20mn - m^2 - 4nm$

4-04 Multiplying terms

15 Simplify each expression.

    a $-2a \times (-3b)$ **b** $-4x^2 \times 6xy$ **c** $-12uv \times (-6vw)$

16 Find a simplified algebraic expression for the area of this rectangle.

4-05 Dividing terms

17 Simplify each expression.

    a $24ab \div (-6b)$ **b** $\dfrac{48x^2}{8x}$ **c** $-14bc \div 28ac$

4-06 Expanding expressions

18 Is each statement true or false?

    a $4(2a - 1) = 8a - 1$

    b $-6(2x + 4) = -12x - 24$

    c $3a(a - 2) = 3a^2 - 2a$

19 Expand and simplify each expression.

    a $3(2m - 4) - (m + 2)$

    b $a(3a + 1) - 2a(3a + 4)$

4-07 Factorising algebraic terms

20 Find the highest common factor for each pair of terms.

    a $9m$ and $18mn$ **b** $8x^2$ and $12xy$ **c** $48m^2n$ and $16\ mn^2$

4-08 Factorising expressions

21 Factorise each expression.

    a $6a - 12b$ **b** $14x - 16xy$

    c $6u^2 - 28uv$ **d** $28ab^2 - 7a^2b$

4-09 Factorising with negative terms

22 Factorise each expression.

    a $-6m - 24n$ **b** $-24x + 16xy$

    c $-15u^2 - 45uvw$ **d** $-48cb^2 - 8a^2b$

TRIGONOMETRY

5

IN THIS CHAPTER YOU WILL:

- label the sides of a right-angled triangle: opposite, adjacent, hypotenuse
- learn the trigonometric ratios for right-angled triangles: sine (sin), cosine (cos), tangent (tan)
- investigate the tangent ratio of angles by measuring right-angled triangles
- use the calculator to evaluate trigonometric ratios and expressions
- use trigonometric ratios to find unknown sides in right-angled triangles
- use trigonometric ratios to find unknown angles in right-angled triangles

Shutterstock.com/Luis Louro

WORDBANK

trigonometry The study of the measurement of sides and angles in triangles.

hypotenuse The longest side of a right-angled triangle, the side opposite the right angle.

opposite side The side facing a given angle in a right-angled triangle.

adjacent side The side next to a given angle in a triangle leading to the right angle.

For angle *C* in this right-angled triangle, *AB* is the opposite side, *BC* is the adjacent side and *AC* is the hypotenuse.

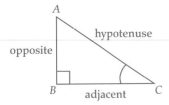

EXAMPLE 1

State the length of the hypotenuse, opposite and adjacent sides for the marked angle in each right-angled triangle.

a

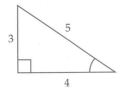

b

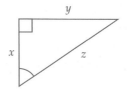

SOLUTION

	Hypotenuse	Opposite side	Adjacent side
a	5	3	4
b	z	y	x

When **labelling triangles**:
- ◾ use capital letters for angles: *P*, *Q*, *R*
- ◾ use small letters for sides: *p*, *q*, *r*
- ◾ use the same letter for sides and angles opposite each other

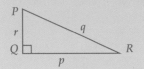

side *p* is opposite angle *P*
side *q* is opposite angle *Q*
side *r* is opposite angle *R*

1   In a right-angled triangle, what is the hypotenuse? Select the correct answer **A**, **B**, **C** or **D**.

   A the shortest side   **B** the side opposite the marked angle

   C the longest side   **D** the side adjacent to the marked angle

2   Which phrase describes the side b in a right-angled triangle? Select **A**, **B**, **C** or **D**.

   A opposite the right angle   **B** opposite angle B

   C adjacent to angle B **D** opposite the hypotenuse

3   Copy and complete each statement for any $\triangle PQR$.

   a Side p is opposite angle ___   **b** Side q is opposite angle ___

   c Angle R is opposite side ___   **d** Angle P is opposite side ___

4   Copy each triangle and complete the labelling of each side and angle.

   a   **b**   **c**

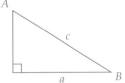 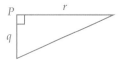

5   For each triangle in question **4**, state the length of the hypotenuse.

6   For each triangle, state the length of the hypotenuse, opposite and adjacent sides in relation to the marked angle.

   a   **b**

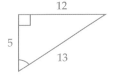

   c   **d**

   e   **f**

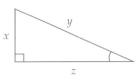

7   Draw a right-angled triangle and label all sides and angles using your choice of letters.

WORDBANK

trigonometric ratio A ratio such as sine, cosine and tangent that compares two sides of a right-angled triangle.

theta, θ A Greek letter often used to represent angles

There are three basic trigonometric ratios for right-angled triangles: **sine**, **cosine** and **tangent**, which are abbreviated as **sin**, **cos** and **tan** respectively.

$$\sin\theta = \frac{\text{opposite}}{\text{hypotenuse}} \qquad \cos\theta = \frac{\text{adjacent}}{\text{hypotenuse}} \qquad \tan\theta = \frac{\text{opposite}}{\text{adjacent}}$$

The following phrase may help you to remember the rules:

Super   Old   Heroes   Can't   Always   Hide   Their   Own   Age

The initials in bold, **SOH CAH TOA**, give the initials of the sin, cos and tan ratios.

EXAMPLE 2

Write the value of sin A, cos A and tan A.

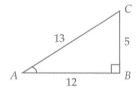

SOLUTION

For angle A, opposite = 5, adjacent = 12 and hypotenuse = 13, so:

$$\sin A = \frac{\text{opposite}}{\text{hypotenuse}} \qquad \cos A = \frac{\text{adjacent}}{\text{hypotenuse}} \qquad \tan A = \frac{\text{opposite}}{\text{adjacent}}$$

$$= \frac{5}{13} \qquad\qquad\qquad = \frac{12}{13} \qquad\qquad\qquad = \frac{5}{12}$$

EXAMPLE 3

Write the expressions for sin P, cos P and tan P.

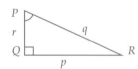

SOLUTION

For angle P, opposite = p, adjacent = r and hypotenuse = q, so:

$$\sin P = \frac{\text{opposite}}{\text{hypotenuse}} \qquad \cos P = \frac{\text{adjacent}}{\text{hypotenuse}} \qquad \tan P = \frac{\text{opposite}}{\text{adjacent}}$$

$$= \frac{p}{q} \qquad\qquad\qquad = \frac{r}{q} \qquad\qquad\qquad = \frac{p}{r}$$

ISBN 9780170351027

1 What is the value of sin B? Select the correct answer **A**, **B**, **C** or **D**.

A $\dfrac{8}{15}$ **B** $\dfrac{15}{17}$ **C** $\dfrac{8}{17}$ **D** $\dfrac{17}{8}$

2 In question **1**, what is the value of sin A? Select **A**, **B**, **C** or **D**.

A $\dfrac{8}{15}$ **B** $\dfrac{15}{17}$ **C** $\dfrac{8}{17}$ **D** $\dfrac{17}{8}$

3 Copy and complete each formula.

a $\sin \theta = \dfrac{\rule{2cm}{0.4pt}}{\text{hypotenuse}}$ **b** $\cos \theta = \dfrac{\rule{2cm}{0.4pt}}{\text{hypotenuse}}$

c $\tan \theta = \dfrac{\text{opposite}}{\rule{2cm}{0.4pt}}$

4 For the triangle XYZ shown, find:

a sin X, cos X, tan X

b sin Y, cos Y and tan Y

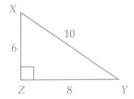

5 For each of the marked angles, find:

a the sin ratio **b** the cos ratio **c** the tan ratio

i

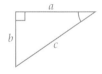

ii

6 Copy and complete each statement for this triangle.

a $\sin F = \dfrac{9}{-}$ **b** $\cos F = \dfrac{}{15}$

c $\tan F = \dfrac{}{12}$ **d** $\sin D = \dfrac{}{15}$

e $\cos D = \dfrac{9}{-}$ **f** $\tan D = \dfrac{}{9}$

7 Draw an appropriate right-angled triangle so that

$\sin A = \dfrac{24}{26}$, $\cos A = \dfrac{10}{26}$ and $\tan A = \dfrac{24}{10}$

EXAMPLE 4

a   Draw a right-angled triangle with an angle of 30°. Make the base of the triangle 6 cm and measure its perpendicular height.

b   Calculate, correct to two decimal places, tan 30° for this triangle.

SOLUTION

a

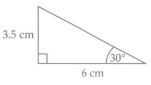

b $\tan 30° = \dfrac{3.5}{6}$ $\dfrac{\text{opposite}}{\text{adjacent}}$

        $= 0.5833\ldots$

        ≈ 0.58

This answer is only approximate because there is always error involved in measurement.

EXAMPLE 5

a   Draw another right-angled triangle with an angle of 30°, but make the base 9 cm and measure its perpendicular height.

b   Calculate, correct to two decimal places, tan 30° for this triangle.

SOLUTION

a

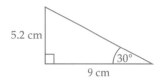

b $\tan 30° = \dfrac{5.2}{9}$ $\dfrac{\text{opposite}}{\text{adjacent}}$

        $= 0.5777\ldots$

        ≈ 0.58

In Examples **5** and **6**, tan 30° has a similar value. In actual fact, the value of tan 30° remains the same no matter what size we draw the triangle. This is the basis of how trigonometry works. Instead of drawing a triangle, we can find tan 30° using a calculator.

Enter tan 30 = to show that tan 30° = 0.577350… ≈ 0.58.

✱ | Make sure that your calculator is in **degrees mode**: a D should appear on the screen.

tan θ has the same value for all right-angled triangles with angle size θ; for example, tan 55° is always 1.4281… This value can be found on the calculator.

1 For each right-angled triangle:

   i draw two large right-angled triangles with the angle size given

   ii for each triangle, measure the opposite side and adjacent side, correct to the nearest 0.1 cm

   iii calculate tan for the angle, correct to two decimal places for both triangles, and see whether the values are similar

   iv use a calculator to calculate tan for the angle, correct to two decimal places, and see whether its value is similar to your measured values in part **iii**

   a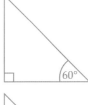

   b

   c

   d

   e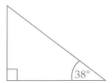

Courtesy Monco Brock Studio

EXAMPLE 6

Evaluate, correct to 3 decimal places where necessary, each trigonometric ratio.

a $\tan 30°$ **b** $\sin 30°$ **c** $\cos 30°$

SOLUTION

a $\tan 30° = 0.57735\ldots$ ←——— On a calculator, enter `tan` 30 `=`

        ≈ 0.577

b $\sin 30° = 0.5$ ←——— On a calculator, enter `sin` 30 `=`

c $\cos 30° = 0.866\,02\ldots$ ←——— On a calculator, enter `cos` 30 `=`

        ≈ 0.866

> To find an unknown angle in a trigonometric ratio, for example, $\sin \theta = 0.73$, use the `SHIFT` or `2ndF` key with the trigonometric key on the calculator.

EXAMPLE 7

Find the size of each angle, correct to the nearest degree.

a $\tan A = 0.6$ **b** $\sin \theta = \dfrac{2}{3}$ **c** $\cos B = 0.1297$

SOLUTION

a $\tan A = 0.6$

        $A = 30.9637\ldots$ ←——— On calculator, enter `SHIFT` `tan` 0.6 `=`

        $A \approx 31°$ ←——— This means $\tan 31° \approx 0.6$

b $\sin \theta = \dfrac{2}{3}$

        $\theta = 41.8103\ldots$ ←——— On calculator, enter `SHIFT` `sin` 2 ⊟ 3 `=`

        $\theta \approx 42°$ ←——— This means $\sin 42° \approx \dfrac{2}{3}$

c $\cos B = 0.1297$

        $B = 82.5477\ldots$ ←——— On calculator, enter `SHIFT` `cos` 0.1297 `=`

        $B \approx 83°$

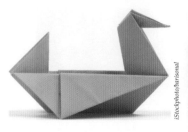

iStockphoto/barisonal

Shutterstock.com/korinoxe

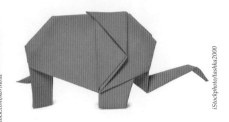

iStockphoto/tashka2000

ISBN 9780170351027

1 Evaluate tan 20° correct to 3 decimal places. Select the correct answer **A**, **B**, **C** or **D**.

    A 0.363   **B** 0.36   **C** 0.364   **D** 0.37

2 Find the size of angle A if sin A = 0.6321. Select **A**, **B**, **C** or **D**.

    A 0.01°   **B** 39°   **C** 40°   **D** 0.011°

3 Evaluate each trigonometric ratio correct to 3 decimal places.

    a tan 48°   **b** sin 48°   **c** cos 48°

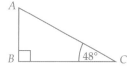

4 **a** What is the size of angle A in question **3**?

    b Evaluate tan A, sin A and cos A.

5 Evaluate each trigonometric ratio correct to 3 decimal places.

    a cos 30°   **b** tan 60°   **c** sin 45°   **d** cos 50°

    e sin 82°   **f** tan 46°   **g** sin 63°   **h** cos 21°

    i tan 75°   **j** cos 14°

6 Evaluate each trigonometric ratio correct to 4 decimal places.

    a sin 38°   **b** tan 64°   **c** cos 84°   **d** tan 25°

    e cos 83°   **f** tan 18°   **g** sin 46°   **h** cos 0°

    i sin 90°   **j** tan 58°

7 Evaluate each trigonometric ratio correct to 1 decimal place.

    a cos 40°   **b** sin 60°   **c** tan 80°   **d** sin 23°

    e tan 29°   **f** cos 23°   **g** tan 74°   **h** sin 32°

    i tan 18°   **j** sin 0°

8 Find the size of each angle correct to the nearest degree.

    a tan A = 0.8   **b** $\sin \theta = \dfrac{3}{4}$ **c** cos B = 0.1286

    d cos A = 0.3   **e** $\tan \theta = \dfrac{1}{5}$ **f** sin B = 0.6319

    g sin A = 0.7   **h** $\cos \theta = \dfrac{3}{8}$ **i** tan B = 1.435

9 Find θ correct to the nearest degree.

    a sin θ = 0.6824   **b** cos θ = 0.5328   **c** tan θ = 0.478   **d** cos θ = 0.5

    e tan θ = 1.624   **f** sin θ = 0.2584   **g** tan θ = 3.6452   **h** sin θ = 0.4531

    i cos θ = 0.8   **j** sin θ = 1   **k** tan θ = 0.98   **l** cos θ = 0

    m tan θ = 0.75   **n** cos θ = 1   **o** sin θ = 0.6   **p** sin θ = 0.5

    q tan θ = 0.9856   **r** cos θ = 0.5

10 Use the sin ratio to find the size of angle B correct to the nearest degree.

The tangent ratio can be used to find unknown sides in right-angled triangles.

$$\tan\theta = \frac{\text{opposite}}{\text{adjacent}}$$

EXAMPLE 8

Find the value of x correct to 2 decimal places.

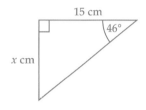

SOLUTION

$\tan 46° = \dfrac{\text{opposite}}{\text{adjacent}}$

$\tan 46° = \dfrac{x}{15}$ ⟵ x is opposite to 46° and 15 is adjacent.

$15 \tan 46° = x$ ⟵ Multiplying both sides by 15

$x = 15.5329\ldots$ ⟵ On calculator, enter: 15 × tan 46 =

$x \approx 15.53$

✱ From the diagram, $x \approx$ 15.53 cm looks reasonable.

EXAMPLE 9

Find the value of a correct to 1 decimal place.

SOLUTION

$\tan 25° = \dfrac{\text{opposite}}{\text{adjacent}}$

$\tan 25° = \dfrac{12}{a}$ ⟵ 12 is opposite 25° and a is adjacent.

$a \tan 25° = 12$ ⟵ Multiplying both sides by a

$a = \dfrac{12}{\tan 25°}$ ⟵ Dividing both sides by tan 25°

$a = 25.7340\ldots$ ⟵ On calculator, enter: 12 ÷ tan 25 =

$a \approx 25.7$

✱ From the diagram, $a \approx$ 25.7 m looks reasonable.

Note that we can jump from the 2nd to 4th lines of working by swapping the places of tan 25° and a:

$$\tan 25° = \frac{12}{a} \to a = \frac{12}{\tan 25°}.$$

1   Which is the correct ratio for this triangle? Select the correct answer **A**, **B**, **C** or **D**.

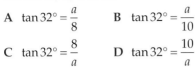

   A $\tan 32° = \dfrac{a}{8}$ **B** $\tan 32° = \dfrac{a}{10}$

   C $\tan 32° = \dfrac{8}{a}$ **D** $\tan 32° = \dfrac{10}{a}$

2   Evaluate $26 \tan 16°$ correct to 2 decimal places. Select **A**, **B**, **C** or **D**.

   A 7.45   **B** 90.67   **C** 7.46   **D** 4.59

3   Copy and complete this solution to find m correct to two decimal places.

   $\tan 28° = \dfrac{m}{5}$

   ___ $\tan 28° = m$

   $m = $ _____

   $m \approx$ _____

4   Find, correct to 2 decimal places, the value of each pronumeral.

   a 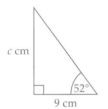   **b** 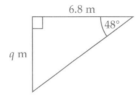   **c**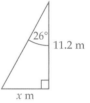

5   Evaluate $\dfrac{16}{\tan 54°}$ correct to 2 decimal places.

6   Find, correct to 1 decimal place, the value of each pronumeral.

   a 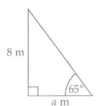   **b** 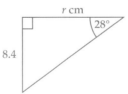   **c**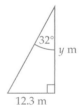

7   Elena is in a boat 25 m from the base of a cliff. She looks up to the top of the cliff through an angle of 29°. How high is the cliff?

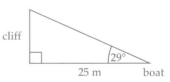

8   A tower is 45 m high. Bianca is standing level with the base of the tower and looks through an angle of 16° to see the top of the tower. How far is Bianca standing from the base of the tower?

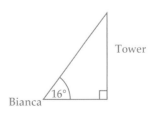

The tangent ratio can be used to find unknown angles in right-angled triangles.

$$\tan \theta = \frac{\text{opposite}}{\text{adjacent}}$$

EXAMPLE 10

Find the size of angle A correct to the nearest degree.

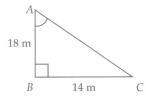

SOLUTION

$$\tan A = \frac{\text{opposite}}{\text{adjacent}}$$

$$\tan A = \frac{14}{18} \quad\longleftarrow\quad \text{14 is opposite to angle } A \text{ and 18 is adjacent.}$$

$$A = 37.8749\ldots \quad\longleftarrow\quad \text{On calculator, enter: } \boxed{\text{SHIFT}} \ \boxed{\text{tan}} \ 14 \ \boxed{\square} \ 18 \ \boxed{=}$$

$$A = 38°$$

✳ From the diagram, $A = 38°$ looks reasonable.

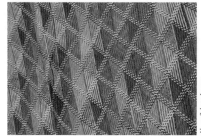

EXAMPLE 11

Find θ correct to the nearest degree.

SOLUTION

$$\tan \theta = \frac{6}{8} \quad\longleftarrow\quad \text{6 is opposite to } \theta \text{ and 8 is adjacent.}$$

$$\theta = 36.8698\ldots \quad\longleftarrow\quad \text{On calculator, enter: } \boxed{\text{SHIFT}} \ \boxed{\text{tan}} \ 6 \ \boxed{\square} \ 8 \ \boxed{=}$$

$$\theta \approx 37°$$

1 Find A if $\tan A = \dfrac{5}{4}$. Select the correct answer **A, B, C** or **D**.

   A 51°   **B** 0.02   **C** 37°   **D** 52°

2 What is $\tan Q$ for this triangle? Select **A, B, C** or **D**.

   A $\dfrac{9}{4}$ **B** $\dfrac{10}{4}$ **C** $\dfrac{4}{10}$ **D** $\dfrac{4}{9}$

3 Calculate the angle, rounding to the nearest degree. Is each statement true or false?

   a $\tan A = \dfrac{3}{5}$ **b** $\tan P = \dfrac{5}{3}$ **c** $\tan Y = \dfrac{1}{12}$ **d** $\tan W = \dfrac{12}{5}$

       $A = 30°$ $P = 59°$ $Y = 5°$ $W = 68°$

4 Find A correct to the nearest degree.

   a   **b**   **c**

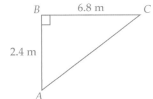
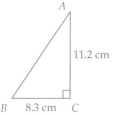

5 Find θ correct to the nearest degree.

   a   **b**   **c**

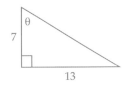
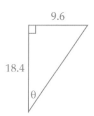
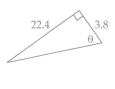

6 Construct $\triangle PQR$, where side p is 5 cm, side q is 12 cm and angle R is 90°. Calculate the size of $\angle P$ correct to the nearest degree, and check your answer by measuring $\angle P$ with your protractor.

7 Ali was rowing out to sea from the base of a cliff that was 28.4 m high. He rowed 12.3 m and then stopped to rest. At what angle must he look up to see the top of the cliff?

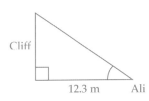

8 Ken was on the roof of a 38 m high building. Sarah was on the ground 18.2 m from the foot of the building. Ken called out to her. Through what angle will she need to look up to see Ken? (Draw a diagram first.)

> **To find an unknown side,** select the correct trigonometric ratio depending on which side and angle is given in the problem:
>
> $$\sin\theta = \frac{\text{opposite}}{\text{hypotenuse}} \qquad \cos\theta = \frac{\text{adjacent}}{\text{hypotenuse}} \qquad \tan\theta = \frac{\text{opposite}}{\text{adjacent}}$$

EXAMPLE 12

a Find b correct to 1 decimal place.

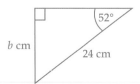

b Find r correct to 2 decimal places.

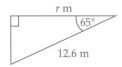

SOLUTION

a b is **opposite** 52° and 24 is the **hypotenuse**, so use **sin**.

$$\sin 52° = \frac{\text{opposite}}{\text{hypotenuse}}$$

$$\sin 52° = \frac{b}{24}$$

$$24 \sin 52° = b$$

$$b = 18.9122\ldots$$

$$b \approx 18.9$$

b r is **adjacent** to 65° and 12.6 is the **hypotenuse**, so use **cos**.

$$\cos 65° = \frac{\text{adjacent}}{\text{hypotenuse}}$$

$$\cos 65° = \frac{r}{12.6}$$

$$12.6 \cos 65° = r$$

$$r = 5.3249\ldots$$

$$r \approx 5.32$$

1   Which trigonometric ratio could be used to find side a?
    Select the correct answer **A, B, C** or **D.**

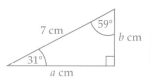

 A sin 31°   **B** sin 59°

 C cos 59°   **D** tan 59°

2   Which trigonometric ratio could be used to find side b in question **1**? Select **A, B, C** or **D.**

 A sin 31°   **B** sin 59°   **C** cos 31°   **D** tan 31°

3   Copy and complete to find x correct to one decimal place.

$$\tan 35° = \frac{}{15.8}$$

15.8 _____ = x

$x =$ _____

$x \approx$ _____

4   Find each pronumeral correct to two decimal places.

 a 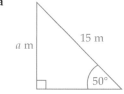 **b**  **c**

 d 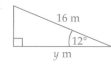 **e** 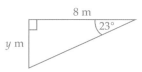 **f**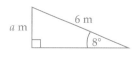

 g  **h**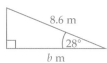

5   Hannah was looking up at a window on the
    third floor of a building through an angle of
    25° to the horizontal. She could see a distance
    of only 5 m in a straight line. How high
    above her was the window, to two decimal
    places?

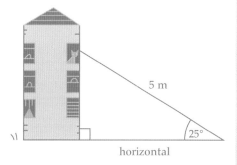

6   Akbar is visiting the Leaning Tower of Pisa
    and is told that it is leaning at 8° to the
    vertical. If the tower is 64 m tall, how high is
    the top of the tower from the ground, to three
    decimal places?

EXAMPLE 13

Find the length of side c in the triangle, correct to 1 decimal place.

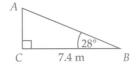

SOLUTION

Side c is opposite the right angle, C.

7.4 is **adjacent**, c is **hypotenuse**, so use **cos**.

$$\cos 28° = \frac{\text{adjacent}}{\text{hypotenuse}}$$

$$\cos 28° = \frac{7.4}{c}$$

$$c = \frac{7.4}{\cos 28°} \quad \longleftarrow \quad c \text{ and } \cos 28° \text{ swap places}$$

$$c = 8.3810\ldots$$

$$c \approx 8.4$$

EXAMPLE 14

Find q correct to 2 decimal places.

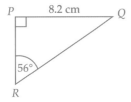

SOLUTION

Side q is opposite angle Q.

q is **adjacent**, 8.2 cm is **opposite**, so use **tan**.

$$\tan 56° = \frac{\text{opposite}}{\text{adjacent}}$$

$$\tan 56° = \frac{8.2}{q}$$

$$q = \frac{8.2}{\tan 56°} \quad \longleftarrow \quad q \text{ and } \tan 56° \text{ swap places.}$$

$$q = 5.5309\ldots$$

$$q \approx 5.53 \text{ cm}$$

1   Which trigonometric ratio can be used to find side c?
    Select the correct answer **A**, **B**, **C** or **D**.

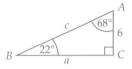

   A $\sin 22°$ **B** $\sin 68°$

   C $\cos 22°$ **D** $\tan 22°$

2   Which trigonometric ratio can be used to find side a in question 1? Select **A**, **B**, **C** or **D**.

   A $\sin 22°$ **B** $\cos 68°$ **C** $\tan 22°$ **D** $\sin 68°$

3   If $\cos 47° = \dfrac{8}{c}$, write a formula for c.

4   Which trigonometric ratio using the marked angle can be used to find each pronumeral?

   a   **b**   **c**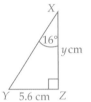

5   Find each pronumeral in question 4 correct to 2 decimal places.

6   Copy and complete to find d correct to one decimal place.

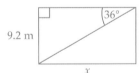

   $\tan 53° = \dfrac{8.3}{d}$

   $d = \dfrac{8.3}{\underline{\quad\quad}}$

   $d = \underline{\quad\quad\quad\quad}$

   $d \approx \underline{\quad\quad}$

7   Find x correct to two decimal places.

   a 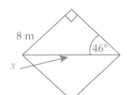   **b**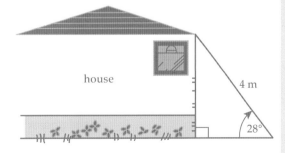

8   Sarila wanted to clean a window
    on the second floor of her house. She
    placed a 4 m ladder at an angle of
    elevation of 28° to the ground. If the
    window is 3.6 m above the ground,
    will the ladder reach it?

EXAMPLE 15

Find A correct to the nearest degree.

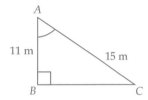

SOLUTION

11 is **adjacent** to angle A and 15 is the **hypotenuse**, so use **cos**.

$$\cos A = \frac{\text{adjacent}}{\text{hypotenuse}}$$

$$\cos A = \frac{11}{15}$$

$A = 42.8334\ldots$ ⟵——————— On a calculator,   11 15 =

$A \approx 43°$

EXAMPLE 16

Find Q correct to the nearest degree.

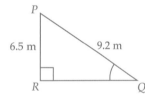

SOLUTION

6.5 is **opposite** angle Q and 9.2 is the **hypotenuse**, so use **sin**.

$$\sin Q = \frac{\text{opposite}}{\text{hypotenuse}}$$

$$\sin Q = \frac{6.5}{9.2}$$

$A = 44.9526\ldots$ ⟵——————— On calculator,   6.5 9.2

$A \approx 45°$

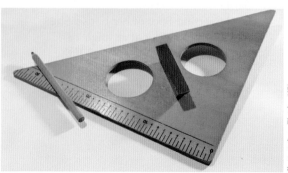

1 Which trigonometric ratio can be used to find angle A?
Select the correct answer **A**, **B**, **C** or **D**.

A cos A **B** tan A

C sin A **D** none of these

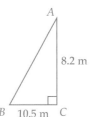

2 Which trigonometric ratio can be used to find angle S?
Select **A**, **B**, **C** or **D**.

A cos S   **B** tan S

C sin S   **D** none of these

3 Copy and complete to find R to the nearest degree.

$\sin R = \dfrac{7}{}$

$R = \underline{\hspace{2cm}}$

$R \approx \underline{\hspace{1.5cm}}$

4 Find A correct to the nearest degree.

a

b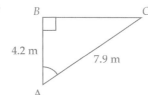

c

5 Find θ correct to the nearest degree.

a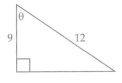

b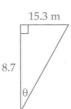

c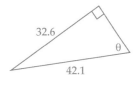

6 **a** Construct $\triangle PQR$, where side p is 7 cm, side r is 16 cm and angle R is 90°. Calculate the size of $\angle Q$ correct to the nearest degree, then check that it is correct by measuring $\angle Q$ with your protractor.

   b Calculate the size of $\angle P$ in $\triangle PQR$.

7 Find the angle between the horizontal and Abbey's roof if it is 2.2 m high in the middle and the distance from the edge to the middle is 5.6 m.

8 Ray placed a 4.5 m long ladder up to a window 3.8 m high. At what angle to the ground (to the nearest degree) should he place the ladder in order for it to reach the window?

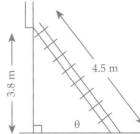

CHAPTER SUMMARY

Copy and complete the chapter summary using words from the list.

degrees	label	opposite	calculator
trigonometry	triangles	hypotenuse	CAH
adjacent	vertices	cosine	small

Trigonometry is the study of measurement in _____. Before we can use _____ we must learn how to _____ triangles. Capital letters are used to name _____ while _____ letters are used to name sides of triangles. The trigonometric ratios are tangent, sine and _____. The tangent ratio is found by putting the opposite side over the _____ side. The sine ratio is _____ side divided by the hypotenuse while the cosine ratio is adjacent side over the_____. The abbreviation to remember is SOH____TOA. A _____ can be used to evaluate the trigonometric ratios. It is important to ensure that your calculator is in _____ mode before evaluating the ratios.

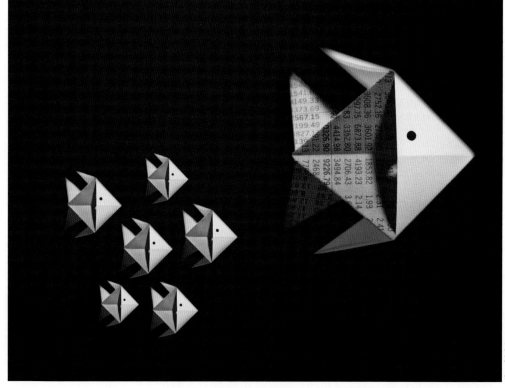

Alamy/Mode Images

Part A General topics

Calculators are not allowed.

1 Decrease $120 by 15%.

2 Expand $-(y + 3)$

3 Solve $\dfrac{x}{6} = 12$

4 Find the perimeter of this shape.

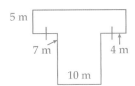

5 Evaluate $\dfrac{2}{3} + \dfrac{1}{2}$

6 Evaluate $\sqrt[3]{64}$

7 Find the median of 20, 14, 4, 17, 12, 11, 18, 13

8 Do the diagonals of a rhombus bisect each other at right angles?

9 Expand $2x(2x - 3)$.

10 What is the probability of rolling a number less than 4 on a die?

Part B Trigonometry

Calculators are allowed.

5-01 The sides of a right-angled triangle

11 For this triangle, name the hypotenuse, opposite and adjacent sides in relation to the marked angle.

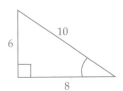

5-02 The trigonometric ratios

12 For the triangle, find the value of cos A. Select the correct answer **A**, **B**, **C** or **D**.

A $\dfrac{8}{17}$ **B** $\dfrac{15}{17}$

C $\dfrac{8}{15}$ **D** $\dfrac{15}{8}$

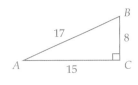

5-04 Trigonometry on a calculator

13 Evaluate each trigonometric ratio correct to 3 decimal places.

  a sin 23° b tan 81° c cos 54°

14 Find θ correct to the nearest degree.

  a sin θ = 0.9864 b tan θ = 1.236 c cos θ = 0.3654

5-05 Using tan to find an unknown side

15 Find b, correct to 2 decimal places.

16 Find x, correct to 1 decimal place.

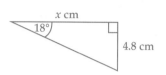

5-06 Using tan to find an unknown angle

17 Find A, correct to the nearest degree.

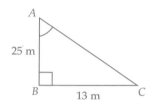

5-07 Finding an unknown side

18 Find q, correct to 1 decimal place.

a

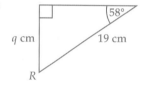

b

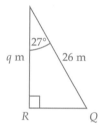

5-08 Finding more unknown sides

19 Find each pronumeral correct to 1 decimal place.

a

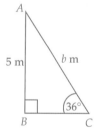

b

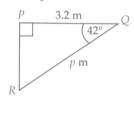

5-09 Finding an angle

20 Find the size of angle θ in each triangle correct to the nearest degree.

a

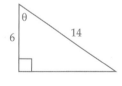

b

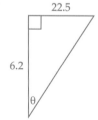

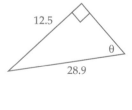

21 Sophie was stranded on the roof of a 28 m high building. She called out to Caitlin who was standing 12 m from the base of the building. Through what angle (to the nearest degree) must Caitlin look up to see Sophie?

PERCENTAGES

6

WHAT'S IN CHAPTER 6?

IN THIS CHAPTER YOU WILL:

- convert between percentages, fractions and decimals
- compare percentages, fractions and decimals
- find a percentage of a number or metric quantity
- express quantities as fractions and percentages of a whole
- calculate percentage increases and decreases
- use the unitary method to find a whole amount given a percentage of it
- solve problems involving profit and loss, cost price and selling price
- solve problems involving discounts and GST
- use the simple interest formula $I = PRN$ to calculate interest, principal and period

Shutterstock.com/Hatchapong Palurtchaivong

A **percentage** is a fraction with a denominator of 100, for example, 30% means 30 out of 100 or $\dfrac{30}{100}$.

EXAMPLE 1

Convert each percentage to a fraction and a decimal.

a 15%

b 72.4%

c $12\dfrac{1}{2}\%$

SOLUTION

a $15\% = \dfrac{15}{100}$

$= \dfrac{3}{20}$

b $72.4\% = \dfrac{72.4}{100}$

$= \dfrac{724}{1000}$

$= \dfrac{181}{250}$

c $12\dfrac{1}{2}\% = \dfrac{12\frac{1}{2}}{100}$

$= \dfrac{25}{200}$

$= \dfrac{1}{8}$

$15\% = 15 \div 100$
$= 0.15$

$72.4\% = 72.4 \div 100$
$= 0.724$

$12\dfrac{1}{2}\% = 12\dfrac{1}{2} \div 100$
$= 0.125$

EXAMPLE 2

Convert each number to a percentage.

a $\dfrac{3}{4}$

b $\dfrac{1}{3}$

c 0.6

d 2.25

SOLUTION

a $\dfrac{3}{4} = \dfrac{3}{4} \times 100\%$

$= 75\%$

b $\dfrac{1}{3} = \dfrac{1}{3} \times 100\%$

$= 33\dfrac{1}{3}\%$

c $0.6 = 0.6 \times 100\%$
$= 60\%$

d $2.25 = 2.25 \times 100\%$
$= 225\%$

TO COMPARE PERCENTAGES, FRACTIONS AND DECIMALS:
- convert them all to decimals
- make them all the same number of decimal places by adding zeros where necessary

EXAMPLE 3

Write $\dfrac{2}{5}$, 25% and 0.456 in ascending order.

SOLUTION

$\dfrac{2}{5} = 2 \div 5 = 0.4$ $25\% = 0.25$ 0.456

Write each decimal with 3 decimal places

$\dfrac{2}{5} = 0.400$ $25\% = 0.250$ $0.456 = 0.456$

From smallest to largest: 0.250, 0.400, 0.456

So 25%, $\dfrac{2}{5}$, 0.456 are in ascending order.

1 Write 20% as a fraction in simplest form. Select the correct answer **A, B, C** or **D**.

   A $\dfrac{20}{1000}$ **B** $\dfrac{20}{100}$ **C** $\dfrac{2}{10}$ **D** $\dfrac{1}{5}$

2 Write 6.5% as a decimal. Select **A, B, C** or **D**.

   A 0.65   **B** 0.065   **C** 6.50   **D** 0.0065

3 Convert each percentage to a simplified fraction.

   a 1%   b 3%   c 10%   d 15%   e 20%

   f 50%   g 62%   h 34%   i 60%   j 40%

   k 12%   l 24%   m 75%   n 82%   o 68%

4 Convert each percentage to a decimal.

   a 70%   b 90%   c 140%   d 18%   e 35%

   f 6%   g 24%   h 96%   i 108%   j 243%

   k 8%   l 7.5%   m 220%   n 4%   o 44%

   p 12%   q 12.5%   r 14.5%   s $12\dfrac{1}{2}\%$ t $2\dfrac{1}{2}\%$

5 Convert each fraction to a percentage.

   a $\dfrac{2}{5}$ b $\dfrac{3}{10}$ c $\dfrac{7}{8}$ d $\dfrac{1}{4}$ e $\dfrac{1}{20}$

   f $\dfrac{4}{5}$ g $\dfrac{3}{40}$ h $\dfrac{5}{16}$ i $\dfrac{12}{25}$ j $\dfrac{17}{40}$

   k $\dfrac{13}{8}$ l $\dfrac{21}{4}$ m $\dfrac{27}{20}$ n $\dfrac{11}{12}$ o $1\dfrac{1}{8}$

6 Convert each decimal to a percentage.

   a 0.02   b 0.06   c 0.05   d 0.09   e 0.08

   f 0.17   g 0.13   h 0.31   i 0.45   j 0.47

   k 0.92   l 0.3   m 0.7   n 4.5   o 1.125

7 Copy and complete this table.

Fraction		$\dfrac{1}{4}$				$\dfrac{3}{4}$		
Decimal		0.25		0.4		0.6		0.8		1
Percentage	10%		30%		50%			90%	

8 Write each set of numbers in ascending order by converting them to decimals first.

   a $\dfrac{3}{5}$, 0.655, 68.5%   b $\dfrac{3}{8}$, 38.2%, 0.38   c 82.5%, $\dfrac{5}{6}$, 0.805

This table lists some commonly-used percentages for mental calculation.

Percentage	Fraction	Decimal
10%	$\frac{1}{10}$	0.1
12.5%	$\frac{1}{8}$	0.125
25%	$\frac{1}{4}$	0.25
$33\frac{1}{3}\%$	$\frac{1}{3}$	$0.\dot{3}$
50%	$\frac{1}{2}$	0.5
$66\frac{2}{3}\%$	$\frac{2}{3}$	$0.\dot{6}$
75%	$\frac{3}{4}$	0.75
80%	$\frac{4}{5}$	0.8
100%	1	1.0

EXAMPLE 4

Use mental calculation to find each quantity.

a   25% of $24

b $66\frac{2}{3}\%$ of 12 days

c   12.5% of 56 m

SOLUTION

a   25% of $24 $= \frac{1}{4} \times 24$
$= \$6$

b $66\frac{2}{3}\%$ of 12 days $= \frac{2}{3} \times 12$
$= 8$ days

c   12.5% of 56 m $= \frac{1}{8} \times 56$
$= 7$ m

EXAMPLE 5

Find each quantity.

a   32% of $440   b   85% of 5 km   c   7.5% of 5 years

SOLUTION

a   32% of $440 = 0.32 × $440   ⟵ or enter 32 [%] [×] 440 [=] on calculator
$= \$140.80$

b   85% of 15 km = 0.85 × 15 km   ⟵ or enter 85 [%] [×] 15 [=] on calculator
$= 12.75$ km

c   7.5% of 5 years = 0.075 × 60 months   ⟵ or enter 7.5 [%] [×] 60 [=] on calculator
$= 4.5$ months   ✱ Convert to smaller metric units if needed

1   Write 30% as a simplified fraction. Select the correct answer **A**, **B**, **C** or **D**.

   A $\dfrac{30}{100}$ **B** $\dfrac{30}{1000}$ **C** $\dfrac{3}{10}$ **D** $\dfrac{1}{30}$

2   Write 45% as a simplified fraction. Select **A**, **B**, **C** or **D**.

   A $\dfrac{45}{100}$ **B** $\dfrac{9}{20}$ **C** $\dfrac{4.5}{10}$ **D** $\dfrac{45}{1000}$

3   Write each percentage as a simplified fraction.

   a 25%   **b** 12.5%   **c** 80%   **d** 50%   **e** $33\frac{1}{3}\%$

   f 100%   **g** 10%   **h** $66\frac{2}{3}\%$ **i** 75%   **j** 40%

4   Copy and complete:

   a $10\% \text{ of } \$200 = \dfrac{1}{-} \times 200$ **b** $75\% \text{ of } 80 \text{ cm} = \dfrac{-}{4} \times 80$

   $\qquad\qquad\qquad = \$ \underline{\quad}$ $\qquad\qquad\qquad = \underline{\quad} \text{ cm}$

5   Use mental calculation to find each quantity.

   a 10% of $7   **b** 10% of 60 mm   **c** 25% of 84c

   d 25% of $7.20   **e** $33\frac{1}{3}\%$ of $96   **f** $33\frac{1}{3}\%$ of 75 km

   g $12\frac{1}{2}\%$ of $840   **h** 75% of $44   **i** 75% of 84 km

   j 50% of $120   **k** 50% of $1.20   **l** 20% of $150

   m 20% of $1.50   **n** $66\frac{2}{3}\%$ of $42   **o** $66\frac{2}{3}\%$ of $4.20

   p 80% of $900   **q** 100% of 6 years   **r** 150% of 24 km

6   Find each quantity.

   a 5% of $250   **b** 27% of 6 m   **c** 40% of 15 m

   d 2.5% of 8 years   **e** 54% of 120 cm   **f** 65% of 9 years

   g 61% of 8 L   **h** 15% of 1 day   **i** 84% of 1 km

   j 51% of $1280   **k** 9.5% of 25 days   **l** 120% of $2000

7   There were 56 000 people shopping on Christmas Eve at a shopping centre and 45% had only one present left to buy.

   a How many people had only one present left to buy?

   b How many people had more than one present left to buy?

8   If a small water tank holds 1800 litres of water and is only 12.5% full, how many litres of water is in the tank?

9   Susanne spent 78% of her holiday savings while overseas. If she had saved $6400, how much money did she have left?

Expressing quantities as fractions and percentages

To express an amount as a fraction of a whole amount, write the fraction as $\dfrac{\text{amount}}{\text{whole amount}}$ and simplify it if possible.

EXAMPLE 6

Express each amount as a fraction.

a   56 marks out of 88

b   18 minutes out of 1 hour

SOLUTION

a   56 marks out of 88 $= \dfrac{56}{88}$

$= \dfrac{7}{11}$

b   18 minutes out of 1 hour $= \dfrac{18}{60}$ ⟵ 1 h = 60 min

$= \dfrac{3}{10}$

To express an amount as a percentage of a whole amount, calculate $\dfrac{\text{amount}}{\text{whole amount}} \times 100\%$

EXAMPLE 7

Express each amount as a percentage.

a   120 marks out of 200

b   15 cm out of 2 m

SOLUTION

a   120 marks out of 200 $= \dfrac{120}{200} \times 100\%$

$= 60\%$

b   15 cm out of 2 m $= \dfrac{15}{200} \times 100\%$ ⟵ 2 m = 200 cm

$= 7.5\%$

When expressing amounts as fractions or percentages, all units must be the same.

iStockphoto/4FR

Developmental Mathematics Book 3

ISBN 9780170351027

1   Express 45 out of 60 as a simplified fraction. Select the correct answer **A, B, C** or **D**.

    A $\dfrac{45}{60}$ **B** $\dfrac{9}{12}$ **C** $\dfrac{15}{20}$ **D** $\dfrac{3}{4}$

2   Express 24 out of 80 as a percentage. Select **A, B, C** or **D**.

    A 3%   **B** 24%   **C** 30%   **D** 80%

3   Copy and complete each statement to write the answer as a fraction.

    a 12 out of 30 = $\dfrac{12}{\ \ }$ = ——   **b** 75 out of 120 = $\dfrac{\ \ }{120}$ = ——

4   Express each amount as a fraction.

    a 20 out of 50   **b** 15 mins out of 1 hour

    c 18 cm out of 1 m   **d** \$2 out of \$10

    e 44 mL out of 1 L   **f** 50c out of \$4

    g 4 months out of 1 year   **h** 12 m out of 2 km

    i 6 hours out of 2 days

5   Express each amount in question 4 as a percentage.

6   Copy and complete each statement to write the answer as a percentage.

    a 15 girls out of 20 students = $\dfrac{15}{\ \ } \times 100\%$

                           = ____%

    b 27 out of 81 = $\dfrac{\ \ }{81} \times 100\%$

              = ____%

7   Express each amount as a percentage.

    a 13 out of 52   **b** 12 out of 96   **c** 14 out of 112

    d 23 out of 115   **e** 5c out of 100c   **f** 10c out of \$2

    g 50c out of \$2.50 **h** 50c out of \$20   **i** 5 min out of 1 h

    j 6 h out of 1 day   **k** 15 s out of 1 min **l** 20 cm out of l m

8   Express each mark as a percentage to determine who performed the best.

    a Kaoru scored 34 out of 50 in English.

    b Gina scored 18 out of 20 in Geography.

    c Barak scored 60 out of 75 in Maths.

9   Jenny scored 7 goals in a netball match. If her team finished the game with a total of 20 goals, what percentage of these did Jenny score?

10 In a class of 25 students, 8 brought their lunch from home, 12 bought their lunch from the canteen and the rest forgot their lunch.

    a What percentage of students bought their lunch from the canteen?

    b What fraction of students forgot their lunch?

    c What percentage of students had a lunch from home?

WORDBANK

increase To make larger by adding.

decrease To make smaller by subtracting.

To increase an amount by a percentage, find the percentage of the amount and add it to the amount.
To decrease an amount by a percentage, find the percentage of the amount and subtract it from the amount.

EXAMPLE 8

a Increase $50 by 15%.

b Decrease 20 L by 4%.

SOLUTION

a Increase = 15% of $50
        = $7.50

   Increased amount = $50 + $7.50
                  = $57.50

b Decrease = 4% of 20 L
        = 0.8 L

   Decreased amount = 20 L – 0.8 L
                  = 19.2 L

EXAMPLE 9

At a sale, a shirt is reduced by 30% from its recommended retail price of $48.

What is the sale price of the shirt?

SOLUTION

Decrease = 30% of $48
       = $14.40

Sale price = $48 – $14.40
        = $33.60

iStockphoto/traity1

1 Increase $200 by 40%. Select the correct answer **A**, **B**, **C** or **D**.

    A $80   **B** $120   **C** $240   **D** $280

2 Decrease 160 m by 75%. Select **A**, **B**, **C** or **D**.

    A 40 m   **B** 120 m   **C** 280 m   **D** 85 m

3 Copy and complete these sentences.

    To **increase** an amount by a percentage, find the percentage first and then _____ it to the existing amount. To **decrease** an amount by a percentage, find the percentage first and then _____ it from the existing amount.

4 Increase:

    a $50 by 10%   **b** 700 m by 30%   **c** $5000 by 25%

    d $20 000 by 70%   **e** 2500 kg by 65%   **f** 80 t by $12\frac{1}{2}$%

5 Decrease:

    a $200 by 15%   **b** 4000 m by 50%   **c** $80 000 by 40%

    d $1 000 000 by 75%   **e** 300 ha by 20%   **f** 6000 kg by $33\frac{1}{3}$%

6 Find the sale price of each item below if the discount is 35% for all items in Jolly John's store.

    a DVD player $520   **b** iPhone 5S $850

    c computer $1280

7 Jamiela earns a wage of $560 for working at a jeans store. If her wage increases by 25%, how much is her new wage?

8 Aaron works for a company that allows him to leave work early to pick up his children from school if he takes a pay cut of 12% off his normal wage of $820 per week. What would be his new weekly wage?

9 Kirsty gave her sales staff an increase of 15% due to extra profits. If the usual combined staff wage was $14 500, what is the new combined staff wage after the increase?

10 Jules had just opened a new shop and decided to add 20% profit to the cost price of all the shoes in her shop. After 2 weeks, she had not sold much stock so she then decided to reduce the price of each pair of shoes by taking 20% off the marked price. If a pair of shoes cost Jules $120, how much would they now be priced at? Is this the same price as she originally bought them for?

iStockphoto/magnez2

WORDBANK

unit Unit means one or each. $6 per unit means $6 for one.

unitary method A method to find a unit amount and then use this amount to find the total amount.

TO USE THE UNITARY METHOD:
- ■ use the given amount to find 1%
- ■ multiply 1% by 100 to find the total amount (100%)

EXAMPLE 10

Vikas donates 12% of his Christmas bonus to charity each year. How much was his Christmas bonus if he donated $216 to charity?

SOLUTION

12% of his Christmas bonus = $216

1% of his Christmas bonus = $216 ÷ 12   ◄——— Calculating 1% first
                         = $18

100% of his Christmas bonus = $18 × 100   ◄——— Multiplying by 100
                           = $1800

So Vikas' Christmas bonus was $1800.

Check: 12% × $1800 = $216

EXAMPLE 11

Jasmine pays 18.5% of her wage in tax. If she pays $224.20 per week in tax, how much does she earn per week before her tax is paid?

SOLUTION

18.5% of Jasmine's wage = $224.20

1% of Jasmine's wage = $224.20 ÷ 18.5   ◄——— Calculating 1% first
                    = $12.1189…

100% of Jasmine's wage = $12.118 91… × 100   ◄——— Multiplying by 100
                      = $1211.891…
                      ≈ $1211.89

So Jasmine's weekly wage is $1211.89.

ISBN 9780170351027

1   If 20% of an amount is $55, what is the amount? Select the correct answer **A**, **B**, **C** or **D**.

   A $1100   **B** $550   **C** $250   **D** $275

2   If 4% of an amount is $160, what is the amount? Select **A**, **B**, **C** or **D**.

   A $1600   **B** $4000   **C** $400   **D** $160

3   Is each statement true or false?

   a   If 10% = $250, then 1% = $25   **b**   If 45% = $900, then 1% = $2

   c   If 82% = $246, then 1% = $30

4   Copy and complete each solution.

   a   15% of an amount = $450   **b**   60% of an amount = $960

   $\qquad$ 1% of the amount = $450 ÷ ___   1% of the amount = $960 ÷ ___

   $\qquad\qquad\qquad\qquad$ = $____ $\qquad\qquad\qquad\quad$ = $____

   $\qquad$ 100% of the amount = $____ × 100   100% of the amount = $____ × 100

   $\qquad\qquad\qquad\qquad\qquad$ = $____ $\qquad\qquad\qquad\qquad\quad$ = $____

5   Find each total amount if:

   a   5% of it is $80   **b**   10% of it is $40   **c**   25% of it is $200

   d   12% of it is 240 m   **e**   65% of it is 1300 L   **f**   90% of it is 27 m

   g   25% of it is 2 hours   **h**   9% of it is $1800   **i**   22% of it is 660 kg

6   Michael scored 88 runs in a game of cricket. This was 20% of his team's runs. How many runs did Michael's team score?

7   Anthea had her gold chain valued and was charged $150 for the value. This was 2.5% of the chain's value. How much is her gold chain worth?

8   The town of Timbuktu has 280 people who cannot speak English. This is 8% of the town's population. How many people live in Timbuktu?

9   Grace was working for a law firm and paid 28.6% tax each week.

   a   If her tax was $328.60, how much was Grace's weekly wage before her tax was paid?

   b   How can you check if this amount is correct?

10 Tom was at a carnival and found that he had spent 40% of his savings by midday by going on all of the rides. If he had spent $64,

   a   how much had he saved for the carnival?

   b   how much money did Tom have left for the rest of the day?

iStockphoto/timhughes

WORDBANK

cost price The price at which an item was bought by a retailer.

selling price The price at which an item was sold by a retailer.

profit To sell an item at a higher price than it was bought.

loss To sell an item at a lower price than it was bought.

GST 10% goods and services tax charged by the government on most goods and services bought.

EXAMPLE 12

A DVD is bought for $28 and sold for $32.

a Find the profit.   **b** Calculate the profit as a percentage of the selling price.

SOLUTION

a Profit = $32 – $28 = $4

b Profit as a percentage of the selling price = $\dfrac{\text{profit}}{\text{selling price}} \times 100\%$

$= \dfrac{4}{32} \times 100\%$ ⟵ Selling price = $32

$= 12.5\%$

EXAMPLE 13

A pair of shoes was bought for $54 and sold for $36.

a Find the loss.   **b** Calculate the loss as a percentage of the cost price.

SOLUTION

a Loss = $54 – $36 = $18

b Loss as a percentage of the cost price = $\dfrac{\text{loss}}{\text{cost price}} \times 100\%$

$= \dfrac{18}{54} \times 100\%$ ⟵ Cost price = $54

$= 33\dfrac{1}{3}\%$

EXAMPLE 14

George bought a cordless drill worth $280 at a sale. 10% GST was added to the price but he was then given a 30% discount. How much did George pay for the drill?

SOLUTION

Price of drill including GST = $280 + 10% of $280 ⟵ Add the GST
$= \$308$

Discount = 30% of $308 = $92.40

Price paid = $308 – $92.40 = $215.60

1   A watch was bought for $35 and sold for $42. What was the profit or loss? Select the correct answer **A, B, C** or **D**.

   A profit $7   **B** loss $7   **C** profit $8   **D** loss $8

2   A tent was bought for $215 and sold for $208. What was the profit or loss? Select **A, B, C** or **D**.

   A profit $7   **B** loss $7   **C** profit $8   **D** loss $8

3   Copy and complete this table.

Cost price	Selling price	Profit or loss
$12.00	$15.00	$3.00 profit
$25.00	$30.00	
$60.00	$20.00	
$250.00		$50.00 loss
	$780.00	$45.00 profit
$1590.90		$120.80 loss
	$450.50	$25.75 profit
$6550.80	$6250.35	

4   A book costs $15. It was sold for $20.

   a   Find the profit.

   b   What was the profit as a percentage of the cost price?

   c   What was the profit as a percentage of the selling price?

5   Owen's new car cost him $55 500. When he sold it a year later, he lost $5000.

   a   What price did he sell his car for?

   b   Calculate his loss as a percentage of the cost price.

6   A tin of coffee cost Joe $12.00. He sells it for $16.50. What is the profit as a percentage of:

   a   the cost price?   **b**   the selling price?

7   If a retailer sells an article for $75.50 and makes a profit of $8.80, what was the article's cost to the retailer?

8   Jenny bought a new washing machine, which cost her $1385. When she moved house, she sold it for $850. How much did she lose?

9   The following items can be bought from an online store, but 10% GST has to be added to find the selling price. Calculate the GST and the selling price for each item.

   a   Dress $158   **b**   Shirt $45   **c**   Trousers $84   **d**   Shoes $120

   e   Belt $36   **f**   Scarf $28   **g**   Necklace $52   **h**   Tie $26

10 The online store is having a sale and all goods are reduced by 40%. Find the sales price for each item in question **9**.

6-07 | Simple interest

WORDBANK

interest Income earned from investing money or a charge spent for borrowing money.

principal The amount invested or borrowed. Interest is calculated on this amount.

simple interest Interest calculated on the original principal, also called **flat-rate interest**.

THE SIMPLE INTEREST FORMULA

Interest = principal × interest rate per period (as a decimal) × number of time periods

$I = PRN$

EXAMPLE 15

Jenny borrowed $10 500 to buy a car. She is charged simple interest at a rate of 6% p.a. for 3 years by the bank. Calculate the simple interest charged.

SOLUTION

$P = \$10\,500$, $R = 6\%$ p.a. = 0.06 as a decimal, $N = 3$ years

✱ Make sure that R and N are given in the same time units, for example, both in years (or p.a. = per annum, which is 'per year') or both in months (or per month).

$I = PRN = \$10\,500 \times 0.06 \times 3 = \$1890$

Jenny is charged $1890 simple interest over 3 years.

EXAMPLE 16

Find the simple interest earned on $5000 at 12% p.a. for 3 months.

SOLUTION

$P = \$5000$, $N = 3$ months

$R = 12\%$ p.a. = 0.12 p.a. = 0.12 ÷ 12 per month = 0.01 per month

✱ R and N are both in months.

$I = PRN = \$5000 \times 0.01 \times 3 = \$150$

EXAMPLE 17

If the simple interest on an investment of $2000 is $220 over 4 years, find the rate of interest p.a.

SOLUTION

$P = \$2000$, $I = \$220$, $N = 4$ years, $R = ?$

$I = PRN$

$\$220 = \$2000 \times R \times 4$

$\$220 = \$8000R$

$\dfrac{220}{8000} = R$

$R = 0.0275 = 0.0275 \times 100\% = 2.75\%$

The interest rate is 2.75%.

1   Calculate the simple interest on $400 at 5% p.a. for 2 years. Select the correct answer **A, B, C** or **D**.

   A $50   **B** $80   **C** $20   **D** $40

2   Calculate the simple interest on $6000 at 8% p.a. for 3 years. Select **A, B, C** or **D**.

   A $1440   **B** $144   **C** $480   **D** $800

3   Phuong invested $5000 in a credit union at 5% p.a. simple interest.

   a   How much interest did she receive in the first year?

   b   How much interest did she receive after 10 years?

4   Copy and complete this solution.

   To find the simple interest on $450 for 2 years at 5% p.a.:

   $P = \$450, R = \_\_\% = 0.05$ as a decimal, $N = \_\_$ years

   $I = PRN$
   $= \$450 \times \_\_\_\_ \times 2$
   $= \$ \_\_\_\_$

5   Calculate the simple interest on each investment.

   a $500 for 2 years at 6% p.a.   **b** $750 for 8 years at 3% p.a.

   c $650 for 10 years at 2.5% p.a.   **d** $1340 for 5 years at 7.5% p.a.

   e $345 for 9 years at 18% p.a.   **f** $659 for 7 years at 13% p.a.

6   Copy and complete this solution.

   To find the simple interest on $2500 for 8 months at 9% p.a.:

   $P = \$2500, R = \_\_\%$ p.a. $= 0.09$ p.a. $= \_\_\_\_$ per month, $N = \_\_$ months

   $I = PRN$
   $= \$2500 \times \_\_\_\_ \times 8$
   $= \$ \_\_\_\_$

7   Calculate the simple interest on each investment.

   a $1000 at 9% p.a. for 6 months   **b** $4000 at 6% p.a. for 9 months

   c $8500 at 2% per month for 2 years   **d** $12 000 at 4% p.a. for 5 months

   e $50 000 at 4% p.a. for 18 months   **f** $16 800 at 9% p.a. for 7 months

8   Maria invested $1400 in government bonds paying 6.5% p.a.

   a   What is the simple interest for 10 years?

   b   What is her investment worth after 10 years?

9   If the simple interest on an investment of $8000 is $2240 at 7% p.a., how many years was the investment earning interest?

10 If the simple interest on an investment of $12 000 is $800 over 6 years, calculate the interest rate p.a. as a percentage correct to 1 decimal place.

FIND-A-WORD PUZZLE

Make a copy of this page, then find all the words listed below in this grid of letters.

C	R	E	G	A	T	N	E	C	R	E	P
O	E	E	C	I	R	P	C	O	S	T	N
M	T	S	E	V	W	C	O	N	L	A	O
M	A	Y	S	I	T	B	U	V	P	M	I
I	I	T	A	L	D	I	T	E	C	O	T
S	N	I	E	S	E	P	W	R	T	U	A
I	E	T	R	T	C	A	N	T	O	N	C
O	R	N	C	P	R	O	F	I	T	T	I
N	R	A	N	Y	E	B	E	S	S	O	L
E	Q	U	I	V	A	L	E	N	T	O	P
W	A	Q	U	E	S	T	R	A	D	E	P
S	L	A	P	S	E	L	L	I	N	G	A

AMOUNT	APPLICATION	COMMISSION	CONVERT
COST	DECREASE	EQUIVALENT	INCREASE
PERCENTAGE	LOSS	PRICE	PROFIT
QUANTITY	RETAINER	SELLING	TRADE

Part A General topics

Calculators are not allowed.

1 Find 15% of $40.

2 Simplify $xy - px + 5xy + xp$.

3 How many hours are in 1 week?

4 Find x.

5 Convert $\dfrac{36}{8}$ to a mixed numeral.

6 Write in descending order:

    –5   7   0   6   3,   –2

7 Evaluate 8.4 – 4.25.

8 Find the value of $3x^2$ if $x = -2$.

9 Expand $-6(2x + 5)$

10 How many hours and minutes are there between 10:35 a.m. to 4:20 p.m.?

Part B Percentages

Calculators are allowed.

6–01 Percentages, fractions and decimals

11 Convert each percentage to a fraction and a decimal.

    a 12%   **b** 70%   **c** 65%

12 Convert $\dfrac{1}{8}$ to a percentage. Select **A**, **B**, **C** or **D**.

    A 8%   **B** 18%   **C** 12.5%   **D** 25%

6–02 Percentage of a quantity

13 Find 22% of $1200. Select **A**, **B**, **C** or **D**.

    A $22   **B** $26.40   **C** $240   **D** $264

14 Find each quantity.

    a 25% of 24 m   **b** $33\dfrac{1}{3}$% of 9 hours   **c** 75% of 64 L

6–03 Expressing quantities as fractions and percentages

15 Write each amount as a fraction.

    a 26 marks out of 30

    b 12 minutes out of 1 hour

    c 3 cm out of 12 m

16 Write each amount in question **15** as a percentage.

6–04 Percentage increase and decrease

17 Increase $250 by 20%.

18 At a sale, a TV is reduced by 40% from its selling price of $1250. Calculate its discounted price.

6-05 The unitary method

19 Gemma was working for a company and paid 18.5% tax on her weekly wage. If her tax was $186.80 per week, what is her weekly wage?

6-06 Profit, loss and discounts

20 Sam was working at a new business and bought an electric bike for $1250. He later sold it for $950. Find:

a the loss

b the loss as a percentage of the sales price (correct to one decimal place).

21 Add 10% GST to a car priced at $25 588.

6-07 Simple interest

22 Find the simple interest on:

a $500 at 6% p.a. for 3 years

b $8500 at 4.8% p.a. for 7 months

23 Find the interest rate p.a. if $12 600 earns $2520 interest in 4 years.

INDICES

7

IN THIS CHAPTER YOU WILL:

- multiply and divide terms with the same base
- find a power of a power
- use zero and negative indices
- round numbers to significant figures
- understand and use scientific notation for large and small numbers
- use a calculator to evaluate expressions with scientific notation

Shutterstock.com/Sdubi

WORDBANK

base The main number that is raised to a power, for example, in 4^3 the base is 4.

power The number at the top right corner of the base that represents repeated multiplication by itself, for example, 4^3 means $4 \times 4 \times 4$ and the power is 3.

index Another word for power.

indices The plural of index, pronounced 'in-da-sees'.

When multiplying terms with the same base, such as $4^3 \times 4^5$, we can add powers because:
$4^3 \times 4^5 = (4 \times 4 \times 4) \times (4 \times 4 \times 4 \times 4 \times 4) = 4^8$ $\quad$ $(3 + 5 = 8)$
$\qquad$ 3 times $\qquad\qquad$ 5 times

To multiply terms with the same base, add the indices.
$a^m \times a^n = a^{m+n}$

EXAMPLE 1

Simplify each expression, writing the answer in index notation.

a $5^3 \times 5^4$ $\qquad$ **b** $x^5 \times x^6$ $\qquad$ **c** $w^2 \times w \times w^4$ $\qquad$ **d** $m^6 \times n^3 \times 3^2$

SOLUTION

a $5^3 \times 5^4 = 5^7$ $\qquad$ **b** $x^5 \times x^6 = x^{11}$ $\qquad$ **c** $w^2 \times w \times w^4 = w^7$ $\qquad$ **d** $m^6 \times n^3 \times 3^2 = 9m^6n^3$

 Add the indices if the bases are the same $\qquad$ $w = w^1$ $\qquad$ The bases are different here.

EXAMPLE 2

Simplify each expression.

a $5a^4 \times 3a^2$ $\qquad$ **b** $6m^3 \times (-3m^2)$ $\qquad$ **c** $7e^2f \times (-4e^3f^5)$

SOLUTION

a $5a^4 \times 3a^2 = 5 \times 3 \times a^4 \times a^2$ $\qquad\qquad$ **b** $6m^3 \times (-3m^2) = 6 \times (-3) \times m^3 \times m^2$
$\qquad\qquad\quad = 15a^6$ $\qquad\qquad\qquad\qquad\qquad\qquad = -18m^5$

c $7e^2f \times (-4e^3f^5) = 7 \times (-4) \times e^2 \times e^3 \times f \times f^5$
$\qquad\qquad\qquad = -28e^5f^6$

iStockphoto/richcarey

1 Simplify $2^6 \times 2^3$. Select the correct answer **A**, **B**, **C** or **D**.

   A 4^{18} **B** 2^{18} **C** 4^9 **D** 2^9

2 Simplify $3^2 \times 3^4$. Select **A**, **B**, **C** or **D**.

   A 3^8 **B** 9^6 **C** 3^6 **D** 9^8

3 Copy and complete each statement.

   a 5^3 means _____   **b** The base of 5^3 is _____

   c The index of 5^3 is _____

4 Write each expression using index notation (powers).

   a $3 \times 3 \times 3 \times 3$ **b** $2 \times 2 \times 2 \times 2 \times 2$

   c 7×7 **d** $8 \times 8 \times 8$

   e $x \times x \times x \times x \times x \times x$ **f** $5 \times 5 \times a \times a \times a$

   g $3 \times 3 \times 6 \times 6 \times 6$ **h** $w \times v \times v \times v$

   i $5 \times a \times 5 \times a \times 5$ **j** $m \times n \times m \times n \times n$

   k $u \times 4 \times u \times 4 \times v$ **l** $3 \times 5 \times m \times n$

5 Simplify each expression using index notation.

   a $3^4 \times 3^3$ **b** $5^2 \times 5^6$ **c** $2^4 \times 2^7$ **d** $6^3 \times 6^0$

   e $m^2 \times m^5$ **f** $x \times x^4$ **g** $p \times p^6$ **h** $n^6 \times n^5$

6 Copy and complete each solution.

   a $3m^3 \times 4m^2 = 3 \times \underline{\hspace{1cm}} \times m^3 \times \underline{\hspace{1cm}}$ **b** $5w^4 \times -3w^7 = \underline{\hspace{1cm}} \times (-3) \times w^4 \times \underline{\hspace{1cm}}$

                  $= 12m\text{---}$ $= \underline{\hspace{0.4cm}} w\text{---}$

7 Simplify each expression.

   a $2n^3 \times 5n^4$ **b** $3a^3 \times 8a^4$

   c $12n^2 \times 5n^5$ **d** $6w^3 \times 8w^5$

   e $-2v^3 \times 6v^4$ **f** $21n^3 \times (-3n^4)$

   g $-5a^3 \times (-9a^4)$ **h** $16a^2 \times (-4a^5)$

   i $-3c^4 \times 3c^2 \times 8c^4$ **j** $6a^3b \times 8a^4b^3$

   k $-4m^3n \times 8m^4n^5$ **l** $-3a^3c \times (-8a^4c^4)$

8 Is each statement true or false?

   a $2^5 \times 2^3 = 4^8$ **b** $3^4 \times 3^6 = 3^{10}$

   c $2^3 \times 3^2 = 6^5$ **d** $5^2 \times 5^3 = 5^5$

9 Can $3^4 \times 2^4$ be simplified by adding indices? Use your calculator to evaluate it.

10 Use a calculator to evaluate each expression.

   a $2^6 \times 3^2$ **b** $3^4 \times 2^2$ **c** $5^3 \times 3^2$ **d** $2^7 \times 4^2$

   e $4^3 \times 5^2$ **f** $7^2 \times 6^2$ **g** $(-5)^3 \times 4^2$ **h** $2^5 \times (-3)^2$

Dividing terms with the same base

When dividing terms with the same base, such as $2^6 \div 2^4$, we can subtract powers because:

$$2^6 \div 2^4 = \frac{2 \times 2 \times \cancel{2} \times \cancel{2} \times \cancel{2} \times \cancel{2}}{\cancel{2} \times \cancel{2} \times \cancel{2} \times \cancel{2}} = 2 \times 2 = 2^2 \qquad (6 - 4 = 2)$$

To divide terms with the same base, subtract the indices.

$$a^m \div a^n = \frac{a^m}{a^n} = a^{m-n}$$

EXAMPLE 3

Simplify each expression, writing the answer in index notation.

a $4^8 \div 4^3$ **b** $m^{10} \div m^4$ **c** $\dfrac{7^6}{7^5}$ **d** $\dfrac{x^{12}}{x}$

SOLUTION

a $4^8 \div 4^3 = 4^5$ **b** $m^{10} \div m^4 = m^6$ **c** $\dfrac{7^6}{7^5} = 7^1 = 7$ **d** $\dfrac{x^{12}}{x} = x^{11}$

 ✱ Subtract the indices if the bases are the same $x = x^1$

EXAMPLE 4

Simplify each expression.

a $12w^{16} \div 4w^6$ **b** $28x^{13} \div (-7x^{10})$ **c** $\dfrac{-36m^4n^6}{-9mn^2}$

SOLUTION

Divide coefficients first.

a $12w^{16} \div 4w^6 = \dfrac{12}{4} \times \dfrac{w^{16}}{w^6}$ **b** $28x^{13} \div (-7x^{10}) = \dfrac{28}{-7} \times \dfrac{x^{13}}{x^{10}}$ **c** $\dfrac{-36m^4n^6}{-9mn^2} = \dfrac{-36}{-9} \times \dfrac{m^4}{m} \times \dfrac{n^6}{n^2}$

            $= 3w^{10}$ $= -4x^3$ $= 4m^3n^4$

Shutterstock.com/AnetaPics

ISBN 9780170351027

1 Simplify $3^8 \div 3^4$. Select the correct answer **A, B, C** or **D**.
   A 3^2 **B** 1^2 **C** 3^4 **D** 1^4

2 Simplify $5^{18} \div 5^6$. Select **A, B, C** or **D**.
   A 5^{12} **B** 1^{12} **C** 5^3 **D** 1^3

3 Copy and complete each solution.
   a $2^8 \div 2^3 = 2^{8-\square}$ **b** $x^9 \div x^3 = x^{\square-3}$
               $= 2^{\square}$ $= x^{\square}$
   c $12^6 \div 12^3 = 12^{6-\square}$
                $= 12^{\square}$

4 Simplify each expression.
   a $2^6 \div 2^3$ **b** $4^8 \div 4^3$
   c $\dfrac{3^{12}}{3^5}$ **d** $3^5 \div 3^0$
   e $d^8 \div d^3$ **f** $\dfrac{m^5}{m^4}$
   g $x^6 \div x^2$ **h** $m^8 \div m^5$

5 Simplify each expression.
   a $12n^8 \div 4n^4$ **b** $8a^{12} \div 2a^4$
   c $15n^6 \div 5n^5$ **d** $16w^{15} \div 8w^5$
   e $-18v^8 \div 6v^4$ **f** $21n^9 \div (-3n^4)$
   g $-45a^7 \div (-9a^4)$ **h** $16a^{12} \div (-4a^5)$
   i $-32c^4 \div 8c^3$ **j** $64a^8b^6 \div 8a^4b^3$
   k $-24m^7n^6 \div 6m^4n^5$ **l** $-16a^8c^4 \div (-2a^4c)$

6 Is each statement true or false?
   a $3^6 \div 3^3 = 3^2$ **b** $x^{12} \div x^6 = x^2$
   c $\dfrac{7^8}{7^3} = 7^5$ **d** $\dfrac{m^{14}}{m^6} = m^8$

7 Can $4^5 \div 3^3$ be simplified by subtracting indices? Use your calculator to evaluate it as a fraction.

8 Use a calculator to evaluate each expression correct to 2 decimal places where appropriate.
   a $8^6 \div 6^4$ **b** $12^9 \div 8^5$
   c $\dfrac{15^4}{11^3}$ **d** $\dfrac{24^6}{18^3}$

When finding a power of a power with the same base, such as $(4^3)^2$, we can multiply powers because:

$$(4^3)^2 = 4^3 \times 4^3 = (4 \times 4 \times 4) \times (4 \times 4 \times 4) = 4^6$$
$$\underbrace{}_{3 \text{ times}} \quad \underbrace{}_{3 \text{ times}}$$

To find a power of a power, multiply the indices.

$(a^m)^n = a^{m \times n} = a^{mn}$

EXAMPLE 5

Simplify each expression, giving each answer in index notation.

a $(3^5)^4$ **b** $(n^2)^5$ **c** $(x^6)^0$

SOLUTION

a $(3^5)^4 = 3^{5 \times 4}$ **b** $(n^2)^5 = n^{2 \times 5}$ **c** $(x^6)^0 = x^{6 \times 0}$
$\qquad = 3^{20} \qquad\qquad\qquad = n^{10} \qquad\qquad\qquad = x^0$

 Multiply the indices

To find a power of ab or $\dfrac{a}{b}$, raise a and b to the power separately.

$(ab)^n = a^n b^n$ and $\left(\dfrac{a}{b}\right)^n = \dfrac{a^n}{b^n}$

EXAMPLE 6

Simplify each expression.

a $(2x^3)^4$ **b** $\left(\dfrac{m^5}{4}\right)^3$ **c** $\left(\dfrac{3a}{b^4}\right)^2$

SOLUTION

Always raise each part in the bracket to the power separately.

a $(2x^3)^4 = 2^4 \times (x^3)^4$ **b** $\left(\dfrac{m^5}{4}\right)^3 = \dfrac{(m^5)^3}{4^3}$

$\qquad\quad = 16x^{12} \qquad\qquad\qquad\qquad = \dfrac{m^{15}}{64}$

c $\left(\dfrac{3a}{b^4}\right)^2 = \dfrac{3^2 a^2}{(b^4)^2}$

$\qquad\quad = \dfrac{9a^2}{b^8}$

Shutterstock.com/Pakhnyushcha

1 Simplify $(2^3)^2$. Select the correct answer **A, B, C** or **D**.

    A 2^5 **B** 8^1 **C** 2^6 **D** 2^1

2 Simplify $(3a^3)^2$. Select **A, B, C** or **D**.

    A $3a^6$ **B** $9a^6$ **C** $9a^5$ **D** $4a^6$

3 Copy and complete.

    a $(5^3)^2 = 5^{3 \times -} = 5^{-}$ **b** $(x^4)^3 = x^{- \times 3} = \_^{12}$

4 Simplify each expression, giving each answer in index notation.

    a $(2^5)^3$ **b** $(6^2)^4$ **c** $(3^4)^3$

    d $(5^6)^3$ **e** $(7^8)^5$ **f** $(x^3)^4$

    g $(n^3)^6$ **h** $(m^8)^2$ **i** $(w^4)^5$

    j $(a^6)^3$ **k** $(9^5)^3$ **l** $(x^2)^4$

    m $(5^4)^6$ **n** $(q^6)^4$ **o** $(p^8)^7$

5 Is each statement true or false?

    a $(2a^4)^3 = 2a^{12}$ **b** $(3x^2)^3 = 27x^6$ **c** $(5n^4)^2 = 25n^8$

6 Simplify each expression.

    a $(2x^3)^2$ **b** $(3a^4)^3$ **c** $(5n^6)^2$

    d $(4m^8)^3$ **e** $(7c^5)^3$ **f** $(2w^8)^5$

    g $(6b^4)^3$ **h** $(9t^4)^2$ **i** $(-2a^6)^5$

    j $(4w^8)^3$ **k** $(-3c^7)^3$ **l** $(-2q^9)^6$

7 Is each statement true or false?

    a $\left(\dfrac{x^2}{3}\right)^3 = \dfrac{x^6}{27}$ **b** $\left(\dfrac{5}{w^4}\right)^2 = \dfrac{5}{w^8}$ **c** $\left(\dfrac{2a^3}{3}\right)^4 = \dfrac{2a^{12}}{81}$

8 Simplify each expression.

    a $\left(\dfrac{a^4}{2}\right)^5$ **b** $\left(\dfrac{m^2}{3}\right)^2$ **c** $\left(\dfrac{3}{w^6}\right)^2$ **d** $\left(\dfrac{5}{m^6}\right)^2$

    e $\left(\dfrac{x^8}{4}\right)^3$ **f** $\left(\dfrac{w^7}{-2}\right)^3$ **g** $\left(\dfrac{4}{c^8}\right)^3$ **h** $\left(\dfrac{-3}{n^6}\right)^4$

9 Is each statement true or false?

    a $\left(\dfrac{2a^3}{-5}\right)^2 = \dfrac{4a^6}{25}$ **b** $\left(\dfrac{a^5}{b^3}\right)^4 = \dfrac{a^{20}}{b^{12}}$

10 Simplify each expression.

    a $\left(\dfrac{3x^6}{2}\right)^4$ **b** $\left(\dfrac{m^6}{n^3}\right)^4$ **c** $\left(\dfrac{3^2}{2a^5}\right)^3$ **d** $\left(\dfrac{a^8}{b^{12}}\right)^5$

What does $3^5 \div 3^5$ equal?

$$3^5 \div 3^5 = \frac{3 \times 3 \times 3 \times 3 \times 3}{3 \times 3 \times 3 \times 3 \times 3} = 1$$

But also, when dividing terms with the same base, we subtract indices:

$3^5 \div 3^5 = 3^{5-5} = 3^0$

So $3^0 = 1$.

What does $3^4 \div 3^6$ equal?

$$3^4 \div 3^6 = \frac{3 \times 3 \times 3 \times 3}{3 \times 3 \times 3 \times 3 \times 3 \times 3} = \frac{1}{3^2}$$

But also, when dividing terms with the same base, we subtract indices:

$3^4 \div 3^6 = 3^{4-6} = 3^{-2}$

So $3^{-2} = \dfrac{1}{3^2}$.

Any term raised to the **power of 0** is 1.

$a^0 = 1$

A term raised to a **negative power** gives a fraction with a numerator of 1 and a denominator that is the same term raised to a positive power.

$a^{-n} = \dfrac{1}{a^n}$

EXAMPLE 7

Simplify each expression.

a 5^0 **b** $3a^0$ **c** $(6n)^0$ **d** $(-4w)^0$

SOLUTION

a $5^0 = 1$ **b** $3a^0 = 3 \times 1$
                            $= 3$ **c** $(6n)^0 = 1$ **d** $(-4w)^0 = 1$

EXAMPLE 8

Simplify each expression.

a 3^{-2} **b** a^{-3} **c** $2x^{-5}$ **d** $\dfrac{1}{2}m^{-4}$

SOLUTION

a $3^{-2} = \dfrac{1}{3^2}$ **b** $a^{-3} = \dfrac{1}{a^3}$

        $= \dfrac{1}{9}$

c $2x^{-5} = 2 \times \dfrac{1}{x^5}$ **d** $\dfrac{1}{2}m^{-4} = \dfrac{1}{2} \times \dfrac{1}{m^4}$

          $= \dfrac{2}{x^5}$ $= \dfrac{1}{2m^4}$

iStockphoto/tpuerzer

1 Write $2a^{-3}$ with a positive index. Select the correct answer **A**, **B**, **C** or **D**.

   A $\dfrac{1}{2a^3}$ **B** $\dfrac{2}{a^3}$ **C** $\dfrac{8}{a^3}$ **D** $\dfrac{1}{8a^3}$

2 Write $\dfrac{3}{x^2}$ with a negative index. Select **A**, **B**, **C** or **D**.

   A $9x^{-2}$ **B** $3x^{-1}$ **C** $3x^{-2}$ **D** $9x^{-1}$

3 Is each statement true or false?

   a $4^0 = 1$ **b** $7^0 = 7$ **c** $2n^0 = 1$ **d** $(2n)^0 = 1$

4 Simplify each expression.

   a 3^0 **b** 5^0 **c** 10^0 **d** x^0

   e m^0 **f** y^0 **g** 6^0 **h** $2x^0$

   i $(3a)^0$ **j** $6m^0$ **k** $(-4n)^0$ **l** $5a^0 + 6^0$

5 Copy and complete each statement.

   a $4^{-3} = \dfrac{1}{4^\square}$ **b** $w^{-5} = \dfrac{\square}{w^\square}$ **c** $4n^{-3} = \dfrac{4}{n^\square}$

6 Write each term with a positive index.

   a 3^{-2} **b** 2^{-4} **c** 4^{-1}

   d 5^{-2} **e** 3^{-4} **f** 7^{-2}

7 Evaluate each expression.

   a 5^0 **b** 3^{-5} **c** 4^{-2}

   d 6^{-1} **e** 9^0 **f** 2^{-7}

8 True or false?

   a $3^{-1} = \dfrac{3}{1}$ **b** $2^{-4} = \dfrac{1}{2^4}$ **c** $5^{-6} = \dfrac{5}{6}$

   d $3^0 = 0$ **e** $10^{-2} = \dfrac{1}{10}$ **f** $8^{-2} = \dfrac{1}{64}$

9 Simplify each expression.

   a $3x^{-4}$ **b** $6a^{-5}$ **c** $4m^{-7}$

   d $2w^{-3}$ **e** $\dfrac{1}{3}x^{-2}$ **f** $\dfrac{1}{4}n^{-3}$

   g $\dfrac{2}{3}x^{-4}$ **h** $\dfrac{3}{5}a^{-6}$ **i** $\dfrac{3}{2}u^{-5}$

   j $5w^{-6}$

10 Evaluate each expression.

   a $2^{-1} + 3^0$ **b** $5^0 - 3^{-2}$ **c** $4^{-1} + 2^{-1} + 10^0$

   d $10^{-1} + 5^{-1} - 9^0$ **e** $4^{-2} \times 2^3$

This table summarises all the index laws learnt so far. Note that these laws only apply to expressions in which all terms have the **same base**.

When multiplying, add indices	$a^m \times a^n = a^{m+n}$
When dividing, subtract indices	$a^m \div a^n = a^{m-n}$
To find a power of a power, multiply indices	$(a^m)^n = a^{m \times n}$
To raise ab to a power:	$(ab)^n = a^n b^n$
To raise $\dfrac{a}{b}$ to a power:	$\left(\dfrac{a}{b}\right)^n = \dfrac{a^n}{b^n}$
To find a zero index:	$a^0 = 1$
To find a negative index:	$a^{-n} = \dfrac{1}{a^n}$

EXAMPLE 9

Simplify each expression.

a $2a^3b^4 \times 3ab^5$

b $12mn^6 \div (-4mn^{-2})$

c $(2a^4)^3 \times (4a)^0$

SOLUTION

a $2a^3b^4 \times 3ab^5 = 2 \times 3 \times a^3a \times b^4b^5$
$\qquad\qquad\quad = 6a^4b^9$

b $12mn^6 \div (-4mn^{-2}) = \dfrac{12}{-4} \times \dfrac{m}{m} \times \dfrac{n^6}{n^{-2}}$
$\qquad\qquad\qquad\qquad = -3 \times 1 \times n^8$
$\qquad\qquad\qquad\qquad = -3n^8$

c $(2a^4)^3 \times (4a)^0 = 2^3\, a^{12} \times 1$
$\qquad\qquad\qquad\quad = 8a^{12}$

Shutterstock.com/flmor.com

1 When dividing terms with the same base, the indices are what? Select the correct answer **A, B, C** or **D**.

    A added   **B** subtracted   **C** multiplied   **D** divided

2 When finding a power of a power, the indices are what? Select **A, B, C** or **D**.

    A added   **B** subtracted   **C** multiplied   **D** divided

3 Is each statement true or false?

    a $2a^4 \times 4a^5 = 8a^{20}$ **b** $18m^6 \div 6m^3 = 3m^3$

    c $6x^0 = 1$ **d** $(3n^6)^2 = 9n^{12}$

4 Simplify each expression.

    a $4^3 \times 4^{-1}$ **b** $3^{-2} \times 3^5$ **c** $7^{-2} \times 7^{-3}$ **d** $5^{-4} \times 5^0$

    e $w^{-2} \times w^3$ **f** $c^7 \times c^{-4}$ **g** $x^{-3} \times x^{-4}$ **h** $a^{-5} \times a^0$

5 Simplify each expression.

    a $6^4 \div 6^{-2}$ **b** $7^{-3} \div 7^{-1}$ **c** $\dfrac{5^4}{5^{-2}}$ **d** $2^{-4} \div 2^{-3}$

    e $x^{-5} \div x^{-3}$ **f** $\dfrac{c^5}{c^{-4}}$ **g** $w^{-4} \div w^{-3}$ **h** $\dfrac{a^{-4}}{a^{-1}}$

6 Simplify each expression.

    a $(2^4)^2$ **b** $(3^2)^3$ **c** $(4^3)^3$ **d** $(5^2)^4$

    e $(3^{-2})^2$ **f** $(2^{-3})^4$ **g** $(4^3)^{-2}$ **h** $(6^{-3})^0$

    i $(x^3)^3$ **j** $(a^4)^3$ **k** $(w^{12})^4$ **l** $(p^8)^3$

    m $(a^{-2})^3$ **n** $(x^7)^{-2}$ **o** $(w^0)^{-2}$ **p** $(x^{-3})^{-4}$

7 Simplify each expression.

    a $2x^4 \times 3x^5$ **b** $12a^6 \div 3a^4$

    c $(2m^3)^4$ **d** $4a^5 \times 3a^{-2}$

    e $\dfrac{15m^6}{3m^{-2}}$ **f** $(4a^5)^2$

    g $(3x^{-3})^4$ **h** $\dfrac{16w^{-8}}{8w^{-4}}$

    i $8a^6b^7 \times (-4a^0b^{-2})$ **j** $-12m^8n^{-3} \div 4m^{-3}n^{-4}$

    k $5w^0 \times (-2w^4)^3$ **l** $42x^6y^4 \div (-6x^5y^{-2})$

8 Is each statement true or false?

    a $3a^8 \times 4a^4 = 12a^{12}$ **b** $18m^4 \div 6m^1 = 3m^3$

    c $(3x^4)^2 = 6x^6$ **d** $(2a^{-2})^3 = 8a^6$

    e $6w^3 \times 3w^{-2} = 18w$ **f** $\dfrac{24m^{-2}}{4m^{-3}} = 6m^{-5}$

    g $(4x^{-2})^0 = 1$ **h** $4a^{12} \times 2x^3 = 8a^{15}$

We already know how to round numbers to **decimal places**, but we can also round to **significant figures**.

- The significant figures in the number 13 740 000 are 1, 3, 7 and 4 because they indicate the size of the number. The 0s at the end are not significant.
- The significant figures in the number 9 056 300 are 9, 0, 5, 6 and 3, but the two 0s at the end are not significant.
- The significant figures in the decimal 0.0428 are 4, 2 and 8 because they indicate the size of the decimal. The 0s at the start are not significant.
- So only 0s at the end of a whole number or at the start of a decimal are not significant.

EXAMPLE 10

Round each number to 3 significant figures.

a 52 698   **b** 4 252 000   **c** 30 756 000

SOLUTION

a 52 698 ≈ 52 700   ⟵——— The 3rd digit 6 is rounded up to 7 as the next digit 9 > 5.

b 4 252 000 ≈ 4 250 000   ⟵——— The 3rd digit 5 is rounded down as the next digit 2 < 5.

c 30 756 000 ≈ 30 800 000   ⟵——— The 3rd digit 7 is rounded up to 8 as the next digit is 5.

EXAMPLE 11

Round each number to 2 significant figures.

a 0.000 428   **b** 0.002 037 1   **c** 3.662

SOLUTION

Count the first 2 non-zero digits in each number.

a 0.000 428 ≈ 0.000 43 ⟵——— The 2nd significant figure 2 is rounded up to 3 as the next digit 8 > 5.

b 0.002 037 1 ≈ 0.0020 ⟵——— The 2nd significant figure 0 is rounded down as the next digit 3 < 5.

c 3.662 ≈ 3.7   ⟵——— The 2nd significant figure 6 is rounded up to 7 as the next digit 6 > 5.

Shutterstock.com/SARANS

ISBN 9780170351027

1   Round 28 624 to 2 significant figures. Select the correct answer **A**, **B**, **C** or **D**.

    A 28 000   **B** 28   **C** 29   **D** 29 000

2   Round 0.005 024 0 to 3 significant figures. Select **A**, **B**, **C** or **D**.

    A 0.005   **B** 0.005 02   **C** 0.005 24   **D** 0.005 024

3   Round each number to 3 significant figures.

    a 23 628   **b** 586 841   **c** 81 620

    d 3 568   **e** 4 227 000   **f** 45 627

    g 32 541 000   **h** 427 832   **i** 407 820

    j 12 093 000   **k** 60 328 125   **l** 2 098 000

4   Write each number correct to 2 significant figures.

    a 0.034 78   **b** 0.003 81   **c** 0.8261

    d 2.4100   **e** 0.003 070 0   **f** 0.0721

    g 0.000 908   **h** 0.7024   **i** 4.8060

    j 0.050 800   **k** 56.032   **l** 0.708 25

5   At the Australian Open tennis in Melbourne, there were 68 722 people in Rod Laver Arena. Round this number correct to:

    a 1 significant figure

    b 2 significant figures

    c 3 significant figures

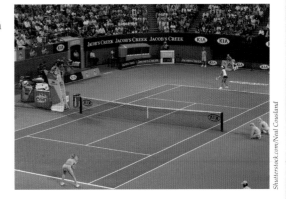

Shutterstock.com/Neal Cousland

6   The concentration of salt in some beach water was 0.0986. Round this number correct to 2 significant figures.

7   The population of Inandarra was 54 728 in the last census.

    a If the population was increasing at a rate of 5% per year, find the population in 1 year's time.

    b Round this number correct to 4 significant figures.

8   Calculate $\dfrac{22.86 - 14.962}{3.48 + 18.624}$ correct to 2 significant figures.

WORDBANK

scientific notation A way of writing very large or small numbers, using a decimal from 1 up to but not including 10 multiplied by a power of 10, such as 2.86×10^7 to write 28 600 000.

A number written in scientific notation has the form $m \times 10^n$, where m is a number between 1 and 10 and n is an integer.

To write a large number in scientific notation:
- use its significant figures to write a decimal between 1 and 10
- for the power of 10, count the number of places in the number after the first digit

EXAMPLE 12

Write each number in scientific notation.

a 724 600   b 48 300 000   c 160 280 000

SOLUTION

a 724 600 = 7.246×10^5 ◄————— 5 places underlined after the first significant figure, 7

b 48 300 000 = 4.83×10^7 ◄————— 7 places underlined after the first significant figure, 4

c 160 280 000 = 1.6028×10^8 ◄————— 8 places underlined after the first significant figure, 1

EXAMPLE 13

Write each number in decimal form.

a 2.6×10^3 b 8.205×10^6 c 6×10^9

SOLUTION

a $2.6 \times 10^3 = 2600$ ◄— Move the decimal point 3 places right or make 3 places after the 2.

b $8.205 \times 10^6 = 8\ 205\ 000$ ◄— Move the decimal point 6 places right or make 6 places after the 8.

c $3 \times 10^9 = 3\ 000\ 000\ 000$ ◄— Move the decimal point 9 places right.

EXAMPLE 14

Write these numbers in ascending order:

3.4×10^4 3×10^7 3.04×10^4

SOLUTION

To compare numbers in scientific notation, first compare the powers of 10.

Both 3.4×10^4 and 3.04×10^4 are definitely smaller than 3×10^7 because $10^4 < 10^7$.

Since 3.4×10^4 and 3.04×10^4 have the same power of 10, we need to compare their decimal parts.

3.04×10^4 is smaller because $3.04 < 3.4$.

So in ascending order, the numbers are 3.04×10^4, 3.4×10^4, 3×10^7.

OR: Convert the numbers to decimal form to compare them.

$3.4 \times 10^4 = 34\ 000$ $3 \times 10^7 = 30\ 000\ 000$ $3.04 \times 10^4 = 30\ 400$

1 Write 685 000 in scientific notation. Select the correct answer **A**, **B**, **C** or **D**.

   A 6.85×10^6 **B** 68.5×10^4 **C** 685×10^3 **D** 6.85×10^5

2 Write 4.02×10^4 in decimal form. Select **A**, **B**, **C** or **D**.

   A 40 200   **B** 402 000   **C** 4020   **D** 402

3 Copy and complete each statement.

   a $33\,000 = 3.3 \times 10^{\square}$ **b** $600\,000 = 6 \times 10^{\square}$

   c $2800 = \square \times 10^3$ **d** $824\,000 = \square \times 10^5$

   e $90\,200\,000 = \square \times 10^{\square}$ **f** $260\,000\,000 = \square \times 10^{\square}$

4 Write each number in scientific notation.

   a 12 000   **b** 456 000   **c** 120 000 000

   d 45 000 000   **e** 6 803 000   **f** 34 500 000

   g 80 000 000 000   **h** 520 000 000

5 Copy and complete each statement.

   a $7.2 \times 10^4 = 72...$ **b** $8.24 \times 10^5 = 824...$ **c** $4.1 \times 10^8 = 41...$

   d $6.03 \times 10^2 = 603...$ **e** $1.98 \times 10^6 = 198...$ **f** $3.728 \times 10^3 = 37...$

6 Write each number as a basic numeral.

   a 5×10^4 **b** 3.8×10^3 **c** 1.68×10^6 **d** 2.45×10^7

   e 4×10^3 **f** 9.2×10^1 **g** 3.004×10^8 **h** 1.09×10^9

7 **a** Write the numbers below in ascending order.

       2.8×10^6 7×10^4 2.006×10^7 6.11×10^6

   b List these same numbers in descending order.

8 Write each number in scientific notation.

   a The dam has the capacity to hold 456 821 megalitres of water.

   b There were 68 000 Australians attending the Anzac Day service in Gallipoli.

   c The population of Australia was approximately 22 900 000.

   d The distance from Earth to the Sun is about 152 000 000 kilometres.

9 Write your answers to each problem in scientific notation.

   a Chen enters a walkathon to raise money for charity. He walks 15 km each day for 82 days. How far has he walked altogether?

   b Hannah flies to Rome in a plane travelling at 800 km every hour. The flight is non-stop and takes 18 hours. How far did she fly?

   c Liam is in training and swims 22 km every day for 48 weeks. How many kilometres has he swum in this time?

   d Sophie is an author. She types 26 000 words every day. How many words will she type in a year?

   e Rajan delivers leaflets in mailboxes every day for 8 weeks. If he is able to deliver 450 leaflets per day, how many leaflets are delivered in 8 weeks?

10 On the Internet investigate how many kilobytes are in a gigabyte and write your answer in scientific notation.

Scientific notation for small numbers

A number written in scientific notation has the form $m \times 10^n$ where m is a number between 1 and 10 and n is an integer.

To write a decimal in scientific notation:
- use its significant figures to write a decimal between 1 and 10
- for the **negative** power of 10, count the number of places in the number up to and including the first significant digit (or count the number of 0s)

EXAMPLE 15

Write each number in scientific notation.

a 0.002 34   **b** 0.000 009   **c** 0.000 010 5

SOLUTION

a 0.002 34 $= 2.34 \times 10^{-3}$ ⟵——3 places underlined up to the first significant figure, 2 (or 3 0s)

b 0.000 009 $= 9 \times 10^{-6}$ ⟵——6 places underlined up to the first significant figure, 9 (or 6 0s)

c 0.000 010 5 $= 1.05 \times 10^{-5}$⟵——5 places underlined up to the first significant figure, 1 (or 5 0s)

EXAMPLE 16

Write each number in decimal form.

a 5.2×10^{-4} **b** 8.02×10^{-7} **c** 9×10^{-6}

SOLUTION

a $5.2 \times 10^{-4} = 0.000\ 52$ ⟵——Move the decimal point 4 places left or insert 4 0s in front.

b $8.02 \times 10^{-7} = 0.000\ 000\ 802$ ⟵——Move the decimal point 7 places left or insert 7 0s in front.

c $9 \times 10^{-6} = 0.000\ 009$ ⟵——Move the decimal point 6 places left or insert 6 0s in front.

EXAMPLE 17

Write the numbers below in descending order.

   4.2×10^{-5} 9.8×10^{-4} 1.35×10^{-4}

SOLUTION

First compare the powers of 10.

Both 9.8×10^{-4} and 1.35×10^{-4} are definitely larger than 4.2×10^{-5} because $10^{-4} > 10^{-5}$.

Since 9.8×10^{-4} and 1.35×10^{-4} have the same power of 10, we need to compare their decimal parts.

9.8×10^{-4} is larger because $9.8 > 1.35$.

So in descending order, the numbers are 9.8×10^{-4}, 1.35×10^{-4}, 4.2×10^{-5}.

OR: Convert the numbers to decimal form to compare them.

$4.2 \times 10^{-5} = 0.000\ 042$ $9.8 \times 10^{-4} = 0.000\ 98$ $1.35 \times 10^{-4} = 0.000\ 135$

1   Write 0.009 12 in scientific notation. Select the correct answer **A**, **B**, **C** or **D**.

    A 9.12×10^{-2} **B** 91.2×10^{-4}

    C 9.12×10^{-3} **D** 912×10^{-5}

2   Write 3.62×10^{-4} in decimal form. Select **A**, **B**, **C** or **D**.

    A 0.000 362   **B** 0.003 62

    C 0.000 0362   **D** 0.0362

3   Copy and complete each statement.

    a $0.005 = 5 \times 10^{\square}$ **b** $0.0012 = 1.2 \times 10^{\square}$

    c $0.000\ 683 = \square \times 10^{-4}$ **d** $0.000\ 07 = \square \times 10^{-5}$

    e $0.000\ 042 = \square \times 10^{\square}$ **f** $0.000\ 000\ 18 = \square \times 10^{\square}$

4   Write each decimal in scientific notation.

    a 0.009   **b** 0.000 12

    c 0.000 097 2   **d** 0.000 000 4

    e 0.05   **f** 0.000 750 8

    g 0.000 721 4   **h** 0.0008

5   Copy and complete each statement.

    a $8 \times 10^{-3} = 0.\underline{\quad}$ **b** $6.24 \times 10^{-4} = 0.\underline{\quad}$

    c $5.01 \times 10^{-3} = 0.\underline{\quad}$ **d** $7.2 \times 10^{-7} = 0.\underline{\quad}$

    e $9.42 \times 10^{-1} = 0.\underline{\quad}$ **f** $2.834 \times 10^{-5} = 0.\underline{\quad}$

6   Write each number as a decimal.

    a 1.2×10^{-2} **b** 3.5×10^{-3} **c** 6×10^{-5} **d** 4.82×10^{-4}

    e 9×10^{-1} **f** 7.6×10^{-6} **g** 1.082×10^{-5} **h** 4.06×10^{-9}

7   **a** Write the numbers below in descending order.

        1.8×10^{-5} 5×10^{-5} 3.129×10^{-7} 6.1×10^{-4}

    b List these same numbers in ascending order.

8   Write each number in scientific notation.

    a The thickness of a human hair is 0.000 08 metres.

    b The length of an ant is 0.007 m.

    c The amount of poison in a substance is 0.000 004 8 grams.

    d The scale factor of a drawing is 0.0015.

    e The wavelength of an electron is 0.000 000 000 001 m.

9   On the Internet, investigate the size of a **micrometre** and a **nanometre**.

To enter a number in scientific notation on a calculator, use the $\boxed{\times 10^x}$ or $\boxed{\text{EXP}}$ key.

EXAMPLE 18

Use a calculator to write each number in decimal form.

a 4.28×10^9 **b** 1.705×10^{-2}

SOLUTION

a $4.28 \times 10^9 = 4\ 280\ 000\ 000$ ⟵ On a calculator, enter: 4.28 $\boxed{\times 10^x}$ 9 $\boxed{=}$

b $1.705 \times 10^{-2} = 0.017\ 05$ ⟵ On a calculator, enter: 1.705 $\boxed{\times 10^x}$ $\boxed{(-)}$ 2 $\boxed{=}$

EXAMPLE 19

Evaluate each expression in scientific notation correct to two significant figures.

a $(2.5 \times 10^5) \times (6.28 \times 10^8)$ **b** $\dfrac{7.12 \times 10^9}{3.8 \times 10^{-6}}$

SOLUTION

a On a calculator, enter: 2.5 $\boxed{\times 10^x}$ 5 $\boxed{\times}$ 6.28 $\boxed{\times 10^x}$ 8 $\boxed{=}$

$(2.5 \times 10^5) \times (6.28 \times 10^8) = 1.57 \times 10^{14}$

$\approx 1.6 \times 10^{14}$

b On a calculator, enter: 7.12 $\boxed{\times 10^x}$ 9 $\boxed{\div}$ 3.8 $\boxed{\times 10^x}$ $\boxed{(-)}$ 6 $\boxed{=}$

$\dfrac{7.12 \times 10^9}{3.8 \times 10^{-6}} = 1.8736 \ldots \times 10^{15}$

$\approx 1.9 \times 10^{15}$

iStockphoto/cemagraphics

1 Use a calculator to write 2.45×10^6 in decimal form. Select the correct answer **A, B, C** or **D**.

   A 245 000   **B** 2 450 000   **C** 24 500   **D** 24 500 000

2 Use a calculator to evaluate $(1.8 \times 10^{12}) \times (7.3 \times 10^{-3})$ in decimal form. Select **A, B, C** or **D**.

   A 131 400 000   **B** 1 314 000 000   **C** 13 140 000 000   **D** 13 140 000

3 Use a calculator to write each number in decimal form.

   a 1.8×10^3 **b** 2.4×10^6 **c** 7×10^9 **d** 2.0056×10^8

   e 4.2×10^{-1} **f** 8.9×10^0 **g** 5×10^7 **h** 3.009×10^{-2}

4 Evaluate each expression, correct to three significant figures where necessary.

   a $3.8 \times 10^5 \times (8.2 \times 10^7)$ **b** $4 \times 10^6 \times (6.1 \times 10^8)$

   c $2.46 \times 10^5 \div (3.6 \times 10^2)$ **d** $\dfrac{9.2 \times 10^9}{1.78 \times 10^3}$

   e $7.5 \times 10^8 \times (7 \times 10^{-4})$ **f** $1.45 \times 10^{-3} \times (5.2 \times 10^{-3})$

   g $9.2 \times 10^5 \div (3 \times 10^{-5})$ **h** $8 \times 10^{-5} \div (5.802 \times 10^{-2})$

   i $5.05 \times 10^{-2} \times (7.1 \times 10^{-6})$ **j** $\dfrac{4.2 \times 10^{15}}{3.78 \times 10^{-4}}$

5 Evaluate each expression, correct to two significant figures.

   a $\dfrac{4.2 \times 10^8}{1.9 \times 10^4}$ **b** $\dfrac{7.24 \times 10^{12}}{1.6 \times 10^3}$ **c** $\dfrac{9.26 \times 10^{12}}{7 \times 10^{-3}}$ **d** $\dfrac{5.03 \times 10^{18}}{8 \times 10^{-4}}$

   e $\dfrac{6.1 \times 10^{16}}{3 \times 10^{-4}}$ **f** $\dfrac{4 \times 10^{-8}}{1.5 \times 10^4}$ **g** $\dfrac{3.24 \times 10^{12}}{1.7 \times 10^{-2}}$ **h** $\dfrac{7 \times 10^{-12}}{2.1 \times 10^3}$

6 Which number is larger?

   a 4.2×10^{-4} or 4.2×10^4 **b** 5.6×10^4 or 5.6×10^3

   c 8×10^3 or 8×10^{-3} **d** 8.6×10^4 or 4.2×10^3

   e 9.6×10^{-3} or 5.7×10^5 **f** 4.8×10^{-2} or 8.9×10^3

   g 1.2×10^{-4} or 7.3×10^{-2} **h** 8.4×10^{-2} or 3.5×10^{-1}

7 Evaluate each expression, correct to 4 significant figures.

   a $(3.4 \times 10^4)^2$ **b** $(6.8 \times 10^5)^3$ **c** $\sqrt{4.3 \times 10^6}$ **d** $\sqrt[3]{6 \times 10^9}$

8 **a** If the mass of the planet Mercury is 3.30×10^{21} kg and the mass of Venus is 4.87×10^{24} kg, which is heavier and how many times heavier is it?

   b If the mass of Jupiter is 1.90×10^{27} kg and the mass of Saturn is 5.69×10^{26} kg, which is heavier and how many times heavier is it?

   c If the mass of Earth is 5.97×10^{24} kg and the mass of Neptune is 1.03×10^{26} kg, which is heavier and how many times is it heavier?

   d If the mass of Mars is 6.42×10^{21} kg and the mass of Uranus is 8.66×10^{21} kg, which is heavier and how many times is it heavier?

   e From your answers in parts **a** to **d** above, order the planets from lightest to heaviest.

MIX AND MATCH

Match each expression on the left with its simplified expression on the right.

1   $a^2 \times a^3$ **A** $2a$

2   $x^6 \div x^3$ **D** x^8

3   $(a^2)^4$ **E** $8a^6$

4   $x \times x^7$ **I** a^8

5   $(x^3)^3$ **M** $6x^5$

6   $\dfrac{12a^4}{6a^3}$ **N** a^5

7   $3x^2 \times 2x^3$ **P** x^9

8   $(2a^2)^3$ **T** $9x^8$

9   $(3x^4)^2$ **X** x^3

Match question numbers with answer letters to decode the answer to this riddle:

What is always in a hurry and is found in a high position?

6-1   3-7-5-6-9-3-8-1-9   3-1-4-8-2

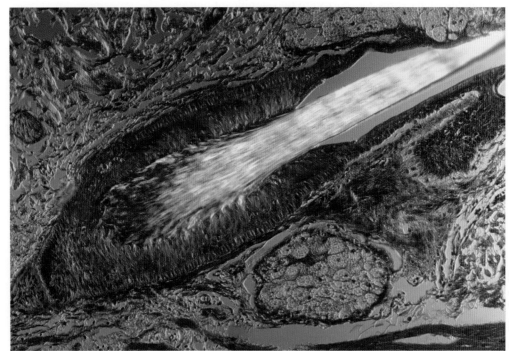

Alamy/Cultura Creative (RF)

Part A General topics

Calculators are not allowed.

1   Decrease $60 by 15%.

2   If $y = -2$, evaluate $8 - 3y$.

3   Find the average of -8 and 6.

4   Copy this diagram and mark two vertically opposite angles.

5   Evaluate $\dfrac{4}{5} - \dfrac{3}{10}$.

6   What percentage of $50 is $15?

7   Convert 1925 to 12-hour time.

8   Factorise $27p^2q - 15pq^2$.

9   Round 82.6854 to two decimal places.

10 What is the probability of selecting a hearts card from a deck of playing cards?

Part B Indices

Calculators are allowed.

7-01 Multiplying terms with the same base

11 Simplify each expression.

   a $a^5 \times a^8$ **b** $3^7 \times 3^4$ **c** $x^6 \times 2 \times x^3 \times 6$ **d** $5n^3 \times 3n^4$

7-02 Dividing terms with the same base

12 Simplify each expression.

   a $15n^{12} \div 3n^4$ **b** $\dfrac{-48x^{18}}{-8x^6}$ **c** $20x^8 \div 5x^4$

7-03 Power of a power

13 Simplify each expression using index notation.

   a $(2^5)^4$ **b** $(m^3)^6$ **c** $(3x^3)^4$ **d** $\left(\dfrac{2^3}{n}\right)^5$

7-04 Zero and negative indices

14 Simplify each expression.

   a 4^0 **b** x^0 **c** $5x^0$

15 Write each expression with a positive index.

   a 7^{-2} **b** $4a^{-3}$ **c** $\dfrac{1}{2}x^{-4}$

7-05 Index laws review

16 Simplify each expression.

   a $2a^2b \times (-5ab^2)$

   b $-32m^9n^4 \div (-8m^6n)$

   c $(3a^3)^2 \times 4a^0$

7-06 Significant figures

17 Round each number to 3 significant figures.

   a 521 670

   b 6 038 214

   c 0.002 087 6

7-07 Scientific notation for large numbers

18 Write each number in scientific notation.

   a 387 200

   b 6 720 000

   c 19 000 000 000

7-08 Scientific notation for small numbers

19 Write each number in scientific notation.

   a 0.0186

   b 0.000 405

   c 0.000 000 004

20 Write these numbers in ascending order.

   3.8×10^6 8×10^{-4} 6.2×10^9

7-09 Scientific notation on a calculator

21 Evaluate each expression.

   a $6.15 \times 10^7 \times 3.08 \times 10^{-3}$

   b $\dfrac{8.64 \times 10^{14}}{9.2 \times 10^{-6}}$

GEOMETRY

8

WHAT'S IN CHAPTER 8?

IN THIS CHAPTER YOU WILL:

- name, measure and classify angles
- solve geometry problems involving right angles, angles on a straight line, angles at a point, vertically opposite angles and angles formed by parallel lines
- prove that two lines are parallel
- name and classify triangles and their properties
- name and classify quadrilaterals and their side, angle and diagonal properties
- solve geometry problems involving the properties and angle sums of triangles and quadrilaterals
- identify congruent figures and their properties
- use the '≡' symbol
- identify the four tests for congruent triangles: SSS, SAS, AAS, RHS

An **angle** measures how much an object turns or spins, and is measured in **degrees** (°).

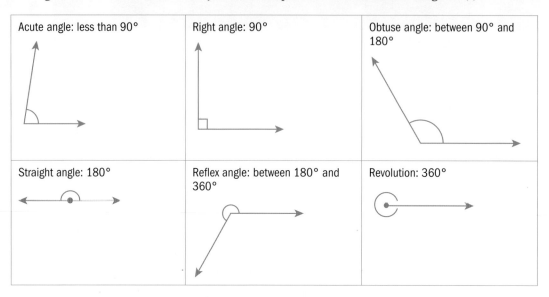

| Acute angle: less than 90° | Right angle: 90° | Obtuse angle: between 90° and 180° |
| Straight angle: 180° | Reflex angle: between 180° and 360° | Revolution: 360° |

An angle is usually named using three letters, with its vertex being the middle letter. This angle is named ∠*PGH* or ∠*HGP*.

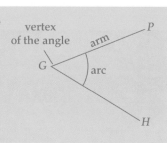

vertex of the angle

arm

arc

EXAMPLE 1

For each angle:

i classify the angle   **ii** name the angle   **iii** measure the angle

a

b

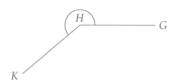

Developmental Mathematics Book 3

ISBN 9780170351027

SOLUTION

a i Obtuse angle.

ii ∠PMQ or ∠QMP.

iii

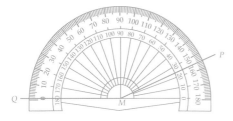

- Place the centre of the protractor on the angle's vertex
- Line up the 0° line of the protractor with one arm of the angle
- Read off the size of the angle using the other arm of the angle

∠PMQ = 155°

b i Reflex angle.

ii ∠KHG or ∠GHK.

iii As the protractor measures up to 180° only, turn it upside down to measure the smaller angle underneath first.

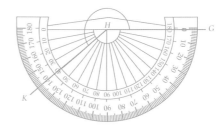

'Upside-down' ∠KHG = 140°
Reflex ∠KHG = 360° − 140° = 220°

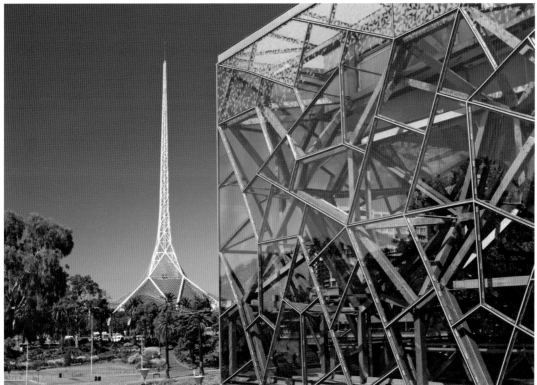

Alamy/William Caram

1   What type of angle is 89°? Select the correct answer **A**, **B**, **C** or **D**.

    A reflex   **B** obtuse   **C** right   **D** acute

2   Name the angle drawn. Select **A**, **B**, **C** or **D**.

    A ∠PQR   **B** ∠RQP

    C ∠QRP   **D** ∠RPQ

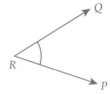

3   Name each angle using 3 letters.

    a

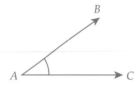

    b

    c

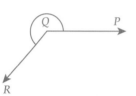

    d

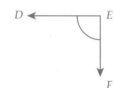

    e

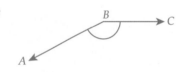

    f

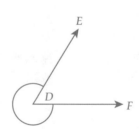

4   Use a protractor to measure the size of each angle in question **3**.

5   Classify each angle in question **3**.

6   Draw an example of each type of angle.

    a right angle   **b** revolution   **c** straight angle

    d obtuse angle   **e** acute angle   **f** reflex angle

7   Use a protractor to draw each angle below.

    a ∠ABC = 42°   **b** ∠BAC = 103°   **c** ∠XYZ = 85°

    d ∠EFD = 225°   **e** ∠PQR = 161°   **f** ∠UVW = 300°

8   Classify each angle below.

    a 58°   **b** 200°   **c** 112°   **d** 90°   **e** 305°

    f 135°   **g** 27°   **h** 180°   **i** 280°   **j** 360°

9   Write an example of each type of angle from question **6** that you can see in everyday life.

WORDBANK

adjacent angles Angles that are next to each other. They share a common arm.

complementary angles Two angles that add to 90°.

supplementary angles Two angles that add to 180°.

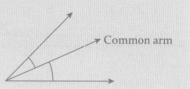
Common arm

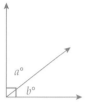

Angles in a right angle are complementary (add up to 90°)
$a + b = 90$

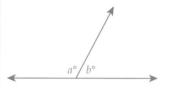

Angles on a straight line are supplementary (add up to 180°)
$a + b = 180$

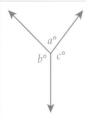

Angles at a point (in a revolution) add up to 360°
$a + b + c = 360$

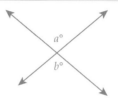

Vertically opposite angles are equal.
$a = b$

EXAMPLE 2

Find the value of each pronumeral, giving reasons.

a

b

c

SOLUTION

a $x + 63 = 90$ (angles in a right angle)
$$x = 90 - 63$$
$$x = 27$$

b $w = 85$ (vertically opposite angles)

c $120 + 90 + a + a = 360$ (angles at a point)
$$2a = 360 - 120 - 90$$
$$= 150$$
$$a = \frac{150}{2}$$
$$= 75$$

1   Which two angles are complementary? Select the correct answer **A, B, C** or **D**.

   A 20° and 50°   **B** 50° and 130°   **C** 53° and 37°   **D** None of these

2   What is the supplement of 62°? Select **A, B, C** or **D**.

   A 118°   **B** 28°   **C** 128°   **D** 38°

3   Calculate the angle size that is complementary to each angle.

   a 40°   **b** 75°   **c** 45°   **d** 34°   **e** 53°

4   Calculate the angle size that is supplementary to each angle.

   a 60°   **b** 35°   **c** 145°   **d** 106°   **e** 56°

5   Sketch the types of angles described below.

   a angles on a straight line

   b adjacent angles

   c angles at a point

   d vertically opposite angles

   e complementary angles that are not adjacent.

6   Copy and complete each solution to find the pronumeral(s).

   a
   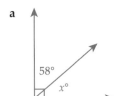

   $x + 58 = 90$ (_____ angles)
   $x = 90 -$ ___
   $x =$ ___

   b
   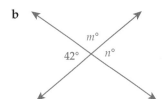

   $n = 42$ (_____ angles)
   $m + 42 = 180$ (_____ angles)
   $m = 180 -$ ___
   $m =$ ___

7   Find the value of each pronumeral, giving reasons.

a

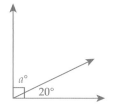

b

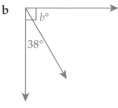

c

d

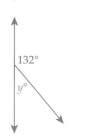

e

f

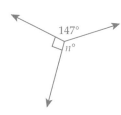

g

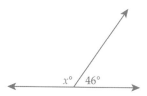

h

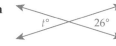

i

j

k

l

8   Draw one diagram that includes a pair of complementary adjacent angles, a straight angle and angles at a point.

WORDBANK

transversal A line that cuts across two or more lines.

corresponding angles Angles on the same side of the transversal and in the same position on the parallel lines. Corresponding angles form the letter F.

alternate angles Angles on opposite sides of the transversal and in between the parallel lines. Alternate angles form the letter Z.

co-interior angles Angles on the same side of the transversal and in between the parallel lines. They form the letter C.

Corresponding angles on parallel lines are equal.	Alternate angles on parallel lines are equal.	Co-interior angles on parallel lines are supplementary (add up to 180°).
		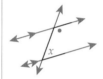

EXAMPLE 3

Find the value of each pronumeral, giving reasons.

a

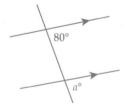

b

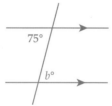

c

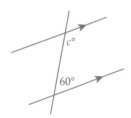

d

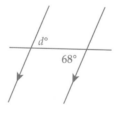

SOLUTION

a $a = 80$ (corresponding angles on parallel lines)

b $b = 75$ (alternate angles on parallel lines)

c $c + 60 = 180$ (co-interior angles on parallel lines)
$$c = 180 - 60$$
$$= 120$$

d $d = 68$ (alternate angles on parallel lines)

8-03 | Angles on parallel lines

Are the lines parallel in the diagram? Give a reason.

SOLUTION

The labelled angles are corresponding angles but they are not equal.
So the lines are not parallel.

EXERCISE 8-03

1   What is a line that crosses two other lines? Select the correct answer **A**, **B**, **C** or **D**.

    A co-interior   **B** alternate   **C** transversal   **D** corresponding

2   What type of angles are in matching positions on parallel lines? Select **A**, **B**, **C** or **D**.

    A co-interior   **B** alternate   **C** transversal   **D** corresponding

3   Copy this diagram and mark all 4 pairs of corresponding angles.
    Mark each pair with a different symbol.

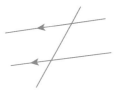

4   Copy the diagram from question **3** again and mark both pairs of alternate angles. Mark
    each pair with a different symbol.

5   Copy the diagram from question **3** again and mark both pairs of co-interior angles. Mark
    each pair with a different symbol.

6   State what type of angles are marked in each diagram.

    a

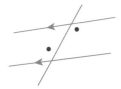

    b

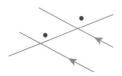

    c

7 Find the value of each pronumeral, giving reasons.

a

70°
$a°$

b

110°
$b°$

c

130°
$c°$

d

115°
$x°$

e

$y°$
68°

f

121°
$d°$

g

125°
$e°$

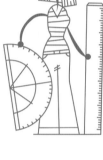

h

82°
$x°$

i
115°
$y°$

j

$d°$
133°

k

120°
$x°$
$y°$
$z°$

l

88°
$a°$
$b°$
$c°$

8 Are the lines parallel in each diagram below? Give reasons for your answer.

a

36°
34°

b
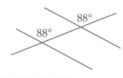
88°
88°

c
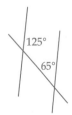
125°
65°

iStockphoto/prgzmat

Triangles can be classified in two different ways, according to their **sides**, or according to their **angles**.

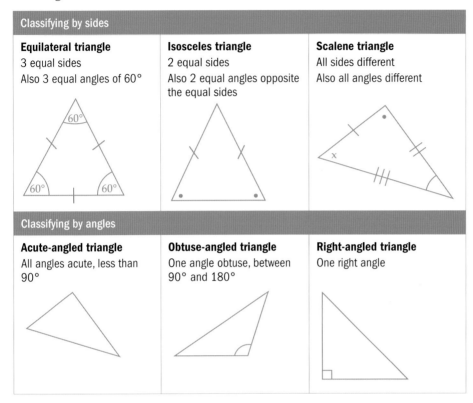

Classifying by sides		
Equilateral triangle 3 equal sides Also 3 equal angles of 60°	**Isosceles triangle** 2 equal sides Also 2 equal angles opposite the equal sides	**Scalene triangle** All sides different Also all angles different
Classifying by angles		
Acute-angled triangle All angles acute, less than 90°	**Obtuse-angled triangle** One angle obtuse, between 90° and 180°	**Right-angled triangle** One right angle

EXAMPLE 5

Classify each triangle according to

   i   sides   ii angles.

a  **b**

SOLUTION

a **i**   Two sides equal, so it is an isosceles triangle.

     ii All angles are acute, so it is an acute-angled triangle.

b **i**   All sides are different, so it is a scalene triangle.

     ii One obtuse angle, so it is an obtuse-angled triangle.

EXAMPLE 6

Sketch a triangle that is:

a   right-angled and isosceles

b   equilateral and acute-angled

SOLUTION

a

b

 An equilateral triangle is always acute-angled, as each angle is 60°

EXERCISE 8–04

1   What type of triangle has no equal sides? Select the correct answer **A, B, C** or **D**.
   A equilateral   **B** right-angled   **C** scalene   **D** isosceles

2   Which of the following is true about obtuse-angled triangles? Select **A, B, C** or **D**.
   A 1 angle obtuse   **B** 2 angles obtuse   **C** no angles acute   **D** all angles acute

3   Sketch a triangle that is:
   a   scalene   **b**   equilateral   **c**   isosceles
   d   obtuse-angled   **e**   right-angled   **f**   acute-angled

4   Classify each triangle according to its sides.

   a 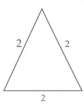   **b**    **c**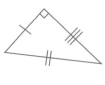

   d    **e**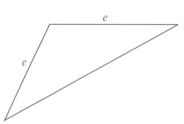

5   Classify each triangle in question **4** according to its angles.

6 Classify each triangle according to its angles.

a 　b 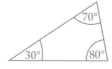　c

d 　e

7 Classify each triangle in question **6** according to its sides.

8 Copy and complete the table below.

Type of triangle	Number of equal sides	Number of equal angles
Scalene		
Isosceles		
	3	

9 Can an isosceles triangle be:

a acute-angled?　b obtuse-angled?　c right-angled?

10 Sketch each triangle described below.

a Scalene and right-angled　b Equilateral and acute-angled

c Isosceles and obtuse-angled　d Right-angled and isosceles

Shutterstock.com/Africa Studio

WORDBANK

exterior angle An outside angle formed by extending one side of a figure.

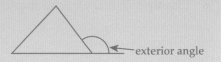

The angle sum of any triangle is 180°.

$a + b + c = 180$

The exterior angle of a triangle is the sum of the two interior opposite angles.

$x = a + b$

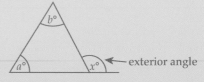

Each angle of an equilateral triangle is 60°.
The two angles opposite the two equal sides of an isosceles triangle are equal.

EXAMPLE 7

Find the value of each pronumeral, giving reasons.

a

b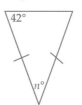

c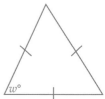

d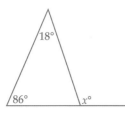

e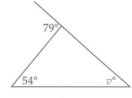

SOLUTION

a $72 + 35 + x = 180$ (angle sum of a triangle)
$$x = 180 - 72 - 35$$
$$x = 73$$

b $42 + 42 + n = 180$ (angle sum of an isosceles triangle)
$$n = 180 - 42 - 42$$
$$n = 96$$

c $w = 60$ (equilateral triangle)

d $x = 86 + 18$ (exterior angle of triangle)
$$x = 104$$

e $v + 54 = 79$ (exterior angle of triangle)
$$v = 79 - 54$$
$$v = 25$$

1 Which angles are equal in an isosceles triangle? Select the correct answer **A**, **B**, **C** or **D**.

    A All three angles   **B** At the base of the triangle

    C The exterior angles   **D** The angles opposite the equal sides

2 What is the exterior angle of a triangle equal to? Select **A**, **B**, **C** or **D**.

    A The sum of the two interior adjacent angles

    B The sum of the two interior opposite angles

    C The sum of the three interior angles

    D The sum of the opposite exterior angle

3 Classify each triangle by sides and angles, and state which angles are equal.

    a   **b**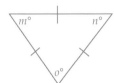

    c 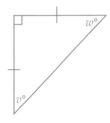   

Shutterstock.com/wolfmaster13

4 For each triangle, name:

    i the exterior angle   **ii** the two interior opposite angles

    a 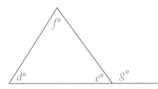   **b**

5 Copy and complete each solution to find the pronumeral.

    a 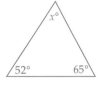   **b**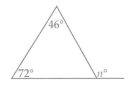

    $52 + 65 +$ ___ $= 180$ (___ of $\triangle$) $n = 72 +$ ___ (exterior ___)

                  $x = 180 - 52 -$ ___ $n =$ ___

                         $x =$ ___

6 Find the value of each pronumeral, giving reasons.

a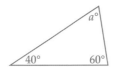

b

c

d

e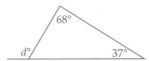

f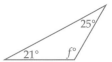

g

h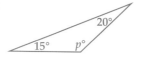

i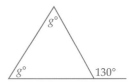

j

k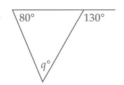

l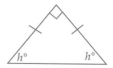

Alamy/William Caram

Convex quadrilateral
All vertices point outwards.
All diagonals lie inside the shape.

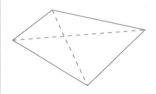

Non-convex quadrilateral
One vertex points inwards.
One diagonal lies outside the shape.
One angle is more than 180°.

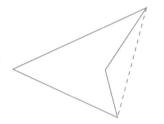

Parallelogram
Two pairs of parallel sides.
Opposite sides are equal.
Opposite angles are equal.
The diagonals bisect each other.

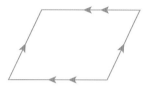

Trapezium
One pair of parallel sides.

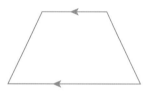

Rectangle
A parallelogram with all angles equal to 90°.
The diagonals are equal.

Rhombus
A parallelogram with all sides equal.
The diagonals bisect each other at right angles.
The diagonals bisect the angles of the rhombus.

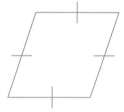

Square
A rectangle with all sides equal.
A rhombus with all angles equal to 90°.

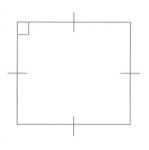

Kite
Two pairs of adjacent sides equal.
One pair of opposite angles equal.
One diagonal bisects the other at right angles.

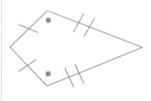

EXAMPLE 8

Classify each quadrilateral.

a 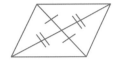   b    c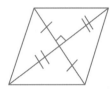

SOLUTION

a The diagonals bisect each other.
So it is a parallelogram.

b All sides are equal and it has a right angle.
So it is a square.

c The diagonals bisect each other at right
angles.
So it is a rhombus.

EXERCISE 8-06

1 Which two of these quadrilaterals have diagonals of equal length? Select two correct
answers **A**, **B**, **C** or **D**.

   A rhombus   **B** rectangle   **C** square   **D** trapezium

2 Draw a sketch of two different:

   a convex quadrilaterals   b non-convex quadrilaterals

3 Draw a sketch of each quadrilateral.

   a trapezium   b kite   c parallelogram   d rhombus

4 a Accurately draw a rectangle using geometrical instruments.

   b Measure the lengths of its sides. Are the opposite sides equal?

   c Measure the size of each angle. Are they all 90°?

   d Draw diagonals on your rectangle and measure each diagonal. Are they both equal?

5 Accurately draw a square and measure its sides, angles and diagonals to check that each
property of the square is true.

6 Classify each quadrilateral.

a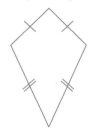

b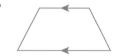

c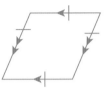

d

e

f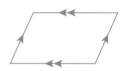

g

h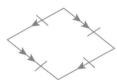

i

j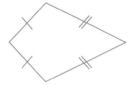

The angle sum of any quadrilateral is 360°.

$a + b + c + d = 360$

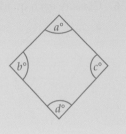

To demonstrate this, draw a quadrilateral and cut it out.
Tear off the 4 angles as shown.

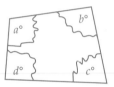

Place the torn-off angles next to each other and they will always form a
revolution. This means that the angle sum of a quadrilateral is 360°.

When finding unknown angles in quadrilaterals, remember:
- the opposite angles of a parallelogram are equal
- one pair of opposite angles of a kite are equal
- the diagonals of a rhombus, square and kite cross at 90°

EXAMPLE 9

Find the value of each pronumeral, giving reasons.

a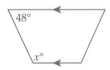

b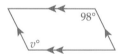

c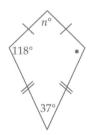

SOLUTION

a $x + 48 = 180$ (co-interior angles on parallel lines)
$$x = 180 - 48$$
$$x = 132$$

b $v = 98$ (opposite angles of a parallelogram)

c • = 118   (opposite angles of a kite)
$118 + 118 + 37 + n = 360$ (angle sum of a quadrilateral)
$$n = 360 - 118 - 118 - 37$$
$$= 87$$

1   What is the size of each angle in a rectangle? Select the correct answer **A, B, C** or **D**.

   A 180°   **B** 90°   **C** 360°   **D** 60°

2   Which word describes the pairs of angles between the parallel sides of a trapezium? Select **A, B, C** or **D**.

   A supplementary   **B** alternate   **C** equal   **D** corresponding

3   Copy and complete this table.

Quadrilateral	Sides	Angles	Diagonals
Parallelogram	Opposite sides equal and parallel.		Diagonals bisect each other.
Trapezium		All angles different.	
Rhombus	All sides equal. Opposite sides parallel.		
Rectangle		All angles are 90°.	
Square		All angles are 90°.	

4   Copy and complete for the parallelogram.

$x + \underline{} = 180$ $(\underline{}$ angles on parallel lines)

$x = 180 - \underline{}$

$x = \underline{}$

5   Copy and complete for the quadrilateral.

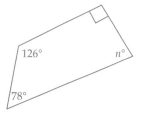

$n + 126 + \underline{} + 78 = \underline{}$ (angle sum of quadrilateral)

$n = 360 - \underline{} - \underline{} - \underline{}$

$n = \underline{}$

6   Find the value of each pronumeral, giving reasons.

a
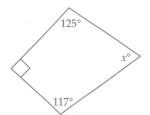

b
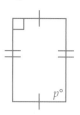

c

d
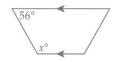

e
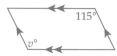

f
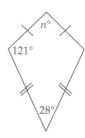

g

h

i

ISBN 9780170351027

Congruent figures

Congruent figures are shapes that are identical in every way, with the same shape and the same size.
- Matching sides are equal
- Matching angles are equal

These two triangles are congruent.

 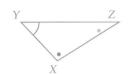

$\angle A$ matches $\angle X$, $\angle B$ matches $\angle Z$, $\angle C$ matches $\angle Y$.
We can write '$\triangle ABC \equiv \triangle XZY$' in matching order of the vertices, where '$\equiv$' stands for 'is congruent to'.
Congruent figures can be formed by **translation**, **reflection**, **rotation** or a combination of them.

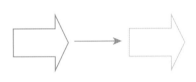   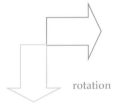

translation   reflection   rotation

EXAMPLE 10

These two triangles are congruent.

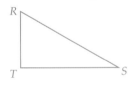 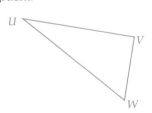

a Which transformation turns the first triangle into the second triangle?

b List the three pairs of matching angles.

c Copy and complete this congruence statement:
$\triangle RST \equiv \triangle$___

SOLUTION

a Rotation

b $\angle R = \angle W$, $\angle S = \angle U$, $\angle T = \angle V$

c $\triangle RST \equiv \triangle WUV$ ←———— matching order of vertices

1 If these two trapeziums are congruent, which angle in *PQRS* matches ∠*A*?

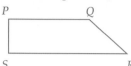

2 For question **1**, which is the correct order of vertices to complete the congruence statement '*ABCD* ≡ _____'? Select the correct answer **A**, **B**, **C** or **D**.

   A *PQRS*   **B** *QSRP*   **C** *RSPQ*   **D** *SRQP*

3 In each pair of congruent figures, which transformation turns the first shape into the second shape?

   a

   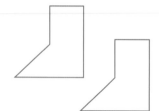

   b

   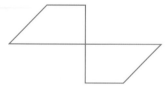

   c

   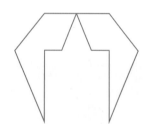

   d

4 Triangle *PQR* is rotated about *Q* to create triangle *P'Q'R*.

   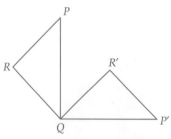

   a Are the triangles congruent?

   b Is △*P'QR'* ≡ △*PQR*?

5   These two parallelograms have sides of length 5 cm and 3 cm.

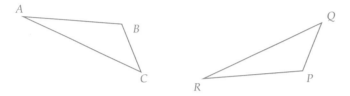

   a   Are they congruent?

   b   What else must you know to draw two congruent parallelograms?

6   If △ABC ≡ △RPQ, list all pairs of matching sides and matching angles.

7   The Dutch artist Maurits Cornelius Escher used congruent shapes to draw amazing works of art. Describe the congruent shapes in this Escher drawing. Which transformation has he used?

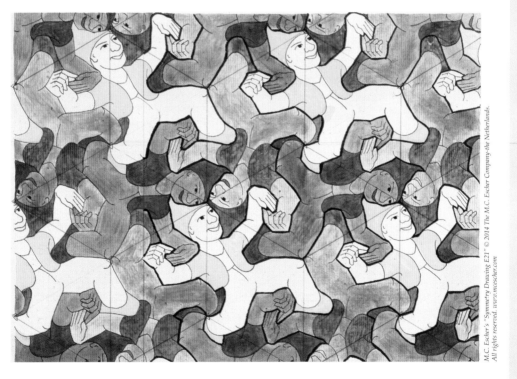

There are 4 tests that can be used to prove that two triangles are congruent.

1   SSS: Three sides of one triangle are equal to three sides of the other triangle.

2   SAS: Two sides and the **included** angle of one triangle are equal to two sides and the **included angle** of the other triangle.

The included angle is the angle between the two sides.

3   AAS: Two angles and one side of one triangle are equal to two angles and the matching side of the other triangle.

4   RHS: A right angle, hypotenuse and a side equal to the right angle, hypotenuse and the matching side of the other triangle.

EXAMPLE 11

Which two triangles are congruent? Which congruence test proves this?

SOLUTION

$\triangle ACB \equiv \triangle ZYX$ Test used was AAS.

Note that $\triangle PQR$ is not congruent to these triangles because PR is not the matching side.

1 Which test proves that these triangles are congruent? Select the correct answer **A, B, C** or **D**.

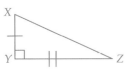

A SAS **B** RHS **C** SSS **D** AAS

2 For question **1**, $\triangle XYZ \equiv$ _____? Select the correct answer **A, B, C** or **D**.
   A $\triangle PQR$ **B** $\triangle RQP$ **C** $\triangle RPQ$ **D** $\triangle PRQ$

3 These two triangles are congruent.

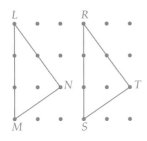

   a Name pairs of matching sides
   b Copy and complete: $\triangle LMN \equiv \triangle$ _____
   c Which test proves this?

4 Which two triangles are congruent?

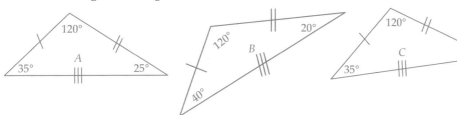

5 **a** Which two triangles are congruent? Remember to state the vertices in the correct order.

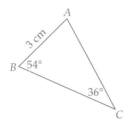 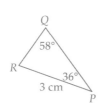 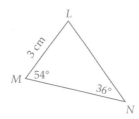

   b Which test did you use?

6 Select all congruent triangles from this set.

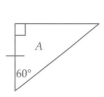 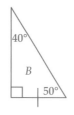 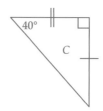 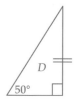

7 Find three pairs of congruent triangles and state which test you used.

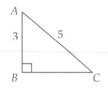 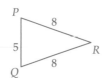 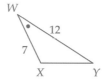

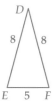 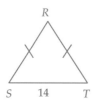

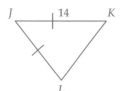 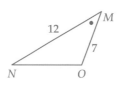

8 Draw two congruent triangles that follow the AAS test.

Corbis/Gollings John/Arcaid

MIX AND MATCH

Make a copy of this page, find the missing angle(s) in each diagram and match the diagram letter to the angle size to decode the rhyme below.

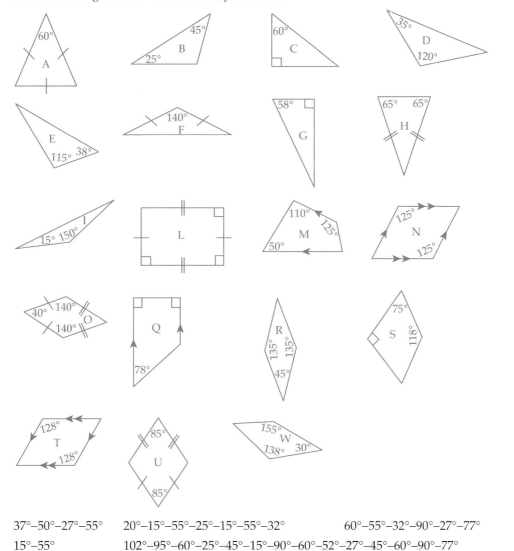

37°–50°–27°–55°   20°–15°–55°–25°–15°–55°–32°   60°–55°–32°–90°–27°–77°

15°–55°   102°–95°–60°–25°–45°–15°–90°–60°–52°–27°–45°–60°–90°–77°

60°–55°–25°   52°–45°–15°–60°–55°–32°–90°–27°–77°,

45°–27°–75°–27°–75°–110°–27°–45°   60°–55°–32°–90°–27°   77°–95°–75°

60°–55°–25°   52°–50°–27°   60°–55°–77°–37°–27°–45°   37°–15°–90°–90°   30°–40°–75°–27°

PRACTICE TEST 8

Part A General topics
Calculators are not allowed.

1 Increase $160 by 5%.

2 Simplify $\dfrac{2k^2}{4k}$.

3 Evaluate $7.40 + $12.60.

4 Find the volume of this cube.

4 cm

5 Copy and complete: $\dfrac{2}{5} = \dfrac{}{25}$

6 Convert 28% to a simple fraction.

7 Simplify 35 : 14

8 Expand and simplify $2(4x-4)+6(x+2)$.

9 Evaluate $18 + 4 \times 5$

10 How many hours and minutes are there from 11:16 a.m. to 5:56 p.m.?

Part B Geometry
Calculators are allowed.

8–01 Types of angles

11 What type of angle is 106°? Select the correct answer **A**, **B**, **C** or **D**.
  A obtuse **B** reflex **C** straight **D** acute

12 Name the angle drawn. Select **A**, **B**, **C** or **D**.

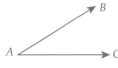

  A $\angle ABC$ **B** $\angle ACB$ **C** $\angle BAC$ **D** $\angle CBA$

8-02 Angle geometry

13 Find the value of each pronumeral, giving reasons.

a

b

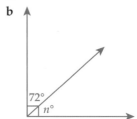

8-03 Angles on parallel lines

14 Copy this diagram and mark two sets of alternate angles on it.

15 Find the value of the pronumeral.

8-04 Types of triangles

16 Classify each triangle according to its sides.

a

b

17 Classify each triangle in question **16** according to its angles.

8-05 Triangle geometry

18 Find the value of each pronumeral, giving reasons.

a

b

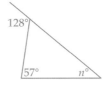

8-06 Types of quadrilaterals

19 Name each quadrilateral.

a

b

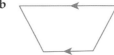

8-07 Quadrilateral geometry

20 Find the value of each pronumeral, giving reasons.

a

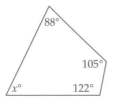

b

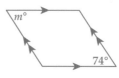

8-08 Congruent figures

21 For these two congruent triangles, complete this statement: $\triangle ACB = \triangle$ _____.

a

b

8-09 Tests for congruent triangles

22 Which test proves that these two triangles are congruent?

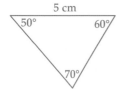

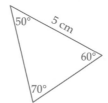

EQUATIONS

WHAT'S IN CHAPTER 9?

IN THIS CHAPTER YOU WILL:

- solve one-step and two-step equations
- solve equations with variables on both sides
- expand and solve equations with brackets
- solve simple quadratic equations
- use equations to solve practical problems

Shutterstock.com/Vladitto

WORDBANK

equation A number sentence that contains algebraic terms, numbers and an equals sign; for example, $x - 3 = 8$.

solve an equation To find the value of the variable that makes the equation true.

inverse operation The 'opposite' process; for example, the inverse operation to adding (+) is subtracting (–).

EXAMPLE 1

Use 'guess-and-check' to solve each equation.

a $x + 4 = 12$ **b** $n - 6 = 18$ **c** $4a = 20$

SOLUTION

✱ Find the value that will make each equation true.

a $x + 4 = 12$ **b** $n - 6 = 18$ **c** $4a = 20$

$8 + 4 = 12$ $24 - 6 = 18$ $4 \times 5 = 20$

So $x = 8$ So $n = 24$ So $a = 5$

EXAMPLE 2

Use inverse operations to solve each equation.

a $w + 6 = 10$ **b** $7a = -42$ **c** $\dfrac{m}{8} = 5$

SOLUTION

✱ Do the same inverse operation to both sides of the equation

a $w + 6 = 10$

$w + 6 - 6 = 10 - 6$

$w = 4$

(Check: $4 + 6 = 10$)

b $7a = -42$

$\dfrac{7a}{7} = \dfrac{-42}{7}$

$a = -6$

(Check: $7 \times (-6) = -42$)

c $\dfrac{m}{8} = 5$

$\dfrac{m}{8} \times 8 = 5 \times 8$

$m = 40$

(Check: $\dfrac{40}{8} = 5$)

✱ These are called one-step equations because they require only one step to solve.

TO SOLVE AN EQUATION:

■ do the same to both sides of the equation: this will keep it balanced

■ use inverse operations to simplify the equation
(+ and − are inverse operations, × and ÷ are inverse operations)

■ write the **solution** (answer) as: x = a number

1 What is the solution to $x - 8 = 21$? Select the correct answer **A**, **B**, **C** or **D**.

   A $x = 13$ **B** $x = 29$ **C** $x = 14$ **D** $x = 28$

2 Find the solution to $n - 9 = 15$.

   A $n = 6$ **B** $n = 135$ **C** $n = -9$ **D** $n = 24$

3 Use 'guess-and-check' to solve each equation.

   a $x + 3 = 6$ **b** $m + 4 = 5$

   c $p + 6 = 9$ **d** $a + 7 = 12$

   e $q + 1 = 2$ **f** $x - 3 = 4$

   g $y - 5 = 10$ **h** $m - 4 = 7$

   i $p - 10 = 1$ **j** $q - 3 = 12$

   k $2x = 10$ **l** $3y = 12$

   m $5a = 20$ **n** $6y = 12$

   o $2m = 4$ **p** $\dfrac{x}{2} = 10$

   q $\dfrac{a}{4} = 3$ **r** $\dfrac{m}{2} = 6$

   s $\dfrac{y}{7} = 2$ **t** $\dfrac{a}{6} = 4$

4 Copy and complete each sentence.

   a To solve $m + 4 = 7$, subtract ___ from both sides.

   b To solve $x - 8 = 17$, _____ 8 to both sides.

   c To solve $5n = 9$, _____ both sides by ___.

   d To solve $\dfrac{a}{4} = 8$, _____ both sides by ___.

5 Solve each equation using inverse operations.

   a $x + 10 = 12$ **b** $y + 12 = 19$ **c** $x + 20 = 30$ **d** $y + 1 = 17$

   e $m + 19 = 21$ **f** $x - 12 = 1$ **g** $a - 4 = 2$ **h** $b - 3 = 6$

   i $y - 7 = 2$ **j** $x - 5 = 3$ **k** $3p = 9$ **l** $5n = 15$

   m $2x = 22$ **n** $7b = 21$ **o** $4t = 16$ **p** $\dfrac{x}{3} = 9$

   q $\dfrac{t}{2} = 11$ **r** $\dfrac{n}{5} = 2$ **s** $\dfrac{p}{8} = 2$ **t** $\dfrac{q}{4} = 8$

6 Solve each equation. The solutions are either negative or fractions.

   a $m + 8 = 4$ **b** $x - 7 = -12$

   c $7e = 18$ **d** $\dfrac{v}{5} = -4$

   e $8w = -9$ **f** $x + 16 = -5$

   g $\dfrac{s}{-4} = 20$ **h** $n - 9 = -30$

EXAMPLE 3

Use inverse operations to solve each equation.

a $2x + 3 = 9$ **b** $3n - 5 = 10$

> ✱ These are called two-step equations because they require two steps to solve.

SOLUTION

a
$$2x + 3 = 9$$
$$2x + 3 - 3 = 9 - 3 \quad \longleftarrow \text{Step 1: } -3 \text{ from both sides}$$
$$2x = 6$$
$$\frac{2x}{2} = \frac{6}{2} \quad \longleftarrow \text{Step 2: } \div \text{ both sides by 2}$$
$$x = 3$$

(Check: $2 \times \mathbf{3} + 3 = 9$)

b
$$3n - 5 = 10$$
$$3n - 5 + 5 = 10 + 5 \quad \longleftarrow \text{Step 1: } + 5 \text{ to both sides}$$
$$3n = 15$$
$$\frac{3n}{3} = \frac{15}{3} \quad \longleftarrow \text{Step 2: } \div \text{ both sides by 3}$$
$$n = 5$$

(Check: $3 \times \mathbf{5} - 5 = 10$)

EXAMPLE 4

Solve each equation.

a $\dfrac{x - 3}{4} = 8$ **b** $26 - 3w = 35$

SOLUTION

a
$$\frac{x - 3}{4} \times 4 = 8 \times 4 \quad \longleftarrow \times \text{ both sides by 4}$$
$$x - 3 = 32$$
$$x - 3 + 3 = 32 + 3 \quad \longleftarrow +3 \text{ to both sides}$$
$$x = 35$$

(Check: $\dfrac{35 - 3}{4} = 8$)

b
$$26 - 3w - 26 = 35 - 26 \quad \longleftarrow -26 \text{ from both sides}$$
$$-3w = 9$$
$$\frac{-3w}{-3} = \frac{9}{-3} \quad \longleftarrow \div \text{ both sides by } (-3)$$
$$w = -3$$

(Check: $26 - 3 \times (\mathbf{-3}) = 35$)

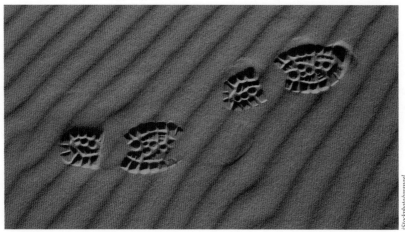

iStockphoto/arangel

1 To solve $5x - 3 = 12$, which operation would you do first? Select the correct answer **A**, **B**, **C** or **D**.

   A $\div 5$ **B** -3 **C** $\times 5$ **D** $+3$

2 To solve $18 - 4n = -16$ which operation would you do first?

   A $+18$ **B** $\div (-4)$ **C** -18 **D** $+4$

3 **a** Make up a one-step equation that has $x = 4$ as its solution.

   b Make up another one-step equation that has $x = 4$ as its solution.

   c Make up a two-step equation that has $x = 4$ as its solution.

4 Copy and complete the solution to each equation.

   a
$$5a - 3 = 17$$
$$5a - 3 + \underline{\ \ } = 17 + \underline{\ \ }$$
$$5a = \underline{\ \ \ }$$
$$\frac{5a}{\ } = \frac{20}{\ }$$
$$a = \underline{\ \ \ }$$

   b
$$2x + 6 = 14$$
$$2x + 6 - \underline{\ \ } = 14 - \underline{\ \ }$$
$$2x = \underline{\ \ \ }$$
$$\frac{\ }{2} = \frac{8}{\ }$$
$$x = \underline{\ \ \ }$$

5 Solve each equation.

   a $2x + 3 = 9$ **b** $2a + 1 = 5$

   c $2b + 2 = 10$ **d** $3m + 1 = 13$

   e $3b + 7 = 1$ **f** $5a + 2 = 22$

   g $2x - 1 = 9$ **h** $2a - 2 = 12$

   i $2b - 1 = 15$ **j** $3y - 1 = 11$

   k $3m - 2 = 13$ **l** $3n - 5 = 22$

6 Check each solution to question **5** by substituting it into the original equation.

7 Solve each equation. The solutions are either negative or fractions.

   a $3x + 4 = 12$ **b** $5n - 4 = 15$

   c $12 - 2a = 16$ **d** $8 + 3x = 15$

   e $15 - 4s = 20$ **f** $6y - 4 = -12$

   g $18 - 3r = -6$ **h** $5w - 3 = -32$

   i $\dfrac{x - 5}{4} = 6$ **j** $\dfrac{n + 6}{3} = -9$

   k $\dfrac{4 - u}{5} = 6$ **l** $\dfrac{-12 - x}{6} = 7$

8 Is each statement true or false?

   a $x = 4$ is the solution to $3x - 8 = 4$

   b $n = -2$ is the solution to $\dfrac{n - 4}{6} = 1$

   c $c = -3$ is the solution to $18 - 2c = 24$

   d $m = 6$ is the solution to $3m + 9 = 25$

TO SOLVE AN EQUATION WITH VARIABLES ON BOTH SIDES:
- ▨ use inverse operations to move all the variables to the left-hand side (LHS) of the equation
- ▨ use inverse operations to move all the numbers to the right-hand side (RHS) of the equation
- ▨ then solve the equation

EXAMPLE 5

Solve each equation.

a $2x - 8 = x + 15$ **b** $5n + 7 = 3n - 15$

SOLUTION

✳ Move all variables to the LHS of the equation

a
$$2x - 8 = x + 15$$
$$2x - 8 - x = x + 15 - x \qquad \longleftarrow \;\; -x \text{ from both sides}$$
$$x - 8 = 15 \qquad \longleftarrow \;\; \text{Simplifying } 2x - x$$
$$x - 8 + 8 = 15 + 8 \qquad \longleftarrow \;\; +8 \text{ to both sides}$$
$$x = 23$$

b
$$5n + 7 = 3n - 15$$
$$5n + 7 - 3n = 3n - 15 - 3n \qquad \longleftarrow \;\; -3n \text{ from both sides}$$
$$2n + 7 = -15 \qquad \longleftarrow \;\; \text{Simplifying } 5n - 3n$$
$$2n + 7 - 7 = -15 - 7 \qquad \longleftarrow \;\; -7 \text{ from both sides}$$
$$2n = -22$$
$$\frac{2n}{2} = \frac{-22}{2} \qquad \longleftarrow \;\; \div \text{ both sides by 2}$$
$$n = -11$$

Shutterstock.com/Karramba Production

EXAMPLE 6

Check the solution to Example **5a** by substituting it into the equation.

SOLUTION

$$2x - 8 = x + 15$$

✳ Check LHS = RHS

Substitute $x = 23$

$$\text{LHS} = 2 \times 23 - 8 \qquad\qquad\qquad \text{RHS} = 23 + 15$$
$$= 38 \qquad\qquad\qquad\qquad\qquad\; = 38$$

LHS = RHS so the solution $x = 23$ is correct.

1 To solve $3x - 4 = x + 16$, which operation would you do first? Select the correct answer **A**, **B**, **C** or **D**.

A $+x$ **B** -4 **C** $-x$ **D** $+16$

2 To solve $4x + 9 = 2x - 7$, which operation would you do first? Select **A**, **B**, **C** or **D**.

A $-2x$ **B** $+4x$ **C** $+2x$ **D** -7

3 Write down the LHS of each equation.

  a $3x - 4 = 2x + 9$ **b** $5a + 6 = 3a - 4$ **c** $8 - 2x = x + 5$

4 Write down the RHS of each equation in question **3**.

5 Solve each equation in question **3** and check your solution as shown in Example **6**.

6 Copy and complete the solution to each equation.

  a
$$2w + 16 = w - 8$$
$$2w + 16 - \underline{} = w - 8 - \underline{}$$
$$w + \underline{} = -8$$
$$w + 16 - \underline{} = -8 - \underline{}$$
$$w = \underline{}$$

  b
$$4x - 15 = 2x + 17$$
$$4x - 15 - \underline{} = 2x + 17 - \underline{}$$
$$2x - \underline{} = 17$$
$$2x - 15 + \underline{} = 17 + \underline{}$$
$$2x = \underline{}$$
$$x = \underline{}$$

7 Solve each equation.

  a $2a - 4 = a + 8$ **b** $3w - 2 = 2w + 7$

  c $5b + 3 = 4b - 5$ **d** $3m + 2 = 2m - 8$

  e $4a + 2 = 3a + 8$ **f** $4x - 8 = 3x + 5$

  g $5a - 4 = 2a + 5$ **h** $4b + 6 = 2b - 8$

  i $6x - 5 = 3x + 13$ **j** $8w - 4 = 2w + 8$

  k $3x - 8 = x + 12$ **l** $5r + 6 = 2r - 9$

  m $9a + 3 = 5a + 11$ **n** $7b - 4 = 5b + 10$

  o $6a + 4 = 2a - 8$

8 Solve each equation. Some answers may be negative or fractions.

  a $5x - 4 = 12 - 3x$ **b** $16 - 3n = 11 + n$

  c $22 + 4x = 3x - 8$ **d** $6d - 8 = 3d$

  e $28 - 4m = -15 + 3m$ **f** $-8w = 9 - 5w$

  g $6 - 2x = 7 + 5x$ **h** $19 - 5n = -2n$

  i $26 - 3x + 7 = 15 - 4x$

An equation with brackets, such as $3(x - 4) = 9$, can be solved by expanding out the brackets first and then solving as before using inverse operations.

EXAMPLE 7

Solve each equation.

a $3(x - 4) = 9$ **b** $5(2a + 6) = 20$

SOLUTION

a
$$3(x - 4) = 9$$
$$3 \times x - 3 \times 4 = 9 \qquad \longleftarrow \text{ Expand}$$
$$3x - 12 = 9 \qquad \longleftarrow \text{ Simplify}$$
$$3x - 12 + 12 = 9 + 12 \qquad \longleftarrow \text{ + 12 to both sides}$$
$$3x = 21$$
$$\frac{3x}{3} = \frac{21}{3} \qquad \longleftarrow \div \text{ both sides by 3}$$
$$x = 7$$

b
$$5(2a + 6) = 20$$
$$5 \times 2a + 5 \times 6 = 20 \qquad \longleftarrow \text{ Expand}$$
$$10a + 30 = 20 \qquad \longleftarrow \text{ Simplify}$$
$$10a + 30 - 30 = 20 - 30 \qquad \longleftarrow \text{ −30 from both sides}$$
$$10a = -10$$
$$\frac{10a}{10} = \frac{-10}{10} \qquad \longleftarrow \div \text{ both sides by 10}$$
$$a = -1$$

EXAMPLE 8

Solve $-4(2x - 1) = 8$ and then check your solution is correct.

SOLUTION

$$-4(2x - 1) = 8$$
$$-4 \times 2x + (-4) \times (-1) = 8 \qquad \longleftarrow \text{ Expand}$$
$$-8x + 4 = 8 \qquad \longleftarrow \text{ Simplify}$$
$$-8x + 4 - 4 = 8 - 4 \qquad \longleftarrow \text{ Subtract 4 from both sides}$$
$$-8x = 4$$
$$\frac{-8x}{-8} = \frac{4}{-8} \qquad \longleftarrow \text{ Divide both sides by (−8)}$$
$$x = -\frac{1}{2}$$

To check the solution, substitute $x = -\frac{1}{2}$:

$$\text{LHS} = -4 \times \left[2 \times \left(-\frac{1}{2} \right) - 1 \right] \qquad\qquad \text{RHS} = 8$$
$$= -4 \times (-2)$$
$$= 8$$

LHS = RHS, so the solution $x = -\frac{1}{2}$ is correct.

ISBN 9780170351027

1   Find the solution to $2(a - 6) = 4$. Select the correct answer **A**, **B**, **C** or **D**.

   A $a = 4$ **B** $a = 8$ **C** $a = -8$ **D** $a = 2$

2   Find the solution to $-3(x + 4) = 6$. Select the correct answer **A**, **B**, **C** or **D**.

   A $x = 6$ **B** $x = 2$ **C** $x = -2$ **D** $x = -6$

3   True or false?

   a $4(a + 2) = 4a + 2$ **b** $-3(w - 2) = -3w - 6$

   c $5(2x + 3) = 10x + 15$ **d** $-2(3n - 4) = -6n + 8$

4   Copy and complete the solution to each equation.

   a $3(w - 4) = 15$ **b** $-2(a + 4) = 12$
               $3w - \_\_\_ = 15$ $-2a - 8 = \_\_\_$
            $3w - 12 + \_\_\_ = 15 + \_\_\_$ $-2a - 8 + \_\_\_ = 12 + \_\_\_$
                     $3w = \_\_\_$ $-2a = \_\_\_$
                      $w = \_\_\_$ $a = \_\_\_$

5   Solve each equation.

   a $2(x - 3) = 8$ **b** $3(a + 4) = 6$

   c $5(b - 3) = 5$ **d** $4(x + 2) = 12$

   e $8(a + 2) = 32$ **f** $2(b - 9) = 6$

   g $9(m + 3) = 36$ **h** $6(a - 4) = 36$

   i $7(v - 3) = 56$ **j** $5(z + 3) = 30$

   k $4(a + 5) = -40$ **l** $8(w - 2) = 80$

6   Check that your solutions to questions **5 a** to **d** are correct.

7   Solve each equation below. The solutions may be negative or fractions.

   a $3(2x - 1) = 9$ **b** $4(3a + 2) = 32$

   c $6(5b + 2) = 62$ **d** $2(3y - 4) = 22$

   e $5(4a + 2) = 50$ **f** $2(4b - 3) = -14$

   g $3(2x + 3) = 21$ **h** $4(2w + 2) = 24$

   i $-3(2a + 4) = 8$ **j** $-5(n + 6) = 7$

   k $-4(3x - 1) = 5$ **l** $-8(2m - 4) = -6$

8   Check your solutions to questions **7 a** to **d**.

9   Write down two different equations with brackets whose solution is $x = 3$.

10 Write down two different equations with brackets whose solution is $x = -2$.

WORDBANK

quadratic equation An equation such as $x^2 = 4$, where the highest power of the variable is 2, that is, the variable squared.

surd A square root whose answer is not an exact decimal number. For example, $\sqrt{8} = 2.8284...$ is a surd because there isn't an exact number squared that is equal to 8.

To solve the quadratic equation $x^2 = 4$, we need to find the number which, when squared, is equal to 4. $2^2 = 4$ but also $(-2)^2 = 4$ as $-2 \times (-2) = 4$.
So the solution to $x^2 = 4$ is $x = 2$ or $x = -2$, which we write as $x = \pm 2$.

EXAMPLE 9

Solve each equation, writing each solution in exact form.

✱ exact form means a number written as a decimal or surd, without rounding

a $x^2 = 9$　　　**b** $x^2 = 5$　　　**c** $a^2 = 17$

SOLUTION

a $x^2 = 9$　　　**b** $x^2 = 5$　　　**c** $a^2 = 17$
　　$x = \pm\sqrt{9}$　　　　$x = \pm\sqrt{5}$　　　　$a = \pm\sqrt{17}$　　　✱ x^2 and $\sqrt{}$ are inverse operations
　　$x = \pm 3$

✱ $\pm\sqrt{5}$ and $\pm\sqrt{17}$ are surds

The solution to $x^2 = c$ is $x = \pm\sqrt{c}$

EXAMPLE 10

Solve $a^2 = 17$ correct to 2 decimal places.

SOLUTION

$a^2 = 17$
$a = \pm\sqrt{17}$
　$= \pm 4.123105...$
　$\approx \pm 4.12$　　　⟵　correct to 2 decimal places

Shutterstock.com/Brian Kinney

1. What is the solution to $x^2 = 16$ in simplest form? Select the correct answer **A**, **B**, **C** or **D**.

    A $x = \pm\sqrt{16}$ **B** $x = \pm4$ **C** $x = \mp\sqrt{16}$ **D** $x = \pm8$

2. What is the solution to $x^2 = 8$, correct to two decimal places?

    A $x = \pm\sqrt{8}$ **B** $x = \pm2.82$ **C** $x = \pm2.83$ **D** $x = \pm4$

3. True or false?

    a   If $x^2 = 25$, then $x = \pm5$ **b**   If $x^2 = 6$, then $x = \pm\sqrt{16}$

    c   If $x^2 = 132$, then $x = \pm12$ **d**   If $x^2 = 47$, then $x = \pm\sqrt{47}$

4. Solve each equation, leaving your answer in exact form.

    a $x^2 = 36$ **b** $x^2 = 81$ **c** $a^2 = 100$ **d** $n^2 = 1$

    e $x^2 = 7$ **f** $x^2 = 11$ **g** $w^2 = 14$ **h** $b^2 = 22$

    i $x^2 = 49$ **j** $x^2 = 38$ **k** $a^2 = 106$ **l** $n^2 = 19$

    m $a^2 = 144$ **n** $x^2 = 68$ **o** $m^2 = 42$ **p** $n^2 = -9$

    q $x^2 = 256$ **r** $x^2 = -25$ **s** $a^2 = 10\,000$ **t** $p^2 = 111$

5. For all solutions in question **4** that are surds, write the solutions correct to one decimal place.

6. Is each statement true or false?

    a $x^2 = c$ has two solutions if c is a positive number.

    b $x^2 = c$ always has two solutions.

    c $x^2 = c$ can have three solutions.

    d $x^2 = 0$ has only one solution.

    e   The solutions to $x^2 = c$ are always whole numbers or fractions.

    f $x^2 = c$ has two solutions if c is a negative number.

7. Solve each quadratic equation correct to 2 decimal places.

    a $x^2 = 182$ **b** $n^2 = 431$ **c** $w^2 = 18.62$

    d $b^2 = 21.6$ **e** $u^2 = 4000$

8. State which equations have surd solutions and find each surd solution.

    a $x^2 = 196$ **b** $n^2 = 28$ **c** $w^2 = 54$ **d** $b^2 = 121$

    e $u^2 = 100\,000$ **f** $a^2 = 49$ **g** $x^2 = 212$ **h** $a^2 = 16$

    i $x^2 = 400$ **j** $v^2 = 900\,000\,000$

EXAMPLE 11

Sally says that if she doubles her favourite number and adds 4, the answer is 18. What is the number?

SOLUTION

Let the number be x.

$2x + 4 = 18$ ←——— Double the number is 2 times the number.

$2x + 4 - 4 = 18 - 4$ ←——— Solve the equation.

$2x = 14$

$\dfrac{2x}{2} = \dfrac{14}{2}$

$x = 7$

∴ Sally's favourite number is 7.

(Check: $2 \times 7 + 4 = 18$)

TO SOLVE A WORD PROBLEM INVOLVING EQUATIONS:
- ▨ let the unknown value be x (or another variable)
- ▨ convert the words into an equation
- ▨ solve the equation
- ▨ answer the problem in words

EXAMPLE 12

Surinder is twice Jenny's age. In 4 years, the sum of their ages will be 26.

How old are Jenny and Surinder now?

SOLUTION

Let Jenny's age now be n. So Surinder's age is $2n$.

In 4 years Jenny will be $n + 4$ and Surinder will be $2n + 4$.

$n + 4 + 2n + 4 = 26$ ←——— Sum of ages will be 26

$3n + 8 = 26$ ←——— Simplifying

$3n + 8 - 8 = 26 - 8$

$3n = 18$

$\dfrac{3n}{3} = \dfrac{18}{3}$

$n = 6$

∴ Jenny is 6 years old now and Surinder is
($2 \times 6 =$) 12 years old.

(Check: In 4 years, the sum of their ages will be $10 + 16 = 26$)

iStockphoto/Vikram Raghuvanshi

1   Which equation says 'Twice a number n plus four is equal to twelve'? Select the correct answer **A, B, C** or **D**.

   A $n + 4 = 12$ **B** $2n - 4 = 12$ **C** $2n + 12 = 4$ **D** $2n + 4 = 12$

2   Convert each worded statement into an equation.

   a   The sum of a number n and 6 is equal to 8.

   b   The difference between a number x and 4 is 18.

   c   The product of a number w and 8 is 24.

   d   The quotient of a number m and 6 is 14.

3   Solve each equation in question **2** to find the number.

4   When n is added to 12, the answer is 32. Find n.

5   When a number (p) is taken from 14, the result is 5. What is the number?

6   When a certain number is divided by 6, the result is 10. What is the number?

7   When Lisa doubles her favourite number and adds 5, the answer is 15. What is the number?

8   I think of a number, multiply it by 3, then subtract 2. The result is 19. What was the number?

9   A side of a square is s cm long. Its perimeter is 32 cm. What is the value of s?

10 The perimeter of this rectangle is 18 cm. What is the value of n?

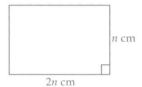

11 The perimeter of this triangle is 30 cm. What is the value of n?

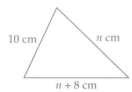

12 Sophie is three times Amy's age. The sum of their ages is 64. How old are Sophie and Amy?

13 Fatima is twice Harry's age. In three years time, the sum of their ages will be 18. How old are Fatima and Harry now?

14 What number is equal to 6 less than 4 times the same number?

15 Anka thinks of a number and divides it by 4. She then adds 5 and multiplies by 6 to get a result of 72. What number did Anka first think of?

16 Twice a number plus 5 is equal to 3 times the number less 9. What is the number?

CODE PUZZLE

Match each equation on the left with its solution on the right.

1	$2x = 28$	**A**	$x = 11$
2	$x - 7 = 5$	**B**	$x = 9$
3	$x + 10 = 12$	**D**	$x = 6$
4	$\dfrac{x}{5} = 4$	**E**	$x = 22$
5	$3x - 6 = 18$	**H**	$x = -6$
6	$2x + 5 = 25$	**I**	$x = 2$
7	$3(x - 4) = 15$	**L**	$x = 4\dfrac{1}{2}$
8	$4(x + 4) = 32$	**O**	$x = 14$
9	$2(3x - 2) = 32$	**Q**	$x = 7$
10	$6(2x + 3) = 72$	**S**	$x = 12$
11	$\dfrac{x - 4}{3} = 6$	**T**	$x = 4$
12	$\dfrac{x + 5}{2} = \dfrac{3}{4}$	**U**	$x = 10$
13	$2x - 4 = x + 7$	**V**	$x = -3\dfrac{1}{2}$
14	$3x + 4 = x - 8$	**W**	$x = 8$
15	$7x - 5 = 3x + 23$	**Y**	$x = 20$

Match the letter of the correct answer with each question number to solve this riddle:

What do an equation and an isosceles triangle have in common?

8–14–11–4   7–1–8–14   14–13–12–11   8–5–1   11–15–6–13–10   2–3–9–11–2!

Alamy/Bill Grant

PRACTICE TEST 9

Part A General topics

Calculators are not allowed.

1 Round $923.674 to the nearest dollar.

2 Expand and simplify $2(x + y) - 3(x - y)$.

3 Convert 9.05% to a decimal.

4 Find the area of this parallelogram.

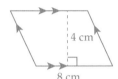

5 Write as a basic numeral 2×10^{-3}.

6 Write 495 000 in scientific notation.

7 Write $\dfrac{874}{1000}$ as a percentage.

8 Evaluate 33.2 − 3.9.

9 Copy and complete: $1 \text{ m}^2 = $ _____ cm^2.

10 Write down an algebraic expression for the perimeter of a rectangle of length l and width w.

Part B Equations

Calculators are allowed.

9-01 One-step equations

11 Solve each equation.

   a $\quad x + 6 = 11$ b $\quad n - 4 = 6$ c $\quad 4x = 24$

12 Solve $a - 8 = -12$. Select the correct answer **A, B, C** or **D**.

   A $\ a = 4$ **B** $\ a = 8$ **C** $\ a = -4$ **D** $\ a = -8$

9-02 Two-step equations

13 Solve $3n + 6 = 21$. Select **A, B, C** or **D**.

   A $\ n = 15$ **B** $\ n = 5$ **C** $\ n = -5$ **D** $\ n = -15$

14 Solve each equation.

   a $\quad 2x - 5 = -13$ b $\quad 22 - 4c = 24$ c $\quad \dfrac{x + 4}{8} = -2$

9-03 Equations with variables on both sides

15 Solve each equation.

   a $\quad 2x - 4 = x + 8$ b $\quad 3a + 8 = 2a - 6$ c $\quad 5n + 6 = 3n - 4$

9-04 Equations with brackets

16 Solve each equation.

   a $\quad 2(x + 3) = 14$ b $\quad -3(a - 5) = 18$ c $\quad 8(2x - 4) = -24$

17 Write down two different equations that have a solution of $x = -2$.

9-05 Simple quadratic equations $x^2 = c$

18 Solve each equation, writing the answer in exact form.

    a $x^2 = 49$ **b** $x^2 = 13$

19 Solve $n^2 = 61$, writing the answer correct to 2 decimal places.

9-06 Equation problems

20 Annika says that if she triples her favourite number and adds 12, the result is 33. Write an equation to find the number and solve it.

21 Find the value of x if the perimeter of this rectangle is 88 cm.

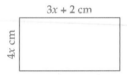

$3x + 2$ cm

$4x$ cm

22 Michael is four times Sunny's age. In 2 years time the sum of their ages will be 84. How old are Michael and Sunny now?

EARNING MONEY

10

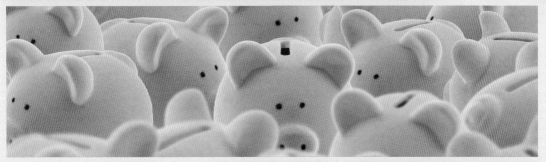

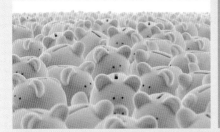

IN THIS CHAPTER YOU WILL:

- calculate wages and salaries
- convert between weekly, fortnightly, monthly and annual incomes
- calculate overtime pay involving time-and-a-half and double time
- calculate commission, piecework and annual leave loading
- calculate income tax based on allowable deductions and taxable income
- read PAYG tax tables to calculate PAYG tax
- calculate net pay from gross pay

Shutterstock.com/Jakub Krechowicz

WORDBANK

wage An income calculated on the number of hours worked, usually paid weekly or fortnightly.

salary A fixed amount of income per year, usually paid weekly, fortnightly or monthly.

per annum / annual Per year.

1 year = 52 weeks = 26 fortnights = 12 months

EXAMPLE 1

The wage paid for a waiter is $25.40 per hour at Delish restaurant. Find Jordan's wage if he works:

a   35 hours   b   28.5 hours

SOLUTION

a   Wage for 35 hours = 35 × $25.40   b   Wage for 28.5 hours = 28.5 × $25.40
                       = $889.00   = $723.90

EXAMPLE 2

Abby is offered a job at Allgood on a salary of $72 000 p.a. and a job at Alright on a fortnightly wage of $2655.50. Which job pays more?

 p.a. = per annum = per year

SOLUTION

To compare incomes, convert to fortnightly amounts (or yearly amounts).

Fortnightly pay for Allgood = $72 000 ÷ 26   ←——————— 1 year = 26 fortnights
                            = $2769.2307…
                            ≈ $2769.23   ←——————— rounded to the nearest cent

This is more than the fortnightly pay at Alright ($2655.50), so Allgood pays more.

EXAMPLE 3

Sean earns $4320 per month as an accountant. Convert this to a weekly pay.

SOLUTION

As there are not an exact number of weeks in a month, we must convert the monthly pay to a yearly pay first, and then the yearly pay to a weekly pay.

Yearly pay = $4320 × 12   ←——————— 1 year = 12 months
           = $51 840

Weekly pay = $51 840 ÷ 52   ←——————— 1 year = 52 weeks
           = $996.92307…
           ≈ $996.92

1   Brock works 19.5 hours per week as a mechanic at $23.51 per hour. How much is he paid? Select the correct answer **A**, **B**, **C** or **D**.

   A $485.44   **B** $45.84

   C $458.45   **D** $4584.45

2   What is Siobhan's fortnightly wage if she works 22 hours one week and 16.4 hours the next at $19.58 per hour? Select **A**, **B**, **C** or **D**.

   A $751.87   **B** $75.18

   C $715.88   **D** $75.19

iStockphoto/Mlenny

3   Calculate the weekly wage for each person listed below.

   a   Julia: 18 hours at $21.60 per hour.

   b   Matt: 38 hours at $19.65 per hour.

   c   George: 22.4 hours at $26.42 per hour.

   d   Paige: 32.7 hours at of $27.93 per hour.

4   Calculate the fortnightly pay for each person in question **3** if they work the same number of hours both weeks.

5   For each annual salary, find:

   i   the monthly pay   **ii** the fortnightly pay   **iii** the weekly pay.

   a $49 920   **b** $71 760   **c** $108 000

6   Calculate the annual income for each person.

   a   Jonah earns a wage of $728.60 per week.

   b   Elise makes $3875.40 per month.

   c   Richard has a fortnightly pay of $2286.21.

7   Clare is paid a salary of $76 000 p.a. and Georgia's wages are $1480 per week.

   a   Who earns more money?

   b   What is the difference in their earnings per fortnight?

8   Convert each monthly pay to a weekly pay.

   a $3896   **b** $5318.16   **c** $4265   **d** $5824.60

9   Convert each weekly pay to a monthly pay.

   a $720   **b** $921.45   **c** $862   **d** $1024.58

10 Who earns more: John on $2828.50 per fortnight or Jack on a salary of $75 600 p.a.?

11 Caitlin is paid a wage of $28.50 per hour for the first 12 weeks while on probation.
    She works 35 hours per week. For the remainder of the year, her wage rises to $32.80 per hour. How much will she earn in her first year?

12 Convert a monthly wage of $5925 to a fortnightly pay.

WORDBANK

overtime Work done outside usual business hours and paid at a higher rate.

time-and-a-half An overtime rate that is 1.5 times the normal hourly rate.

double time An overtime rate that is 2 times the normal hourly rate.

EXAMPLE 4

Amelia earns $27.55 per hour as a store manager and works a 30-hour week from Monday to Friday. Last week, she also worked 4 hours on Saturday at time-and-a-half and 3.5 hours on Sunday at double time. How much was her wage last week?

SOLUTION

Normal wage = $27.55 × 30
= $826.50

Saturday overtime = 4 × 1.5 × $27.55   ←—————— Time-and-a-half (× 1.5)
= $165.30

Sunday overtime = 3.5 × 2 × 27.55   ←—————— Double time (× 2)
= $192.85

Total weekly wage = $826.50 + $165.30 + $192.85
= $1184.65

1 How much is Sam's weekly wage if he works 35 hours at a rate of $18.90 per hour and does 3 hours overtime at time-and-a-half? Select the correct answer **A**, **B**, **C** or **D**.

   A $661.50   **B** $718.20   **C** $746.55   **D** $1077.30

2 How much is Sue's weekly wage if she works 30 hours at a rate of $22.45 per hour and does 2.5 hours overtime at double time? Select the correct answer **A**, **B**, **C** or **D**.

   A $673.50   **B** $729.63   **C** $1459.25   **D** $785.75

3 Find the overtime pay earned for time-and-a-half working:

   a 4 hours at a normal rate of $19.50   **b** 6.5 hours at a normal rate of $21.60

4 Find the overtime pay earned in question **3** if the rate was double time.

5 Copy and complete the wage for a worker who earned $23.60 per hour for a 35-hour week. Overtime was worked at time-and-a-half for 4 hours and double-time for 1.5 hours.

   Normal wage = $23.60 × ___ = $ ____

   Overtime = $23.60 × 1.5 × ___ + $23.60 × 2 × ____

             = $ _____

   Total wage = $ _____ + $ _____

          = $ _____

6 Babar works at a games store and earns a basic wage of $437.50 per week. If he works 4 hours overtime at time-and-a-half, what will his wage for the week be if his normal rate of pay is $21.40 per hour?

7 Jade works at the local fruit market for 20 hours at a rate of $19.25 per hour. She then works overtime for 5 hours on Saturday at time-and-a-half and 3 hours on Sunday at double time. What is Jade's wage for the week?

8 Madeline works at the veterinary hospital and is paid $756 for 30 hours. She works 6 hours overtime in one week and 4 hours the next week, all at double time. What is her wage for the fortnight?

9 Stephanie works at a theme park and is paid $25.80 per hour on weekdays for the first 8 hours and then time-and-a-half for any time after that. On weekends she is paid double time. Find her wage if she works 8 a.m. to 6 p.m. Monday to Friday and 10 a.m. to 4 p.m. on Saturday.

10 Max works at the same theme park. Copy and complete this timesheet, then calculate his wage for the week.

Day	Start	Finish	Normal hours	Time-and-a-half	Double-time
Monday	9:00 a.m.	5:00 p.m.	8	0	0
Tuesday	8:00 a.m.	6:00 p.m.		2	
Wednesday	10:00 a.m.		5	0	0
Thursday	7:30 a.m.	4:30 p.m.			
Friday	8:15 a.m.		8	1.75	0
Saturday	11:00 a.m.	3:30 p.m.	0	0	
Sunday		3:20 p.m.	0		5.5

WORDBANK

commission Income for salespeople and agents, calculated as a percentage of sales.

piecework Income calculated per item made or processed, such as sewing of clothes or delivery of items.

bonus A one-off payment to employees for a job completed on time or to a high standard.

EXAMPLE 5

Jessie sells cosmetics and is paid a retainer of $470 plus 5% commission on her total sales for the week. Find her pay for the week if she sold $450 worth of skin care products, $1880 worth of perfume and $645 worth of lipstick and mascara.

SOLUTION

Total sales = $450 + $1880 + $645 ←——— Adding all her sales first.
     = $2975

Jessie's pay = $470 + 5% × $2975 ←——— retainer + commission
      = $618.75

EXAMPLE 6

Marco charges $1.10 per brick for a feature wall he is asked to build. The wall measures 3 m by 12 m and needs 48 bricks per m^2 to complete the pattern. How much does Marco charge for the job?

SOLUTION

Calculate the number of bricks required first.

Area of the wall = 3 × 12 = 36 m^2

Number of bricks = 48 × 36 = 1728

Job charge = 1728 × $1.10 = $1900.80

EXAMPLE 7

Harry earns a salary of $120 000 plus a bonus of 12.5% if he meets his sales targets for the year. Calculate his bonus and his total annual earnings if he meets his sales targets.

SOLUTION

Bonus = 12.5% of $120 000
   = $15 000

Total annual earnings = $120 000 + $15 000
       = $135 000

ISBN 9780170351027

1   What is Lucy's wage if she sells $2500 worth of phone plans and earns a retainer of $520 plus a commission of 8%? Select the correct answer **A**, **B**, **C** or **D**.

    A $720   **B** $3020   **C** $520   **D** $2541.60

2   What are Ben's annual earnings if he is paid a salary of $115 600 and is given a bonus of 11% of his salary? Select **A**, **B**, **C** or **D**.

    A $115 600   **B** $128 316   **C** $115 611   **D** $12 716

3   Calculate each of the following earnings.

    a A life insurance salesman is paid 35% commission on $8250 worth of premiums sold.

    b A phone salesman is paid 28% commission on phone sales of $4860.

    c A shop assistant is paid 40% commission on sales of $3265.

    d A car salesman is paid a retainer of $240 per week plus 4.5% commission on $25 500 worth of car sales.

4   Calculate each pay earned by piecework.

    a Jack folds 2200 leaflets at 45c per leaflet.   **b** Chloe sews 195 dresses at $4.50 per dress.

    c Sam paints 624 posts at $1.48 per post.   **d** Emilie writes 1256 pages at 62c per page.

5   Calculate each bonus earned by the employees.

    a Kara's bonus is 9% of her $78 000 salary.

    b Ziad earned a 15% bonus of his annual salary, which is paid by monthly amounts of $3650.

    c William's bonus is 12.5% of his annual earnings, which are paid by $685.25 per week.

    d Jacinta's bonus is 15% of her annual earnings, which are paid at a rate of $1750 per fortnight.

6   Natasha works in a shoe shop. Her wages are $350 plus 5% commission. In one week, she sold shoes to the value of $910. What were her earnings that week?

7   A real estate agent sold a new home for $580 000. His commission was 5% on the first $10 000 and 2.5% on the rest. How much was his commission?

8   Kath works at a garment factory where she earns $6.85 for each garment she completes, plus a bonus of $100 if the factory turns out more than 1000 garments during the week. Calculate how much Kath earned this week if she completed 71 garments and the factory produced 1008 garments.

9   David and Paul have a contract to deliver 12 400 telephone books. They are paid 30 cents for each book delivered. What is the total amount they earn?

10   A sheep-shearer earns $150 for every 100 sheep he shears (or $1.50 per sheep). The table below shows his tally for one week.

Monday	Tuesday	Wednesday	Thursday	Friday
120	135	112	92	101

    a What was the total number of sheep shorn that week?

    b How much did the sheep shearer earn?

WORDBANK

annual leave loading Also called holiday loading, this is extra pay to employees for their 4 weeks of annual holidays, calculated at 17.5% of 4 weeks pay.

Annual leave loading = 17.5% × weekly pay × 4
Total holiday pay = 4 weeks pay + annual leave loading

EXAMPLE 8

For Ray's annual holiday, he takes a cruise around the Mediterranean for 4 weeks.

a Calculate Ray's annual leave loading if he earns $6200 per month.

b Find his total holiday pay.

SOLUTION

a Calculate Ray's weekly pay first.
Yearly pay = $6200 × 12
= $74 400

Weekly pay = $74 400 ÷ 52
≈ $1430.77

Annual leave loading = 17.5% × $1430.77 × 4 ⟵ 17.5% of 4 weeks' pay
= $1001.539
≈ $1001.54

b Holiday pay = 4 weeks wage + annual leave loading
= $1430.77 × 4 + $1001.54
= $6724.62

EXERCISE 10-04

1 Jules earns a salary of $68 500 p.a. Find her pay for 4 weeks. Select the correct answer **A**, **B**, **C** or **D**.

   A $5708.33   **B** $5269.23   **C** $1317.31   **D** $2634.62

2 Calculate Jules' annual leave loading from question **1**. Select **A**, **B**, **C** or **D**.

   A $998.96   **B** $922.12   **C** $230.53   **D** $461.06

3   Is each statement true or false?

    a   The amount of annual leave loading depends on the employee's pay.

    b   Holiday pay is the same as annual leave loading.

    c   Annual leave loading is extra money paid at holiday time.

4   Copy and complete the solution to this problem.

    Lucas earns $725.50 per week working at a travel agent. Calculate his total holiday pay.

    4 weeks wage = $725.50 × __ = $_____

    Annual leave loading = 17.5% × ____ = $_____

    Total holiday pay = $ _____ + $ _____ = $_____

5   Riley earns $425.00 per week as a horse trainer.

    a   How much will his holiday loading be if he takes 4 weeks holiday at Easter?

    b   What will be the amount of his holiday pay?

6   Elena earns $24.50 per hour for a 36-hour week. She takes 4 weeks holidays to travel overseas.

    a   What is her normal weekly wage?

    b   What will be her wage for 4 weeks?

    c   How much will her holiday loading be?

    d   What amount of money will Elena be paid altogether for her holiday?

7   Thanh works for a computer company and is paid a salary of $72 000 p.a. He wants to travel home for Christmas and takes 4 weeks holidays.

    a   How much does Thanh earn for 4 weeks?

    b   What will Thanh's holiday loading be?

    c   How much holiday pay will Thanh be paid?

8   A security firm employs eight full-time staff whose earnings are shown below.

    2 senior officers salary: $92 000

    3 officers $27.25 per hour for a 32-hour week

    2 assistants $21.50 per hour for a 35-hour week

    1 trainee $18.20 per hour for a 2- hour week

    The security firm closes for 4 weeks over Christmas. Calculate the total amount of holiday pay for all employees.

9   Sarah earns a salary of $82 500 p.a. and decides to take 4 weeks annual leave plus an additional 3 weeks unpaid leave for a holiday to Canada. Calculate:

    a   her annual leave loading   **b**   her total holiday pay

    c   the amount of money lost by taking 3 weeks unpaid leave.

10 Tony earns $1275 per fortnight working for a building firm. He is paid a 7.5% bonus on his annual wage as well as his 4 weeks annual leave at Easter. Calculate:

    a   his bonus   **b**   his annual leave loading   **c**   his total pay at Easter.

WORDBANK

gross income The total amount earned including bonuses and overtime.

allowable (tax) deductions Any amounts subtracted from a gross income, such as work expenses and donations to charity, that are not taxed.

taxable income Gross income less allowable deductions, the remaining income that is charged tax, rounded down to the nearest dollar.

Income tax is a tax paid to the government for services such as roads, schools and hospitals. The amount of tax paid depends upon the amount of income and is calculated according to this table.

Taxable income (rounded down to the nearest dollar)	Tax on this income
$0 – $18 200	Nil
$18 201– $37 000	19c for each $1 over $18 200
$37 001 – $80 000	$3572 plus 32.5c for each $1 over $37 000
$80 001 – $180 000	$17 547 plus 37c for each $1 over $80 000
$180 001 and over	$54 547 plus 45c for every $1 over $180 000

Taxable income = Gross income – total allowable deductions

EXAMPLE 9

Ryan's salary is $74 600 p.a. He has weekly allowable deductions of $35.80 for union fees and $100 for voluntary superannuation, and annual work-related expenses of $6500. Use the table above to calculate his income tax payable.

 Superannuation is a retirement fund that both employer and employee can pay into

SOLUTION

Total annual deductions = $35.80 × 52 + $100 × 52 + $6500
$$= \$13\,561.60$$

Taxable income = $74 600 – $13 561.60 ⟵———— Gross income – allowable deductions
$$= \$61\,038.40$$
$$\approx \$61\,038 ⟵———————— \text{Rounded down to the nearest dollar}$$

Tax payable = $3572 + 0.325 × ($61 038 – $37 000)

 Use line 3 of the table.
32.5c for each dollar means 32.5% or 0.325

$$= \$11\,384.35$$

1 What is the taxable income if the gross income is $72 800 p.a. and the total deductions are $185 per week? Select the correct answer **A**, **B**, **C** or **D**.

   A $72 615   **B** $63 180   **C** $67 990   **D** $15 207

2 Use the income tax table to find the tax payable on a taxable income of $29 780. Select **A**, **B**, **C** or **D**.

   A $5658.20   **B** $3458   **C** $2200.20   **D** $2200.01

3 Calculate the income tax payable for each taxable income.

   a $4890   **b** $48 780   **c** $32 760

   d $82 000   **e** $205 960

4 **a** Sanjay's salary is $126 000 p.a. Calculate his taxable income and income tax payable if he has annual travel expenses of $8650 and his superannuation contribution is $305 per fortnight.

   b What percentage (correct to one decimal place) of Sanjay's gross income is income tax?

5 Find the total annual deductions for each person below.

   a Alicia: travel costs $58/week, union fees $20.50/fortnight, superannuation $50/week

   b Peter: uniforms $28/week, car expenses $155/week, accommodation $250/month

   c Jack: social club $25/week, charities $200/month, phone calls $280/fortnight

   d Bonnie: travel $128/week, union fees $32.80/month, training $560/quarter

Alamy/Chris Rout

6 Find the taxable income (rounded down to the nearest dollar) and tax payable using the tax table for the employees in question **5** if:

   a Alicia earns $42 500 p.a.   **b** Peter earns $81 500 p.a.

   c Jack earns $155 000 p.a.   **d** Bonnie earns $78 250 p.a.

WORDBANK

PAYG tax Stands for **Pay As You Go** tax, which is income tax taken from your pay every payday to avoid you paying a huge sum of tax at the end of the financial year.

net pay Gross pay less tax and other deductions.

Net pay = Gross pay - tax - other deductions

EXAMPLE 10

Liam worked in an electrical store during the Christmas school holidays and earned $388 per week for 6 weeks.

a Calculate Liam's gross pay for the 6 weeks.

b Use the PAYG tax table to calculate the PAYG tax that was deducted over the 6 weeks.

Weekly Earnings ($)	PAYG tax withheld ($)
384 – 386	87
387 – 389	88
390 – 392	89
393 – 394	90

c Calculate Liam's net pay for the 6 weeks if he also paid $50 per fortnight into his superannuation.

SOLUTION

a Liam's gross pay = $388 × 6 = $2328

b Liam's PAYG tax = $88 × 6 = $528 ◄——————— Using line 2 of the table

c Liam's net pay = $2328 – $528 – 3 × $50 ◄——————— 6 weeks = 3 fortnights
= $1650

EXAMPLE 11

Bronte earns $1255 per fortnight. Her deductions are $125 for superannuation and $45.50 for health insurance. Calculate Bronte's PAYG tax and net pay per fortnight.

Fortnightly earnings ($)	PAYG tax withheld ($)
1240 – 1244	336
1246 – 1250	338
1252 – 1256	340
1258 – 1262	342

SOLUTION

Bronte's PAYG tax = $340 per fortnight ◄——————— Using line 3 of the table

Bronte's net pay = $1255 – $340 – $125 – $45.50 ◄——————— Gross income – deductions
= $744.50

1   Use the weekly PAYG table to find the net pay for an employee earning $391 per week. Select the correct answer **A**, **B**, **C** or **D**.

    A $304   **B** $303   **C** $302   **D** $301

2   Use the fortnightly PAYG table to find the net pay for an employee earning $1248 per fortnight. Select **A**, **B**, **C** or **D**.

    A $912   **B** $910   **C** $908   **D** $906

3   Find the PAYG tax payable on each wage.

    a $394/week for 5 weeks   b $385/week for 3.5 weeks

    c $23/hour for 17 hours/week   d $21.50/hour for 18 hours/week

    e $1243 per fortnight for 2 fortnights   f $1254 per fortnight for 5 fortnights

    g $24.60/hour for 51 hours/fortnight   h $19.80/hour for 63 hours/fortnight

4   Pania's gross pay is $615 a week. If she has deductions of $150.20 for tax and $185 for loan repayments, what is her weekly net pay?

5   Phillip works a 35-hour week and is paid $16.40 an hour.

    a   What is his gross weekly wage?

    b   If taxation of 32% is deducted, what is his net pay?

6   A truck driver earns gross wages of $802 per week. Taxation is 25% of this amount and his house payments of $188 are also deducted each week.

    a   What are his gross wages for one year?

    b   What is his weekly net pay?

7   Copy and complete each pay statement.

Name:	Ali Mc Phee	Date:	20/07/17 to 24/07/17
Hourly rate:	$21.60	Gross pay:	
Hours worked:	31.5	Tax:	
Tax rate:	24.5%	Net pay:	

8   Lachlan worked in a nursery during his holidays and earned $385 per week for 5 weeks. Use the PAYG table to calculate:

    a   Lachlan's gross pay for the 5 weeks

    b   the PAYG tax that was deducted

    c   Lachlan's net pay for the 5 weeks if he contributed $40 per fortnight to his superannuation.

9   Kylie earns a gross income of $1246 per fortnight. Her deductions are $95 for superannuation and $52.80 for private health insurance, as well as PAYG tax.

    a   Calculate Kylie's net pay per fortnight.

    b   Find Kylie's deductions as a percentage of her gross pay, correct to 1 decimal place.

FIND-A-WORD PUZZLE

Make a copy of this page, then find all the words listed below in this grid of letters.

S	O	H	P	B	Y	K	Y	Z	R	C	S	F	Q	I
S	U	A	O	A	O	N	E	E	R	O	S	T	V	S
T	Y	P	V	L	M	N	N	E	Y	M	O	G	M	V
G	H	L	E	O	I	I	U	Q	W	M	R	J	B	N
D	R	G	W	R	A	D	I	S	D	I	G	W	G	W
G	N	G	I	T	A	N	A	E	X	S	Z	E	Q	A
I	C	U	E	N	C	N	D	Y	H	S	K	M	H	G
M	G	R	F	O	T	U	N	X	E	I	E	I	S	E
S	J	R	M	E	C	R	A	U	L	O	S	T	A	F
T	B	E	D	T	R	T	O	D	A	N	N	R	L	S
W	T	U	I	N	Y	D	X	F	Q	T	E	E	A	N
K	R	O	W	E	C	E	I	P	Z	V	I	V	R	T
X	N	L	O	A	D	I	N	G	A	G	Z	O	Y	E
S	M	O	N	T	H	Y	J	E	A	N	D	P	N	N
G	C	N	D	V	L	W	L	A	U	N	N	A	I	T

ANNUAL	BONUS	COMMISSION	DEBT
DEDUCTIONS	FORTNIGHT	GROSS	HOLIDAY
INCOME	LEAVE	LOADING	MONTH
NET	OVERTIME	PAYG	PIECEWORK
REFUND	RETAINER	SALARY	SUPERANNUATION
TAX	WAGE	WEEK	

Part A General topics

Calculators are not allowed.

1 Find 10% of $46.

2 Expand $-2(3a + 9)$.

3 Convert 0.015 to a percentage.

4 Find the area of this triangle.

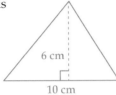

6 cm
10 cm

5 Find $\frac{2}{3}$ of $33.

6 Find the mode of:

6 3 2 6 5 5 6

7 Convert $17\frac{1}{2}\%$ to a decimal.

8 How many fortnights are there in one year?

9 Write a prime number between 10 and 20.

10 How many hours and minutes are there between 8:30 a.m. and 4:45 p.m.?

Part B Earning money

Calculators are allowed.

10-01 Wages and salaries

11 Find the weekly wage for Sarah who earns $18.90/hour for 29 hours work. Select the correct answer **A, B, C** or **D**.

    A $54.81   **B** $5481.00   **C** $548.10   **D** $548.01

12 Find the fortnightly pay for Ewan who earns $84 500 p.a. Select **A, B, C** or **D**.

    A $1625   **B** $3250   **C** $7041.67   **D** $3520.83

10-02 Overtime pay

13 Find Sharelle's weekly pay if she works 32 hours at $21.60 per hour and then does 3 hours overtime at time-and-a-half. Select **A, B, C** or **D**.

    A $788.40   **B** $756   **C** $1134   **D** $691.20

14 Find each weekly wage.

    a Billal earns $18.55/hour and works 35 hours at the normal rate, 4 hours at time-and-a-half, as well as 3.5 hours at double time.

    b Jackie earns $21.80/hour and works a normal week of 30 hours plus 3 hours overtime at time-and-a-half and 2.5 hours at double time.

10-03 Commission and piecework

15 Georgia sells real estate and is paid commission of 5% on the first $10 000 and 2.5% on the remaining sales price of each home sold. Find her commission on a house sold for $625 000.

16 Luis is a bricklayer and is paid $1.05 per brick for a fence that he is asked to build. The fence measures 16 m by 1.4 m and he needs 42 bricks per square metre. How much will Luis earn for building the fence?

17 A sales manager is paid a salary of $115 000 p.a. plus a bonus of 8.5% if he meets his sales target for the year. Find his annual salary if he met his sales target.

10-04 Annual leave loading

18 Connor earns $528.60 per week at his local supermarket. He takes 4 weeks annual leave and is paid his normal wage for 4 weeks plus 17.5% annual leave loading. Calculate:
   a   Connor's annual leave loading   b   his 4 weeks holiday pay.

19 Gillian earns a salary of $75 600 p.a. She is paid a bonus of 9.5% and then decides to take 4 weeks holiday. Calculate:
   a   Gillian's annual leave loading   b   her total 4 weeks holiday pay.

10-05 Income tax

20 Use the income tax table on page 204 to calculate the income tax payable for each employee.
   a   Robert: Salary $68 500 p.a.   Deductions $185/week
   b   Sally: Wage $780/week   Deductions $58.40/week

10-06 PAYG tax and net pay

21 Use the PAYG tax tables on page 206 to calculate the PAYG tax payable for each person.
   a   Jenny: $388/week for 4 weeks   b   Sam: $1249/fortnight for 10 weeks

22 Find each employee's net pay in question 21 if the following deductions for superannuation were made.
   a   Jenny: $28/week   b   Sam: $68/fortnight

INVESTIGATING DATA

11

WHAT'S IN CHAPTER 11?

IN THIS CHAPTER YOU WILL:

- name and classify different types of data
- know the difference between a sample and a census
- find the mean, mode, median and range of a set of numerical data
- draw dot plots and stem-and-leaf plots
- complete a frequency distribution table
- draw frequency histograms and polygons
- identify the shape of a frequency distribution

Shutterstock.com/Adriano Castelli

WORDBANK

data A collection of raw facts or information.

categorical data Data that can be divided into groups or categories. They are usually words, such as favourite colour, or state of Australia.

quantitative data Data that has a numerical value, such as heights or ages of students. Quantitative comes from the word quantity, meaning number of items.

discrete data Quantitative data that is counted and has separate values, such as the number of people in a house. The number could be 5 or 6, but not 5.5. There are no in-between scores.

continuous data Quantitative data that is measured and has a smooth scale of values, such as people's heights. Your height could be 161 cm, 162 cm or 161.3 cm. In-between values are allowed.

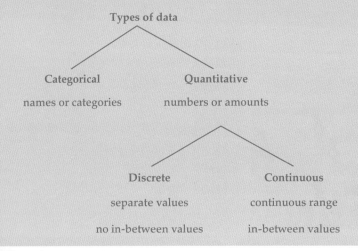

EXAMPLE 1

Classify each type of data as categorical or quantitative. If it is quantitative, state whether it is discrete or continuous.

a   number of chairs in a classroom

b   ice-cream flavour

c   time taken to read this page

d   colour of eyes

SOLUTION

a   quantitative, discrete   ⟵ Number of chairs must be whole numbers

b   categorical   ⟵ Ice-cream flavours are names or categories

c   quantitative, continuous   ⟵ Time has a smooth scale

d   categorical   ⟵ Eye colours are categories

1 What type of data are film classifications such as G, PG and M? Select the correct answer **A**, **B**, **C** or **D**.

   A quantitative   **B** continuous   **C** discrete   **D** categorical

2 Which term below best describes discrete data? Select **A**, **B**, **C** or **D**.

   A categorical

   B continuous

   C numerical with in-between values

   D numerical with no in-between values

3 Classify each type of data as categorical or quantitative. If it is quantitative, state whether it is discrete or continuous.

   a Number of people in the park

   b Arm length of basketballers

   c Hair colour of people

   d Shoe sizes of shoppers

   e Favourite music of teenagers

   f Time taken to swim 100 m

   g Dress sizes of women

   h Quality of food in a restaurant

   i Weight of singers in a band

   j Favourite type of TV show

Shutterstock.com/1000 Words

4 State whether each type of quantitative data is discrete (D) or continuous (C).

   a A football team's score

   b Your school's student population

   c The reaction time of a driver

   d The speed of a car

   e The cost of an ice-cream

   f The circumference of a ball

   g A mobile phone bill

   h The amount of money in a wallet

   i Your age last birthday

   j Your exact age right now

5 Is each statement true or false?

   a The temperature of a stove is continuous quantitative data.

   b The make of a car is categorical data.

   c A school grade of A, B or C is discrete quantitative data.

   d Your postcode is continuous quantitative data.

6 Write down 3 examples of categorical data.

7 **a** Write down 3 examples of quantitative data.

   b State whether this data is continuous or discrete.

WORDBANK

population In statistics, all of the items under investigation. For example, if a survey is done on the cattle on a farm, the population is all the cattle on this farm.

sample A part of the population selected to be surveyed.

census A survey of the entire population.

A **sample** should:
- represent the population fairly
- be **random**, where every item of the population has an equal chance of being selected
- be **not biased**, where one type of item is not favoured over another
- be **large**, because the larger the sample, the more accurately it represents the population

A **census** surveys the whole population but can be expensive and time-consuming to conduct.

EXAMPLE 2

State whether a sample or a census would be the more suitable for each survey.

a Finding the most popular Australian film of last year.

b Collecting data on the number of people living in retirement homes.

c Surveying people's opinions on the salaries of professional football players.

iStockphoto/digitalhallway

SOLUTION

Use the definitions to help you decide.

a Sample, as it would be too expensive to conduct a census of all Australians for this reason.

b Census, as reliable data should be available at each retirement home.

c Sample, as it would be too time-consuming to survey everyone.

1   Which phrase best describes a census? Select the correct answer **A**, **B**, **C** or **D**.

    A quick and easy   **B** biased   **C** accurate   **D** random

2   Which phrase best describes a sample?

    A expensive   **B** random   **C** biased   **D** time-consuming

3   Determine whether a sample or a census would be the more suitable for each survey. Give a reason for your answer.

    a Testing the quality of a restaurant's food.

    b Finding the number of students enrolled in Kindergarten in a country town.

    c Determining Australia's most favourite TV show.

    d Surveying students on the most popular brand of smartphone.

    e Finding the population in a particular suburb.

    f Surveying the best brand of meat pie sold in Victoria.

    g Testing the strength of all males in a weightlifting competition.

    h Determining the most popular potato chip flavour.

4   State whether each of the following methods of selecting a sample of 10 people is random or biased.

    a Mixing the names of people in a barrel and drawing out 10 names.

    b Interviewing the first 10 customers in the store when it opens.

    c Choosing 10 students from the 9A class when surveying students about a new school uniform.

    d Interviewing 10 customers from the same store when investigating the local council rate rises.

5   This table shows the number of students in each year level at Happy Valley High School.

Year	7	8	9	10	11	12
Students	120	98	115	122	180	172

    a What is the school population?

    b What is the ratio of Year 11 students to Year 7 students in simplest form?

    c If I wish to survey a sample of 50 Year 7 and Year 11 students about their sports interests, how many students should I choose from Year 11?

6   Investigate how often a national census is held in Australia. When will the next census be?

7   In your own words, desribe how you would select a random sample of students to investigate the favourite foods sold at your school canteen.

WORDBANK

mean The average of a set of data scores, with symbol $\bar{x}$, calculated by adding the scores and dividing the sum of the scores by the number of scores.

mode The most popular score(s) in a data set, the score that occurs most often.

outlier An extreme score that is much higher or much lower than the other scores in the data set. An outlier affects the value of the mean.

The mean and the mode are called **measures of location** because they measure the central or middle position of a set of data.

EXAMPLE 3

For this set of data, 5 6 5 8 9 4 16 6 5 5 6 find:

a the mean, correct to one decimal place   **b** the mode   **c** any outliers

SOLUTION

a $\text{Mean} = \bar{x} = \dfrac{\text{sum of scores}}{\text{number of scores}}$

$= \dfrac{5+6+5+8+9+4+16+6+5+5+6}{11}$

$= \dfrac{75}{11}$

$= 6.818\,181...$

≈ 6.8

b Mode = 5   ⟵ 5 is the most popular score, occurring 4 times

> ✱ Note that the mean and mode are around the centre of the set of scores

c Outlier = 16   ⟵ 16 is much higher than all the other scores

EXAMPLE 4

The mean of 6 scores is 12. If 5 of the scores are 14, 9, 11, 10 and 8, find the missing score.

SOLUTION

Let the missing score be x.

$\dfrac{14+9+11+10+8+x}{6} = 12$ ⟵ $\text{Mean} = \bar{x} = \dfrac{\text{sum of scores}}{\text{number of scores}}$

$\dfrac{52+x}{6} = 12$

$52 + x = 12 \times 6$

$= 72$

$x = 72 - 52$

$= 20$

So the missing score is 20.

> ✱ Check that the mean of the 6 scores is 12

1   Find the mode of the scores   7   5   6   7   3   6   5   7   8
    Select the correct answer **A**, **B**, **C** or **D**.
    A 4   **B** 5   **C** 6   **D** 7

2   Find the mean of the scores in question **1**.
    A 4   **B** 5   **C** 6   **D** 7

3   Copy and complete this paragraph.

    The mean of a set of scores is found by _____ the scores together and then
    _____ by the _____ of scores. The mode is the score that occurs the most
    _____. It is the most _____ score. There can be one or more modes for
    a set of scores and sometimes there is ___ mode at all. An outlier is a score that is much
    _____ or _____ than the other scores.

4   Find the mean of each set of data, correct to one decimal place if necessary.
    a   3, 3, 3, 4, 5   b   5, 7, 9, 11, 13
    c   2, 2, 2, 7, 8, 3, 2, 6   d   5, 8, 12, 12, 15, 16
    e   15, 15, 17, 13, 16, 17   f   28, 34, 35, 41
    g   6.2, 2.6, 5.7, 6.2, 2.6, 2.6   h   21.5, 13.6, 5.7, 8.9, 11.1, 21.5, 5.7, 11.1
    i   456, 131, 131, 524, 456, 131, 524   j   54.8, 45.7, 32, 76.9, 123

5   Find the mode(s) for each set of data in question **4**.

6   Seven students were asked how much pocket money they were paid each week.
    $25 $28 $19 $20 $50 $26 $25
    a   Find the mean amount of pocket money, correct to the nearest cent.
    b   What is the outlier?
    c   Calculate the mean amount of pocket money without the outlier.
    d   Describe the effect the outlier had on the mean.

7   a   Calculate, correct to two decimal places, the mean for the following assignment marks
        (out of 50).
        42   38   40   48   39   18   44   45   42   41   46
    b   Find the outlier.
    c   If the outlier was removed, would the mean be higher or lower?
    d   Calculate the mean if the outlier is removed.
    e   Is there a mode? If so, write it down.

8   Last week, eight houses were sold in Wilderness Cove. The selling prices were:
    $580 000 $745 000 $824 500 $690 000
    $985 000 $1 220 500 $585 250 $721 000
    a   What is the mode?   b   Find the mean.
    c   Is there an outlier? If so, what is it?
    d   If another house in Wilderness Cove sold last week for $850 000, would the mean
        increase or decrease?

WORDBANK

median The middle score when the scores are ordered from lowest to highest.

range The highest score minus the lowest score.

TO FIND THE MEDIAN:
- order the scores from lowest to highest
- if there are an odd number of scores, **median = middle score**
- if there are an even number of scores, **median = average of the two middle scores**

Range = Highest score – lowest score
The median is another measure of location, while the range is a measure of spread.

EXAMPLE 5

Find the median and range of each set of scores.

a 2.8 3.2 5.6 2.4 7.2 4.9 6.7 1.6 4.5

b 56 76 65 52 66 21 73 68

SOLUTION

a 1.6 2.4 2.8 3.2 (4.5) 4.9 5.6 6.7 7.2

✱ Arrange the scores in order first

Median = 4.5 ⟵ 4.5 is in the middle: there are 4 scores on either side of it

Range = 7.2 – 1.6 ⟵ highest score – lowest score

= 5.6

b 21 52 56 (65) (66) 68 73 76

✱ As there is an even number of scores, there are 2 scores in the middle.

Median = $\dfrac{65+66}{2}$ ⟵ average of 2 middle scores

= 65.5

Range = 76 – 21

= 55

EXAMPLE 6

a Which set of scores in Example 5 has an outlier?

b What effect does the outlier have on the median and the range?

SOLUTION

a The scores in Example 5b have an outlier of 21.

b If the outlier was removed, then the median would be 66 and not be much different. However, the range would be 24 and much smaller. So the outlier does not affect the median much, but it affects the range significantly (making it larger).

1   Find the range for the scores: 11   14   12   18   16   19
    Select the correct answer **A, B, C** or **D**.
    A 8   **B** 15   **C** 14   **D** 16

2   Find the median for the scores in question **1**.
    A 8   **B** 15   **C** 14   **D** 16

3   Copy and complete this paragraph.

    The range for a set of scores is the _____ score minus the _____ score. The median
    is found by ordering the scores from _____ to _____ and finding the _____ score for
    an odd number of _____. If there is an _____ number of scores, the median is the
    _____ of the two _____ scores.

4   Find the median of each set of scores.
    a   4, 8, 9, 5, 6, 7, 2   b   3, 8, 7, 1, 8, 4, 8, 3   c   12, 14, 18, 10, 11, 17
    d   21, 29, 23, 25, 28, 24, 22   e   45, 53, 72, 38, 65, 34, 58, 65

5   Find the range for each set of scores in question **4**.

6   Three teams of swimmers swam a relay race at a school's swimming carnival. The times
    taken (in seconds) for each swimmer in the teams to complete a lap of the pool are shown
    below. Find the median and range for each team's times.

    | **Red:** | 48.6 | 43.7 | 49.5 | 52.1 | 52.6 | 44.8 |
    | **Yellow:** | 46.5 | 51.8 | 44.6 | 48.2 | 58.4 | 42.6 |
    | **Green:** | 41.9 | 55.8 | 53.4 | 46.8 | 52.6 | 51.7 |

7   a   Which team in question **6** should represent the school at the zone swimming carnival,
        if it is the one with the lowest median?
    b   If you want the fastest swimmer in the school in the team, which team would you choose?

8   Find the range and median for each set of scores.
    a   2, 2, 1, 1, 2, 3, 3, 5   b   14, 14, 16, 8, 10, 20
    c   0, 2, 0, 4, 1, 0, 6   d   3, 3, 3, 2, 2, 2, 4, 6, 3
    e   101, 100, 99, 98, 99, 100, 102, 100   f   12, 13, 10, 9, 10, 10, 12, 11
    g   1, 6, 8, 3, 7, 7, 6, 7, 3, 9   h   34, 29, 29, 30, 31, 31, 32, 31, 30
    i   6, 3, 1, 1, 1, 2, 2, 6, 6, 5, 4, 5   j   5, 4, 0, 0, 2, 1, 3, 4, 3, 2, 3, 3, 3

9   Three of the scores are outliers in question **8g** above. If these outliers are ignored, find the
    range and median.

10 In 12 rounds of golf, Greg scored:
    72   76   71   77   75   71   74   72   71   79   73   74
    Find the mode, range, median and mean of these scores.

11 A selector wanted to find the most reliable measure of Greg's golf game. If Greg wants to
    impress him, which measure of location should he show him?

WORDBANK

dot plot A diagram showing frequency of data scores using dots.

stem-and-leaf plot A table listing data scores where the tens or hundreds digits are in the stem and the units digits are in the leaf.

cluster Where scores in a set of data are close together in a group.

EXAMPLE 7

This dot plot shows the daily sales in dollars from a cupcake shop for 2 weeks.

Daily sales ($)

iStockphoto/carterdayne

a What was the range of daily sales?

b What was the mode?

c Find the mean (correct to the nearest cent) and median.

d Write down any clusters or outliers.

SOLUTION

a Range = $1100 − $750   ←—— Highest score − lowest score

= $350

b Mode = $950   ←—— Most common amount

c Mean = $\dfrac{750 + 4 \times 850 + 3 \times 900 + 5 \times 950 + 1100}{14}$ ←—— $\dfrac{\text{sum of scores}}{\text{no. of scores}}$

$= \dfrac{12\,700}{14}$

$= 907.1428\ldots$

$\approx \$907.14$

Median $= \dfrac{900 + 900}{2}$ ←—— 14 scores: the two middle dots (7th and 8th) are both 900

$= \$900$

900

d A cluster of scores occurs from $850 to $950.

There are 2 outliers: $750 and $1100.

EXAMPLE 8

Organise the scores below into a stem-and-leaf plot.

24 26 37 22 48 25 51 60 28 33 42 56
62 45 58 69 43 27 50 47 68 25 37 61

SOLUTION

The scores range from 22 to 69, so we need stems of 2, 3, 4 up to 6.

In the leaf, we write the units digit of each score.

For 24, write a '4' in the leaf next to the 2 stem; for 26, write a '6' in the leaf next to the 2 stem, and so on.

Stem	Leaf						
2	4	6	2	5	8	7	5
3	7	3	7				
4	8	2	5	3	7		
5	1	6	8	0			
6	0	2	9	8	1		

To make things neater, we should write each row of digits in order to make an **ordered** stem-and-leaf plot:

Stem	Leaf						
2	2	4	5	5	6	7	8
3	3	7	7				
4	2	3	5	7	8		
5	0	1	6	8			
6	0	1	2	8	9		

Shutterstock.com/Useng

EXERCISE 11–05

1   In a dot plot, the score with the highest column of dots is what? Select the correct answer **A**, **B**, **C** or **D**.

   A the range   **B** the mean   **C** the median   **D** the mode

2   In a stem-and-leaf plot, the leaf shows the:

   A tens digit   **B** units digit   **C** median   **D** range

3   Copy and complete this paragraph.

   A dot plot is a diagram that shows the distribution of scores using a _____ axis and _____ to represent each score. A stem-and-leaf plot lists each _____ by using the tens or _____ digit as the stem and the units digit as the _____.

4   Draw a dot plot for each set of scores.

   a   3   4   8   7   7   8   5   4   6   7

   b   5   6   9   8   7   8   6   8   9   4

   c   10   11   9   14   11   13   11   12   11   9

   d   22   25   23   24   25   24   22   26   24   24

   e   50   60   58   55   58   52   56   58   57   59

5   For each dot plot drawn in question **4**, find:

   i   the lowest score   ii   the highest score   iii   the mode

6   The amounts below show Simone's weekly part-time wages (in dollars) for 20 weeks. Organise them into a dot plot.

   190   220   280   200   250   240   230   220   180   200

   270   280   260   250   240   250   230   250   180   250

   a   What was the greatest amount that Simone earned for a week?

   b   What was the mode?

   c   If she saved everything she earned above $200/week, how much did she save in this time?

   d   Find Simone's median weekly wage.

   e   Are there any clusters or outliers in Simone's earnings?

7   Organise each set of scores below into an ordered stem-and-leaf-plot.

   a   12   13   26   32   49   15   27   43   35   28

   b   59   67   69   73   82   75   86   52   74   86

   c   36   45   59   64   72   38   42   44   56   58

   d   79   82   95   65   66   72   83   91   60   72

   e   129   137   145   138   156   123   160   134   145   150

8   Calculate the range, median and mean of the scores in question **7d**.

9   The data below shows the daily number of people watching a particular superhero movie at a cinema for 3 weeks.

   36   72   98   45   37   56   62   72   45   34   45

   73   66   72   52   39   55   64   74   62   72

   a   Draw a stem-and-leaf plot displaying this data.

   b   Find the range and mode for this data.

   c   Calculate, correct to the nearest dollar, the mean daily sales takings for the cinema if each person pays $18 to see a movie.

   d   What is the median?

   e   Are there any clusters of scores? If so, where are they?

   f   Is there an outlier? If so, what is it?

WORDBANK

frequency The number of times a score occurs in the data set.

cumulative frequency A running total of frequencies used for finding the median, combining the frequencies of all scores less than or equal to the given score.

frequency table A table that shows the frequency of each score in a data set.

EXAMPLE 9

A group of families were surveyed on the number of tablet devices owned.

4 3 5 3 0 2 4 1 4 5 6 3 2 5 2 1 3 5 6 1 3 2

a Arrange these scores in a frequency table, including columns for cumulative frequency and *fx*.

b For this data set, find:

    i   the range   ii   the mode

    iii the median   iv   the mean (correct to two decimal places)

SOLUTION

a

Score (x)	Tally	Frequency (f)	Cumulative frequency	fx
0	|	1	1	0
1	|||	3	4	3
2	||||	4	8	8
3	⊬⊦	5	13	15
4	|||	3	16	12
5	||||	4	20	20
6	||	2	22	12
	Totals	22		70

> ✱ cumulative frequency is a running total of frequencies
> fx means 'frequency (f) × score (x)'

b i Range = 6 − 0 = 6   ←——— Highest score – lowest score (from the Score column)

  ii Mode = 3   ←——— The score with the highest frequency (5)

  iii There are 22 scores, so the median is the average of the 11th and 12th scores.

From the cumulative frequency table, it can be seen that the 9th to 13th scores are all 3s, so the 11th and 12th scores are both 3.

$$\text{Median} = \frac{3+3}{2} = 3$$

  iv Mean $\bar{x} = \dfrac{\text{sum of scores}}{\text{number of scores}} = \dfrac{\text{sum of } fx}{\text{sum of } f}$

$$= \frac{70}{22} \quad ←——— \text{Use the totals at the bottom of the table}$$

$$\approx 3.18$$

For data in a frequency table:
- ▪ a cumulative frequency column can be included to find the median
- ▪ an fx column can be included to find the mean
- ▪ Mean $\bar{x} = \dfrac{\text{sum of } fx}{\text{sum of } f}$

EXERCISE 11–06

1 Which column do you need to find the median in a frequency table? Select the correct answer **A, B, C** or **D**.

    A score   **B** frequency   **C** fx **D** cumulative frequency

2 Which column do you need to find the mean in a frequency table?

    A score   **B** frequency   **C** fx **D** cumulative frequency

3 A random sample of 50 matchboxes was taken and the contents of each box counted.

50, 49, 52, 50, 51, 50, 52, 49, 48, 53, 48, 52, 49, 50, 49, 48, 47,

51, 52, 51, 51, 51, 49, 48, 47, 52, 51, 50, 50, 49, 47, 50, 50, 48,

49, 53, 52, 50, 48, 50, 50, 50, 49, 51, 53, 48, 47, 50, 50, 52

    a Arrange the results in a frequency table with columns for Score, Tally and Frequency.

    b Find the range and the mode for this data.

4 A random sample of 30 tyres was run continuously until they were worn out. The results below are rounded to the nearest 1000 km.

23 000	26 000	24 000	28 000	27 000	23 000	26 000	28 000	29 000	23 000
22 000	21 000	25 000	23 000	23 000	22 000	21 000	23 000	25 000	26 000
28 000	21 000	23 000	27 000	26 000	23 000	23 000	26 000	24 000	23 000

    a Organise these scores in a frequency table including an fx column.

    b Find the range of wear for the tyres.

    c What is the mean?

5 The manager of a shoe store recorded the sizes of shoes sold over a weekend.

5, 6, 7, 7, 7, 6, 5, 8, 4, 3, 7, 7, 8, 9, 4, 3, 7, 3, 7, 9, 2, 9, 7, 6, 6, 7, 7, 7, 6, 7

    a Arrange the information in a frequency table, including an fx column.

    b Find the range of shoe sizes sold.

    c If you were ordering new stock, which measure of location (mode, mean or median) would be most useful? Find this measure.

    d Find the mean shoe size.

6   The number of spelling errors made by 50 students is shown below.

2 3 4 0 0 2 1 3 3 4 1 3 2

2 2 1 3 3 2 2 2 1 2 2 2 4

3 4 5 3 2 2 2 3 4 3 3 5 2

2 2 1 0 0 0 2 3 3 2 2

a   Express the data in a frequency table, including columns for cumulative frequency and *fx*.

b   Find the mode.

c   Calculate the mean number of spelling errors.

d   Find the median.

7   Copy each frequency table and add columns for cumulative frequency and *fx*. For each data set, find:

a

Score	Frequency
0	3
1	6
2	9
3	4
4	3

b

Score	Frequency
12	4
13	9
14	12
15	7
16	5
17	3

   i   the range   ii   the mode

   iii the median   iv   the mean (correct to two decimal places).

8   A sample of people was surveyed on how often they visited a restaurant last month. Draw up a frequency table for this data set. Find:

2 5 0 2 3 4 6 4 3 1 4 0 3

1 4 8 7 2 4 5 3 4 2 4 5 4

3 4 2 1

a   the range.

b   the mode.

c   the median.

d   the mean amount spent per person on restaurants in a month, at $45 per visit.

123RF/Jean-Marie Guyon

WORDBANK

frequency histogram A column graph that shows the frequency of each score. The columns are joined together.

frequency polygon A line graph that shows the frequency of each score, drawn by joining the middle of the top of each column in a histogram.

EXAMPLE 10

This frequency table shows the results of a survey on the number of bedrooms in each home.

Graph this data on a frequency histogram and a frequency polygon.

Number of bedrooms	Frequency
1	4
2	7
3	8
4	5
5	1

SOLUTION

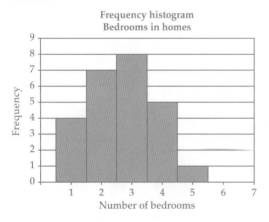

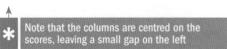

* Note that the columns are centred on the scores, leaving a small gap on the left

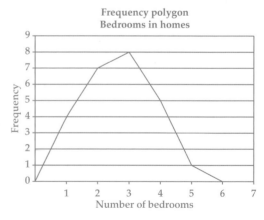

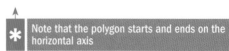

* Note that the polygon starts and ends on the horizontal axis

1 Copy and complete this paragraph.

A frequency histogram is a type of _____ graph, where the columns are _____.
A frequency polygon is a type of _____ graph, where the points are the middle of
the top of each _____ in the histogram.

2 The number of competition points scored by 18 football teams in a season are shown in
the frequency table.

Points scored	Frequency
16	2
18	3
20	7
22	5
24	1

a Draw a frequency histogram and polygon for this data.

b What was the most common score? What is the statistical name for this score?

c How many competition points did the winning team score?

3 The table below shows the number of parcel deliveries to a sample of houses in the last month.

Parcel deliveries	Frequency
1	4
2	5
3	8
4	2
5	1

  a Draw a frequency histogram of this information.

  b Write down the range and mode for this data.

  c In your own words, explain the meaning of the mode for this data.

  d Use the frequency table to calculate the mean number of parcels delivered.

4 A die was thrown 30 times and the results are recorded below.

Number on die	Frequency
1	3
2	5
3	6
4	7
5	4
6	5

Shutterstock.com/Lisa S.

  a Draw a frequency polygon for the table.

  b What was the most likely number to be thrown?

  c How many times was a 5 thrown?

  d What number was least likely to be thrown?

  e Write down the range and write its meaning in words.

  f What percentage of throws were 4 or more? Answer correct to one decimal place.

5 The marks below are the test results for a Year 9 Science class.

  72 76 78 74 72 74 76 78 73 75

  74 73 76 74 78 74 75 72 74 73

  a Draw a frequency table of the examination marks.

  b Graph the results on a frequency histogram.

  c Find the range and mode of the data.

  d Calculate the mean mark.

  e What percentage of students passed if the mean or higher was considered a pass?

ISBN 9780170351027

When looking at histograms, dot plots and stem-and-leaf plots, the shape of the data can be seen by drawing a curve around the graph.

WORDBANK

symmetrical When the scores are evenly spread or balanced around the centre.

skewed When the scores are bunched or clustered at one end, while the other end has a tail.

bimodal (pronounced 'bye-mode-l') When the scores have two peaks.

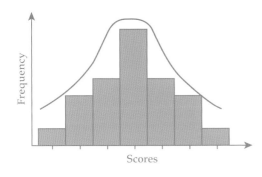

Symmetrical
The scores are clustered around the centre.

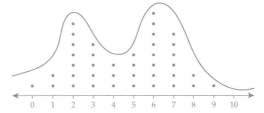

Positively skewed
The scores are clustered to the left and the tail is to the right.

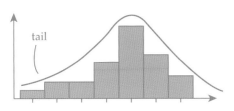

Negatively skewed
The scores are clustered to the right and the tail is to the left.

Bimodal
Two peaks

11-08 | The shape of a distribution

EXAMPLE 11

For each data display:

i describe the shape

ii identify any clusters or outliers

a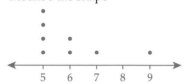

b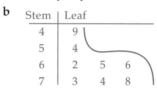

Stem	Leaf		
4	9		
5	4		
6	2	5	6
7	3	4	8

✱ When looking at the shape of a stem-and-leaf plot, turn the plot sideways so that the stem is at the bottom.

SOLUTION

a i The shape is positively skewed.

  ii 9 is an outlier and there is clustering at 5 and 6.

b i The shape is negatively skewed.

  ii There is a cluster at the 60s and 70s stems but no outliers.

EXERCISE 11-08

1 What is a feature of a positively-skewed graph? Select the correct answer **A, B, C** or **D**.

  A the scores are clustered evenly   **B** the tail points to the right

  C the tail points to the left   **D** the scores have two peaks

2 Sketch an example of a symmetrical graph.

3 What is a feature of a symmetrical graph? Select **A, B, C** or **D**.

  A the scores are spread out everywhere

  B the scores are clustered everywhere

  C the scores are evenly spread around the centre

  D the scores have two peaks

4 Copy and complete this paragraph.

  A symmetrical set of scores are clustered around the _____, while a positively-skewed set of scores are clustered to the _____, and a negatively-skewed set of scores are clustered to the _____. Data that is positively-skewed has a tail pointing _____, while data that is negatively-skewed has a tail pointing _____. A bimodal graph has ___ peaks.

Developmental Mathematics Book 3

ISBN 9780170351027

5 For each data display:

    i describe the shape   **ii** identify any clusters or outliers

a

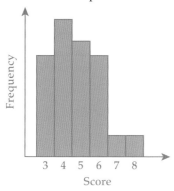

b

c

Stem	Leaf		
12	1	3	6
13	2	8	
14	3		
16	8		

d

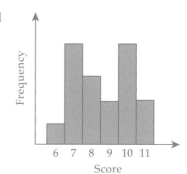

6 The stem-and-leaf plot shows the monthly maximum temperatures in Mt Alpine for 15 months.

Stem	Leaf			
1	0			
1	8	9		
2	3	5		
2	6	8	9	
3	1	3	4	
3	6	8	8	9

    a Describe the shape of the graph.

    b Find the mode and range.

    c Calculate the mean temperature correct to 1 decimal place.

    d What is the median temperature?

    e Where are most of the scores clustered?

7 If 10 is considered an outlier for the scores in question **6**, which measure does it affect the most? Recalculate the mode, range, mean and median without using the outlier to answer this question.

CLUELESS CROSSWORD

Make a copy of this crossword and complete it using words from this chapter.

Part A General topics

Calculators are not allowed.

1   Find 30% of $85.

2   Simplify $\dfrac{5ab^2 \times 10b}{2a^2b}$.

3   Evaluate $\dfrac{3}{4} + \dfrac{1}{3}$.

4   Find the value of x.

5   If $y = -3x - 1$, find y when $x = 2$.

6   Solve $5x - 26 = 3x + 19$.

7   Write 4 numbers that have a range of 9 and a mean of 7.

8   Name this shape.

9   Evaluate $22 + 56 \div 8$.

10 If there is a probability of $\dfrac{3}{10}$ of a hot day tomorrow, what is the probability that it will not be hot tomorrow?

Part B Investigating data

Calculators are allowed.

11-01 Types of data

11 What type of data is the number of goals scored in a basketball match?
    A categorical   B discrete   C continuous   D biased

12 What type of data is the measured height of a student since their birth?
    A categorical   B discrete   C continuous   D biased

11-02 Sample vs. census

13 A random sample surveys …
    A part of the population   B all of the population
    C a particular type of population   D a small area of the population

14 State whether a census or a sample is more appropriate for surveying:
    a   the most popular TV show
    b   the number of Year 9 students in a school district.

11-03 The mean and mode

15 For the scores: 32   37   42   38   41   78   35   37   41   37
    find the mode and write down any outliers.

16 a   Calculate the mean for the scores in question 15.
    b   Does the outlier affect the mean?

11-04 The median and range

17 For the scores: 5.2   6.4   5.8   5.9   2.7   5.4   6.2   5.5   7.3   5.1   7.4   6.8
find the range and write down any outliers.

18 a Calculate the median for the scores in question **17**.
b Does the outlier affect the median?

11-05 Dot plots and stem-and-leaf plots

19 Illustrate this data set 10   25   15   30   20   35   20   15   15   25   20 using:
a a dot plot   **b** a stem-and-leaf plot

11-06 Frequency tables

20 For this frequency table, find the:

a mode
b mean
c median

Score	Frequency
26	4
27	6
28	7
29	3

11-07 Frequency histograms and polygons

21 Draw a frequency table using the frequency polygon below.

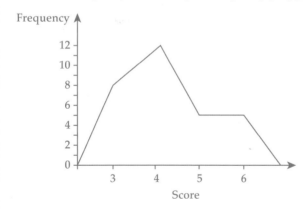

11-08 The shape of a distribution

22 Describe the shape of each distribution.

a

Score

b

Stem	Leaf
4	3 5 8
5	2 6
6	4
7	3
8	1

LENGTH AND TIME

12

WHAT'S IN CHAPTER 12?

IN THIS CHAPTER YOU WILL:

- convert between metric units of length, capacity, mass and time
- find the limits of accuracy of measuring instruments
- calculate the perimeter of shapes, including circles and composite shapes
- calculate with time, including 24-hour time
- read timetables, understand and use international time zones, and solve problems involving them

Shutterstock.com/Amore

WORDBANK

milli- One thousandth $\dfrac{1}{1000}$

centi- One hundredth $\dfrac{1}{100}$

kilo- Thousand (1000)

mega- Million (1 000 000)

The metric system is based on powers of ten, with the above prefixes used in metric units such as metre, litre and gram. The table below shows the metric units for length, capacity, mass and time.

Length		
millimetre (mm)	1000 mm = 1 m	Smallest gap on your ruler
centimetre (cm)	100 cm = 1 m	Width of a pen
metre (m)	base unit	Height of a kitchen bench
kilometre (km)	1000 m = 1 km	Distance between bus stops
Capacity		
millilitre (mL)	1000 mL = 1 L	Dose of medicine for babies
litre (L)	base unit	Carton of milk
kilolitre (kL)	1000 L = 1 kL	Water in 4 bath tubs
megalitre (ML)	1 ML = 1 000 000 L	Water in 2 Olympic swimming pools
Mass		
milligram (mg)	1000 mg = 1 g	A grain of salt
gram (g)	1000 g = 1 kg	A tablet
kilogram (kg)	base unit	A packet of sugar
tonne (t)	1000 kg = 1 t	A small car
Time		
second (s)	60 s = 1 min	One tick of a clock
minute (min)	60 min = 1 h	Time to run 500 m
hour (h)	base unit	Length of most TV shows
day (d)	24 h = 1 day	From midnight to midnight

To convert units, remember these initials:
- **SOLD = Small Over to Large Divide:** to convert from a small unit to a large unit, divide
- **LOSM = Large Over to Small Multiply:** to convert from a large unit to a small unit, multiply

EXAMPLE 1

Convert:

a 500 cm to m   **b** 3 L to kL   **c** 250 kg to g   **d** 4 days to h

SOLUTION

Use the table of units to convert

a 500 cm = 500 ÷ 100 m   ←————   SOLD Small Over to Large Divide, 100 cm = 1 m
           = 5 m

b 3 L = 3 ÷ 1000 kL   ←————   SOLD Small Over to Large Divide, 1000 L = 1 kL
        = 0.003 kL

c 250 kg = 250 × 1000 g   ←————   LOSM Large Over to Small Multiply, 1000 g = 1 kg
          = 250 000 g

d 4 days = 4 × 24h   ←————   LOSM Large Over to Small Multiply, 24 h = 1 day
       = 96 h

EXERCISE 12–01

1 What is the height of a door handle closest to? Select the correct answer **A**, **B**, **C** or **D**.
    A 1 mm   **B** 1 cm   **C** 1 m   **D** 1 km

2 What is the capacity of a kitchen sink closest to? Select **A**, **B**, **C** or **D**.
    A 9 mL   **B** 9 L   **C** 9 kL   **D** 9 ML

3 Choose the most appropriate unit for each measurement.
    a The thickness of a matchstick
    b The distance from Sydney to Melbourne
    c The width of a bathroom
    d The length of a textbook
    e The length of your shoe
    f Your height
    g The amount of water in a pool
    h The mass of an apple
    i The time taken to fly to Singapore
    j The capacity of a cup
    k The weight of a truck
    l The time to swim 50 m

4 Write down an approximate measurement for each item in question **3**.

iStockphoto/triamide

Shutterstock.com/saksawad

5 Copy and complete this table.

Millimetres	Centimetres	Metres	Kilometres
4000			
	6500		
		786	
			95
		85.4	
	790		
654 000			

6 Convert:

a 4000 mL to L   b 150 min to h   c 2500 g to kg

d 6.5 kL to L   e 3.5 h to min   f 7.85 g to mg

g 7 ML to kL   h 4.75 days to h   i 9.6 t to kg

j 25 mg to g   k 84 h to days   l 6.8 g to mg

7 What number does each metric prefix mean?

a milli-   b mega-   c kilo-

d centi-   e micro-   f giga-

8 Computer memory is measured in **bytes**. One **kilobyte (kB)** stores about half a page of text. Research the following questions.

iStockphoto/sweetym

a How many bytes in a kilobyte?

b A **megabyte (MB)** is about the memory size of a novel. How many kB in a MB?

c A **gigabyte (GB)** stores about 600 photos or 7.5 hours of video. How many MB in a GB?

d Convert 2750 kB to MB.

e Convert 1.6 GB to MB.

f What is 1000 GB called?

WORDBANK

precision The size of one unit on a measuring instrument; for example, 1 mm on a ruler.

limit of accuracy The amount of error in reading a measuring instrument. It is ±0.5 of one unit on the instrument's scale.

All measurements are **approximate** because:
* they depend upon the **precision** of the measuring instrument
* there is **human error** in reading the measurement
* there may be **faults** in the measuring instrument

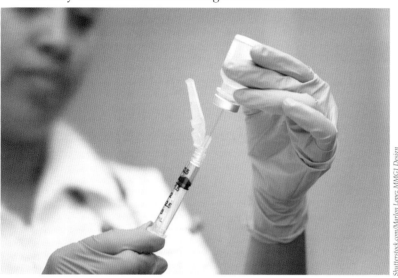

Shutterstock.com/Marion Lopez MMG1 Design

EXAMPLE 2

Find the limits of accuracy of each measuring scale.

a

b

SOLUTION

a Precision = 1 kg ⟵──────── The size of one unit on the scale
   Limits of accuracy = ± 0.5 × 1 kg = ± 0.5 kg
   A measurement of 62 kg on this scale means that the actual mass is 62 ± 0.5 kg
   (62 plus or minus 0.5), that is, between 61.5 and 62.5 kg.

b Precision = 5 mL
   Limits of accuracy = ± 0.5 × 5 mL = ± 2.5mL
   A measurement of 10 mL on this scale means that the actual capacity is 10 ± 2.5 mL,
   that is, between 12.5 and 7.5 mL

1 What is the size of one unit on this measuring cylinder? Select the correct answer **A**, **B**, **C** or **D**.

   A 1 mL   **B** 10 mL   **C** 50 mL   **D** 100 mL

2 What are the limits of accuracy of this measuring cylinder in question **1**? Select **A**, **B**, **C** or **D**.

   A ± 0.5 mL   **B** ± 10 mL   **C** ± 20 mL   **D** ± 5 mL

3 What is the precision (the size of one unit) for each measuring instrument?

a

b

c

d

e

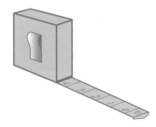

f

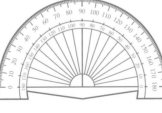

g

h

i    j    k

4 Write down the limits of accuracy for each measuring instrument in question **3**.

5 For each measurement below, find:

  i the limits of accuracy

  ii the range of values that the actual measurement lies between.

    a 75 kg, precision 1 kg   **b** 280 mL, precision 10 mL

    c 220 m, precision 20 m   **d** 450 g, precision 50 g

    e 2.6 cm, precision 0.1 cm   **f** 55 mm, precision 5 mm

    g 2750 mg, precision 50 mg   **h** 11.24 s, precision 0.01 s

Shutterstock.com/Paul Matthew Photography

WORDBANK

perimeter The distance around the edges of a shape.

composite shape A combination of shapes; for example, a rectangle
    and a triangle.

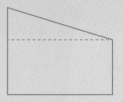

To find the **perimeter** of a shape, add the lengths of its sides.

EXAMPLE 3

Find the perimeter of each shape.

a

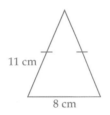

b

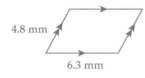

4.8 mm

6.3 mm

c

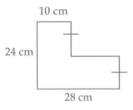

10 cm

24 cm

28 cm

SOLUTION

a   Perimeter = 11 + 11 + 8 ←——————— 2 equal sides in the isosceles triangle
         = 30 cm

b   Perimeter = 2 × 6.3 + 2 × 4.8 ←——————— opposite sides are equal in the parallelogram
         = 22.2 mm

c   First find the lengths of the unknown sides.   10 cm
    Perimeter = 10 + 12 + 18 + 12 + 28 + 24
             = 104 cm $\frac{1}{2} \times 24 = 12$ cm

                                                         28 − 10 = 18 cm

                                          24 cm

                                                         12 cm

                                                28 cm

1 What is the perimeter of a square of side length 6 cm? Select the correct answer **A, B, C** or **D**.

   A 18 cm   **B** 24 cm   **C** 30 cm   **D** 36 cm

2 What is the perimeter of a rectangle with length 8 mm and width 4 mm? Select **A, B, C** or **D**.

   A 24 mm   **B** 12 mm   **C** 16 mm   **D** 48 mm

3 Find the perimeter of each rectangle.

   a    **b**    **c**

4 Copy and complete this table.

Shape	Side lengths	Perimeter
Rectangle	9 cm and 4 cm	
Square	12 mm	
Equilateral triangle	4.2 cm	
Rhombus	8.4 m	
Parallelogram	9 cm and 4 cm	
Kite	3.6 cm and 5.7 cm	

5 Find the perimeter of each shape. All units are in cm.

   a    **b**    **c**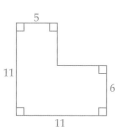

   d 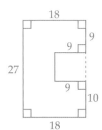   **e** 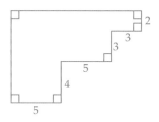   **f**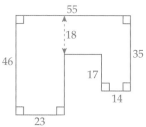

6 The area set aside for a tennis court is 100 m by 50 m.

   a What is the perimeter?

   b If shrubs are to be planted every 10 m around the boundary, how many shrubs will be needed?

   c If each shrub costs $10.50 how much will the shrubs cost?

WORDBANK

circumference The distance around a circle, the perimeter of a circle.

diameter The distance across a circle from one side to another, through the centre.

pi (π) The special number 3.14159... represented by the Greek letter π, which as a decimal has digits that run endlessly, without repeating or stopping.

quadrant quarter of a circle, with a right angle

To find the **circumference of a circle,** multiply π by the circle's diameter.
$C = πd$ means $C = π ×$ diameter
or $C = 2πr$ means $C = 2 × π ×$ radius

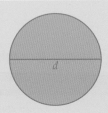

Note that the diameter of a circle is double its radius.
The value of π is stored on your calculator and can be found by pressing the [π] key, which may require the [SHIFT] key to use.

EXAMPLE 4

Find, correct to one decimal place, the circumference of each circle.

a

6 cm

b

4 m

SOLUTION

a $C = πd$ ⟵ diameter = 6 cm
$= π × 6$ ⟵ [π] [×] [6] [=] on calculator
$= 18.8495...$
$≈ 18.8$ cm

b $C = 2πr$ ⟵ radius = 4 m
$= 2 × π × 4$ ⟵ 2 × [π] [×] [4] [=]
$= 25.1327...$
$≈ 25.1$ m

EXAMPLE 5

Calculate the perimeter of this quadrant correct to two decimal places.

SOLUTION

Perimeter $= \dfrac{1}{4} ×$ circumference $+ 3 + 3$ ⟵ quadrant is $\dfrac{1}{4}$ of a circle

$= \dfrac{1}{4} × 2 × π × 3 + 6$ ⟵ radius = 6 cm

$= 10.7123...$
$≈ 10.71$ cm

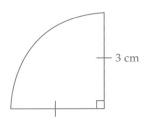

3 cm

EXAMPLE 6

Find the perimeter of this shape:

a correct to one decimal place   b in terms of π.

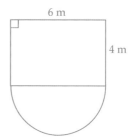

SOLUTION

a Perimeter $= \dfrac{1}{2} \times$ circumference $+ 4 + 4 + 6$

$\qquad = \dfrac{1}{2} \times \pi \times 6 + 14$ ⟵——— diameter = 6 m

$\qquad = 23.4247\ldots$

$\qquad \approx 23.4$ m

b Perimeter $= \dfrac{1}{2} \times \pi \times 6 + 14$ ⟵——— as in part **a**

$\qquad = (3\pi + 14)$ m ⟵——— $\dfrac{1}{2} \times \pi \times 6 = 3\pi$

✱ Leaving an answer in terms of π is more exact because we do not round the answer as a decimal.

EXERCISE **12-04**

1 Find the circumference of a circle with a diameter of 5 cm. Select the correct answer **A, B, C** or **D**.

   A 15.7 cm   **B** 7.9 cm   **C** 31.4 cm   **D** 78.5 cm

2 Find the circumference of a circle with radius 3.6 m. Select **A, B, C** or **D**.

   A 11.3 m   **B** 22.6 m   **C** 5.7 m   **D** 40.7 m

3 Calculate, correct to one decimal place, the circumference of a circle with:

   a diameter 6 cm   b radius 4 mm   c radius 2.8 m

4 Calculate, correct to one decimal place, the perimeter of a semicircle with:

   a diameter 8.2 cm   b diameter 48 mm   c radius 5.4 m

5 Find the perimeter of each shape:

   i correct to two decimal places   **ii** in terms of π.

   a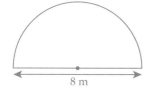

   b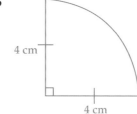

6 Calculate, correct to one decimal place, the perimeter of each shape.

a

16 cm

b

2 m

c

2.6 m

d

3.4 cm

e

11.2 m

f

6.8 m

g

10 m

3 m

h

3 cm

i

2.8 m

8.6 m

7 Sam is about to fence the edge of his semicircular garden with a picket fence. The radius of the garden is 2.3 m.

a How long does the fence need to be, to 1 decimal place?

b Jody buys the palings for the fence in Sam's garden and knows she needs 10 palings for each metre of fencing. How much do all the palings cost her if they cost $3 each?

iStockphoto/swedewah

24-hour time uses 4 digits to describe the time of day and does not require a.m. or p.m.

12-hour time	24-hour time	12-hour time	24-hour time
1:00 a.m.	0100	1:00 p.m.	1300
2:00 a.m.	0200	2:00 p.m.	1400
3:00 a.m.	0300	3:00 p.m.	1500
4:00 a.m.	0400	4:00 p.m.	1600
5:00 a.m.	0500	5:00 p.m.	1700
6:00 a.m.	0600	6:00 p.m.	1800
7:00 a.m.	0700	7:00 p.m.	1900
8:00 a.m.	0800	8:00 p.m.	2000
9:00 a.m.	0900	9:00 p.m.	2100
10:00 a.m.	1000	10:00 p.m.	2200
11:00 a.m.	1100	11:00 p.m.	2300
12:00 midday	1200	12:00 midnight	0000

Shutterstock.com/holbox

EXAMPLE 7

Write each time in 24-hour time.

a 4:16 a.m.　　**b** 9:53 p.m.　　**c** 12:28 a.m.

SOLUTION

a 4:16 a.m. = 0416 in 24-hour time ◄──── After 1 a.m., insert a 0 in front to make 4 digits

b 9:53 p.m. = 2153 in 24-hour time ◄──── After 1 p.m., add 12 to the hour: 9 + 12 = 21

c 12:28 a.m. = 0028 in 24-hour time ◄──── 12 midnight is 00 for the first 2 digits

EXAMPLE 8

Round 4 h 21 min to the nearest hour.

SOLUTION

4 h 21 min ≈ 4 h ◄──── 21 min is less than 30 min (half an hour), so round down

EXAMPLE 9

Convert:

a 146 minutes to hours and minutes

b 3.4 hours to hours and minutes

SOLUTION

a 146 minutes = (146 ÷ 60) h ← ───── 1 h = 60 min

= 2.43333… h

= 2 h 26 min ← ───── Enter ⊙'" or 2ndF DMS on a calculator or calculate 0.43333… × 60 for minutes

b 3.4 hours = 3 hours + 0.4 × 60 min ← ─── or enter 3.4 = ⊙'" or 2ndF DMS on a calculator

= 3 hours 24 min

Shutterstock.com/PomInOz

EXAMPLE 10

Calculate the time difference from 8:35 a.m. to 3:10 p.m.

SOLUTION

Use a number line and 'build bridges' like we did in Chapter 2 for mental subtraction.

```
   25 min          + 6 h           + 10 min
 ⌒             ⌒              ⌒
8:35 a.m.   9 a.m.          3 p.m.   3:10 p.m.
```

Time difference = 25 min + 6 h + 10 min

= 6 h + 35 min

OR: convert to 24-hour time and use the calculator's ⊙'" or DMS keys:

8:35 a.m. = 0835, 3:10 p.m. = 1510, so enter 15 ⊙'" 10 ⊙'" ─ 8 ⊙'" 35 ⊙'" =

So 6 hours 35 minutes is the time difference.

1   Convert 4:36 p.m. to 24-hour time. Select the correct answer **A, B, C** or **D**.

    A 1436   **B** 0436   **C** 1636   **D** 1236

2   Convert 6.6 hours to hours and minutes. Select **A, B, C** or **D**.

    A 6 h 6 min   **B** 6 h 36 min   **C** 6 h 60 min   **D** 6 h 10 min

3   Write each time in 24-hour time.

    a 2:10 a.m.   **b** 5:48 p.m.   **c** 4:39 a.m.   **d** 6:52 p.m.

    e 12:00 midday   **f** 7:55 a.m.   **g** 12:46 a.m.   **h** 9:25 p.m.

4   Convert to 12-hour time.

    a 0500   **b** 2100   **c** 0655   **d** 1848

    e 0850   **f** 1139   **g** 2256   **h** 0054

5   Round each time.

    a 3 h 28 min to the nearest hour

    b 8 h 43 min to the nearest hour

    c 5.6 hours to the nearest hour

    d 7.4 hours to the nearest hour

    e 12 min 10 s to the nearest minute

    f 5 min 35 s to the nearest minute

    g 7.4 min to the nearest minute

    h 18.65 min to the nearest minute

6   Convert each time to hours and minutes.

    a 5.8 h   **b** 186 min   **c** 1258 min

7   Convert each time to minutes and seconds.

    a 17.5 min   **b** 220 s   **c** 521 s

8   Find the time difference from:

    a 3:20 a.m. to 8:45 p.m.   **b** 2:50 a.m. to 11:59 a.m.

    c 9:21 p.m. to 10:56 a.m. next day   **d** 0235 to 1554

    e 1127 to 2148   **f** 1356 to 0817 next day

9   Ashok boarded a train at 0840 and arrived at his destination at 1435. How long was his train journey?

10 Rhonda bakes cupcakes in the oven for 28 minutes. If she placed them in the oven at 6:23 p.m., at what time should she take them out?

11 Find the sum of 3 h 25 min, 7 h 42 min and 5.4 hours.

This is a section of the train timetable from Newcastle to Sydney (Central).

Newcastle and Central Coast line – Newcastle to Central via Strathfield or Chatswood (weekdays)

	am	am	am	am	am	am	am	am	am	am	am	am	am	am	am	am	am
Newcastle				5.22				5.45	6.20		6.29		6.40		7.29		7.45
Civic				5.24				5.47			6.31		6.42		7.31		7.47
Wickham				5.26				5.49					6.44				7.49
Hamilton				5.28				5.51	6.24		6.34		6.46		7.34		7.51
Broadmeadow				5.32				5.55	6.28		6.38		6.50		7.38		7.55
Adamstown				5.35				5.58					6.53				7.57
Kotara				5.37				6.00					6.55				8.00
Cardiff				5.42				6.05	6.35		6.45		7.00		7.45		8.04
Cockle Creek				5.46				6.09					7.04				8.08
Teralba				5.50				6.13					7.08				8.11
Booragul				5.52				6.15					7.10				8.13
Fassifern				5.56				6.19	6.46		6.55		7.14		7.55		8.17
Awaba				6.01				6.24			7.00		7.19				8.21
Dora Creek				6.10				6.33			7.09		7.27				8.30
Morisset				6.14				6.37	7.01		7.13		7.32		8.11		8.35
Wyee				6.21				6.44			7.20		7.39		8.17		
Warnervale				6.28				6.51			7.27		7.46		8.24		
Wyong	6.03			6.35	6.44			6.58		7.21	7.34		7.53		8.31		
Tuggerah	6.05			6.37	6.46			7.00			7.36		7.55		8.33		
Ourimbah	6.12			6.44	6.53			7.07			7.43		8.02		8.40		
Lisarow	6.15			6.47				7.10					8.05				
Niagara Park	6.17			6.49				7.12					8.07				
Narara	6.20			6.52	6.58			7.15					8.10				
Gosford arr.	6.23			6.55	7.01			7.18	7.29	7.35	7.51		8.13		8.48		
Gosford dep.	6.36		6.48	6.57	7.02	7.07		7.20	7.30	7.36	7.52		8.15		8.53		
Point Clare	6.39					7.10		7.23					8.18				
Tascott	6.41					7.12		7.25					8.20				
Koolewong	6.44					7.15		7.28					8.23				
Woy Woy	6.47		6.56	7.06	7.11	7.18		7.31	7.38	7.45	8.00		8.26		9.01		
Wondabyne	6.54							7.38									
Hawkesbury River	7.01					7.31		7.45									
Cowan	7.12					7.42		7.56					8.51				
Berowra	7.17				7.38	7.47		8.01				8.20	8.56	9.00		9.21	
Asquith					7.45							8.30		9.09		9.30	
Hornsby	7.28		7.34	7.44	7.50	7.56	8.01	8.12	8.15	8.20	8.37	8.38	9.07	9.13	9.37	9.43	
Chatswood		7.48	7.53		8.06		8.21			8.36		8.59		9.34		10.04	
St Leonards		7.53	7.58		8.11		8.26			8.41		9.03		9.39		10.09	
North Sydney		8.01	8.04		8.19		8.34			8.49		9.09		9.46		10.16	
Wynyard		8.06	8.09		8.24		8.39			8.54		9.14		9.51		10.21	
Town Hall		8.09	8.12		8.27		8.42			8.57		9.17		9.54		10.24	
Epping	7.37			7.53		8.07		8.27			8.46		9.16		9.46		
Eastwood						8.10		8.30									
West Ryde	7.41							8.32									
Strathfield	7.50			8.05		8.19		8.43	8.35		8.58		9.27		9.57		
Redfern	8.01					8.30					9.08						
Central	8.04	8.11	8.14	8.18	8.29	8.33	8.44	8.56	8.49	8.59	9.11	9.19	9.40	9.56	10.10	10.26	

Source: http://www.cityrailinfo/timetable. Reproduced by permission of RailCorp.

ISBN 9780170351027

EXAMPLE 11

Nina lives in Gosford and her office is a 10-minute walk from Town Hall Station in Sydney.

a   If she needs to be at work by 8:45 a.m., what is the latest train she could catch from Gosford?

b   How long would this train journey take?

c   If she caught the earlier train from Gosford, at what time would she arrive at Town Hall Station?

SOLUTION

a   Working backwards, if she needs to be at work by 8:45 a.m., she needs to reach Town Hall Station by 8:45 a.m. − 10 min = 8:35 a.m.

One train arrives at Town Hall Station at 8:27 a.m., and looking upwards, it departs Gosford at 7:02 a.m. So Nina would need to catch the 7:02 a.m. train from Gosford.

b   The journey lasts from 7:02 a.m. to 8:27 a.m., which is 1 h 25 min.

c   The train before 7:02 a.m. that goes from Gosford to Town Hall is at 6:48 a.m., which arrives at Town Hall at 8:12 a.m.

EXERCISE 12-06

1   If Tony needs to be at Town Hall Station by 8:30, what is the latest train that he can catch from Tuggerah? Select the correct answer **A**, **B**, **C** or **D**.

   A 6:05   **B** 6:48
   C 6:37   **D** 6:46

2   If Lyn needs to be at Woy Woy by 8:06, at what time should she catch the train from Gosford? Select **A**, **B**, **C** or **D**.

   A 7:18   **B** 7:51
   C 7:35   **D** 7:29

3   Find which is the latest train to catch and how long the journey will take if you have to be at:

   a   Central station at 8.15 a.m., travelling from Wyong

   b   Strathfield at 8.40 a.m., travelling from Hamilton

   c   Hornsby at 9.15 a.m., travelling from Cardiff

   d   West Ryde at 7.45 a.m., travelling from Tuggarah

   e   Epping at 8.30 a.m., travelling from Gosford.

4   What would be the quickest train for Sandy to catch from Gosford to Central Station if she has to reach Central Station by 9 a.m.?

5 a Vincent is catching a train from Gosford and needs to be at St Leonards for a job interview at 9 a.m. What is the latest train departing from Gosford that he can catch?

   b How long will the train journey take?

   c How much time will Vincent have to walk from St Leonards station to the interview?

6 Colin is flying from Perth to Hong Kong and the details of his flight are shown here.

Departure city	Departure time	Boarding gate	Arrival time	Destination
Perth	1110 14 /01/15	26	1850 14/01/15	Hong Kong

   a If Colin needs to check-in at the airport $2\frac{1}{2}$ hours before his flight leaves, what time should he arrive at Perth airport? Answer in 12-hour time.

   b How long will the flight take if there are no delays?

   c At what time will Colin arrive in Hong Kong if the flight is 40 minutes late? Answer in 12-hour time.

12-07 International time zones

The world is divided into 24 different time zones, each one representing a 1-hour time difference. World times are measured in relation to the Greenwich Observatory in London, either ahead or behind UTC (Coordinated Universal Time), also known as GMT (Greenwich Mean Time).

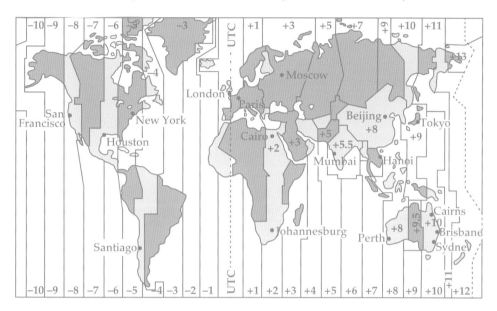

EXAMPLE 12

If it is 8 a.m. in Greenwich, what is the time in:

a   Beijing?   b   Sydney?   c   New York?

SOLUTION

a   Beijing is 8 hours ahead of UTC, so its time is 8 a.m. + 8 h = 4 p.m.

b   Sydney is 10 hours ahead of UTC, so its time is 8 a.m. + 10 h = 6 p.m.

c   New York is 5 hours behind UTC so its time is 8 a.m. – 5 h = 3 a.m.

Shutterstock.com/Sean Pavone

Shutterstock.com/Thorsten Rust

Shutterstock.com/richie81

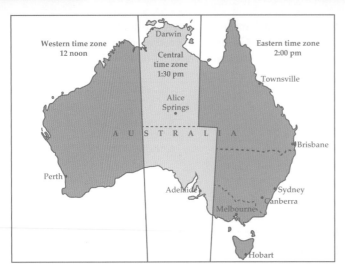

Australia has three time zones: Western (UTC+8), Central (UTC+9.5) and Eastern (UTC+10)

EXAMPLE 13

If it is 11:30 a.m. in Alice Springs, then find the time in:

a Perth   **b** Newcastle   **c** Adelaide

SOLUTION

a Perth is 1.5 hours behind the time in Alice Springs, so its time is
11:30 a.m. − 1.5 h = 10 a.m.

b Newcastle is half an hour ahead of the time in Alice Springs, so its time is
11:30 a.m. + 0.5 h = 12 midday.

c Adelaide is in the same time zone as Alice Springs, so its time is 11:30 a.m.

EXERCISE 12-07

1 If it is 9:30 a.m. in Greenwich, what time will it be in Johannesburg in South Africa?
Select the correct answer **A, B, C** or **D**.

     A 7:30 a.m.   **B** 8:30 a.m.   **C** 10:30 a.m.   **D** 11:30 a.m.

2 If it is 4:30 p.m. in Mumbai in India, what time will it be in London? Select **A, B, C** or **D**.

     A 12:30 p.m.   **B** 11 a.m.   **C** 12 midday   **D** 10 p.m.

3 Given that it is 9 a.m. in London, find the time in each city.

     a Moscow, Russia   **b** Santiago, Chile

     c Mumbai   **d** Tokyo, Japan

     e Houston, USA   **f** Sydney

4   Given that is 12 noon in Johannesburg, find the time in each city.

   a   London   b   Paris, France   c   New York

   d   Perth   e   Beijing   f   Sydney

5   Clare-Louise caught a direct flight from Sydney to Paris. She left Sydney at 11:20 a.m. on Thursday and the flight was 28 hours long. What time and day did she arrive in Paris?

6   Thanh goes to school in Sydney but her grandmother lives in Hanoi in Vietnam. Thanh finishes school at 3.30 p.m. and rings her grandmother when she gets home at 4.30 p.m. What time is it in Hanoi when Thanh rings?

7   If it is 11.00 a.m. in Darwin, find the time in each city.

   a   Brisbane   b   Melbourne   c   Perth   d   Alice Springs

8   If it is 8.00 p.m. in Perth, find the time in each city.

   a   Canberra   b   Adelaide   c   Townsville   d   Hobart

9   Anna lives in Perth and rings her father, who is in Adelaide on business. If she rings at 3 p.m., what time is it in Adelaide?

10 Pramona is keen to watch the Wimbledon final live on TV at his home in Brisbane. The match is played in London at 6 p.m. and lasts 4 hours and 15 minutes. Between what times does Pramona watch the match in Brisbane?

11 Each year, most Australian states change to daylight saving time from October to March to take advantage of the longer hours of daylight. Clocks are turned forward one hour during this period, ahead of standard time. When it is 4 p.m. in the eastern states in February, what time is it in Queensland, where daylight saving does not operate?

12 Vas left Sydney at 10.30 p.m. daylight saving time to travel to Cairns in Queensland on a flight that took 3 hours. What was the local time when he arrived?

FIND-A-WORD PUZZLE

Make a copy of this page, then find all the words listed below in this grid of letters.

D	R	E	T	E	M	A	I	D	Y	R	C	I	E	H
H	R	E	Z	X	A	Y	M	W	M	E	A	N	R	C
K	I	L	O	G	R	A	M	A	N	C	C	T	T	I
D	E	R	T	S	G	Q	R	T	S	T	C	E	E	W
R	G	M	E	H	O	G	I	L	T	A	U	R	M	N
L	M	N	I	T	L	M	Z	Q	Y	N	R	N	O	E
E	L	B	A	T	E	M	I	T	C	G	A	A	L	E
N	D	Q	Y	T	L	M	I	L	I	L	C	T	I	R
G	N	N	R	M	L	C	I	F	R	E	Y	I	K	G
T	O	E	M	Z	A	N	E	R	T	I	L	O	M	V
H	C	N	H	P	R	M	U	K	E	Q	U	N	I	U
W	E	E	A	Q	A	K	X	P	M	P	Y	A	N	S
W	S	C	O	M	P	O	S	I	T	E	U	L	U	S
C	I	R	C	U	M	F	E	R	E	N	C	E	T	A
R	W	R	E	R	T	I	L	I	L	L	I	M	E	M

ACCURACY	CAPACITY	CENTIMETRE	CIRCUMFERENCE
COMPOSITE	DIAMETER	GRAM	GREENWICH
INTERNATIONAL	KILOGRAM	KILOMETRE	LENGTH
LITRE	MASS	METRE	METRIC
MILLILITRE	MINUTE	PARALLELOGRAM	PERIMETER
RECTANGLE	SECOND	TIME	TIMETABLE

Shutterstock.com/Maksim Budnikov

PRACTICE TEST 12

Part A General topics

Calculators are not allowed.

1 Convert 1935 to 12-hour time.

2 Simplify 28 : 16.

3 Evaluate 9^{-2}.

4 Write Pythagoras' theorem for this triangle.

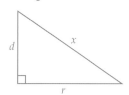

5 Factorise: $10x^2y - 15xy^2$

6 Write an algebraic expression for the number of days in w weeks.

7 Solve $6x - 26 = x + 1$

8 Expand $6(2x - 10)$.

9 Write $\dfrac{2}{3}$, 0.65, 0.6 and 68% in descending order.

10 What is the probability of rolling a number on a die that is a factor of 6?

Part B Length and time

Calculators are allowed.

12–01 The metric system

11 How many millilitres in a litre? Select the correct answer **A, B, C** or **D**.

    A 10   **B** 100   **C** 1000   **D** 1 000 000

12 How many millimetres in a kilometre? Select **A, B, C** or **D**.

    A 10   **B** 1000   **C** 10 000   **D** 1 000 000

12–02 Limits of accuracy of measuring instruments

13 What are the limits of accuracy for this scale?

12–03 Perimeter

14 Find the perimeter of each shape.

    a

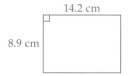

    b

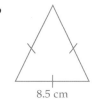

15 A rectangular shape 18.5 m long and 9.6 m wide is set aside for a garden. What is its perimeter?

12-04 Circumference of a circle

16 Find the circumference of each circle, to 1 decimal place.

a
9 cm

b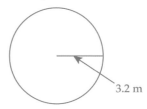
3.2 m

17 Find the perimeter of this shape, to 1 decimal place.

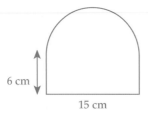
6 cm
15 cm

12-05 Time

18 Find the time difference from:

   a   2:35 p.m. to 3:15 a.m. the next day   b   0954 to 1622

19 Round to the nearest hour.

   a   21 h 42 min   b   9.45 h

12-06 Timetables

20 Use the timetable on page 250 to answer the questions below.

   a   What is the latest train that I can catch from Newcastle if I have to meet a friend in Gosford at 8:15 a.m.?

   b   How long would this train journey take?

12-07 International time zones

21 Use the time zones on page 253 to answer the questions below.

   a   If it is 4:50 a.m. Sunday in Moscow, what time and day is it in San Francisco?

   b   If it is 5:15 p.m. in Townsville, what time is it in Perth?

AREA AND VOLUME

13

WHAT'S IN CHAPTER 13?

IN THIS CHAPTER YOU WILL:

- calculate the areas of triangles, quadrilaterals and composite shapes
- calculate the areas of circles, semicircles, quadrants, sectors, annuluses and composite circular shapes
- calculate the surface areas of cubes, rectangular prisms and triangular prisms
- calculate the surface area of a cylinder
- convert between metric units for volume and capacity
- calculate the volumes of prisms, composite prisms and cylinders

Shutterstock.com/S.Pytel

Area is the amount of surface space occupied by a flat shape. Area is measured in square units such as square centimetres (cm²) or square metres (m²). The table below shows the formulas for finding the areas of four shapes.

Square	Rectangle
	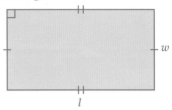
$A = $ (side length)^2	$A = $ length × width
$A = s^2$	$A = lw$
Triangle	**Parallelogram**
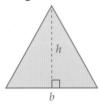	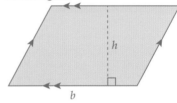
$A = \dfrac{1}{2} \times$ base × height	$A = $ base × height
$A = \dfrac{1}{2}bh$	$A = bh$

EXAMPLE 1

Find the area of each shape.

a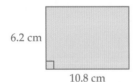

b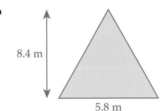

c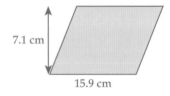

SOLUTION

a $A = lw$

 rectangle

$= 10.8 \times 6.2$

$= 66.96 \text{ cm}^2$

b $A = \dfrac{1}{2}bh$

 triangle

$= \dfrac{1}{2} \times 5.8 \times 8.4$

$= 24.36 \text{ m}^2$

c $A = bh$

 parallelogram

$= 15.9 \times 7.1$

$= 112.89 \text{ cm}^2$

EXAMPLE 2

Find the area of this composite shape.

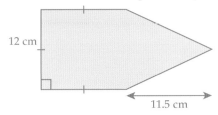

SOLUTION

This is a combined square and triangle.

Area = Area of square + area of triangle

$$= 12^2 + \frac{1}{2} \times 12 \times 11.5$$

$$= 213 \text{ cm}^2$$

EXERCISE 13-01

1 Find the area of a rectangle with length 18 cm and width 11 cm. Select the correct answer **A**, **B**, **C** or **D**.

   A 99 cm²   **B** 49.5 cm²   **C** 198 cm²   **D** 396 cm²

2 Find the area of a triangle with base 15 m and height 12 m. Select **A**, **B**, **C** or **D**.

   A 180 m²   **B** 90 m²   **C** 360 m²   **D** 225 m²

3 Find the area of each shape, to 2 decimal places if required.

a

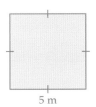

5 m

b

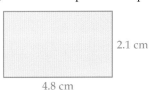

2.1 cm

4.8 cm

c

5.3 cm

1.7 cm

d

3.2 m

e

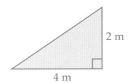

2 m

4 m

f

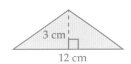

3 cm

12 cm

g
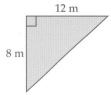
12 m
8 m

h
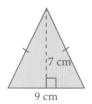
7 cm
9 cm

i
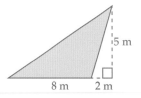
5 m
8 m   2 m

j
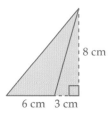
8 cm
6 cm   3 cm

k
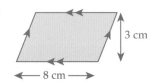
3 cm
8 cm

l
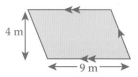
4 m
9 m

m
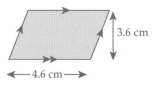
3.6 cm
4.6 cm

n
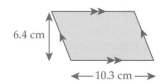
6.4 cm
10.3 cm

4 Find the area of each composite shape.

a
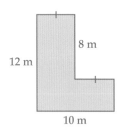
8 m
12 m
10 m

b
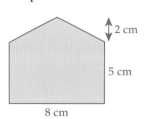
2 cm
5 cm
8 cm

c
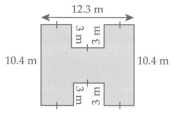
12.3 m
3 m   3 m
10.4 m   10.4 m
3 m   3 m

d
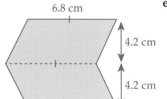
6.8 cm
4.2 cm
4.2 cm

e
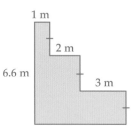
1 m
2 m
6.6 m
3 m

f
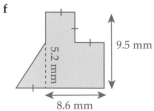
9.5 mm
5.2 mm
8.6 mm

To find the **area of a circle,** multiply π by the circle's radius squared.
$A = \pi r^2$ means $A = \pi \times \text{radius}^2$

If you are given the diameter, halve it to find the radius first.

EXAMPLE 3

Find the area of each circle correct to 1 decimal place.

a

6.2 m

b

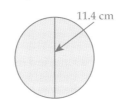

11.4 cm

SOLUTION

a $A = \pi r^2$ ⟵ $r = 6.2$
$= \pi \times 6.2^2$
$= 120.7628...$
$\approx 120.8 \text{ m}^2$

b $A = \pi r^2$ ⟵ $r = \dfrac{1}{2} \times 11.4 = 5.7$
$= \pi \times 5.7^2$
$= 102.0703...$
$\approx 102.1 \text{ cm}^2$

EXAMPLE 4

Find the area of each shape correct to 2 decimal places.

a

2.85 cm

b

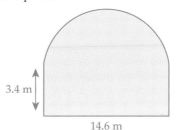

3.4 m

14.6 m

SOLUTION

a Area $= \dfrac{1}{4} \times$ area of circle
$= \dfrac{1}{4} \times \pi \times 2.85^2$
$= 6.37939...$
$\approx 6.38 \text{ cm}^2$

b Area $= \dfrac{1}{2} \times$ area of circle $+ 14.6 \times 3.4$
$= \dfrac{1}{2} \times \pi \times 7.3^2 + 49.64$ ⟵ $r = \dfrac{1}{2} \times 14.6 = 7.3$
$= 133.3477...$
$\approx 133.35 \text{ m}^2$

1   Find the area of a circle with radius 4 cm. Select the correct answer **A**, **B**, **C** or **D**.
    A 25.1 cm^2 **B** 50.3 cm^2 **C** 12.6 cm^2 **D** 100.5 cm^2

2   Find the area of a circle with diameter 6.2 m. Select **A**, **B**, **C** or **D**.
    A 30.2 m^2 **B** 19.5 m^2 **C** 120.8 m^2 **D** 60.4 m^2

3   Find the area of each circle correct to 2 decimal places.

    a   **b**   **c**

4   A circular lawn has a diameter of 20 m. Find its area, correct to the nearest square metre.

5   Find, correct to one decimal place, the area of each shape.

    a   **b**   **c**

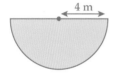

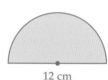

    d   **e**   **f**

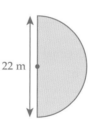

6   A duck pond is built in the shape of a quadrant. How many square metres of land
    (to 2 decimal places) will it occupy if the radius of the pond is 3.6 m?

7 Find the area of each shape, to 2 decimal paces if required.

a

b

c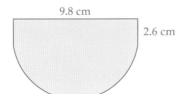

d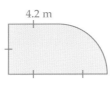

e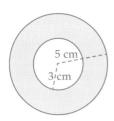

f

8 A business logo is shown below. A sign is needed for an advertisement. How much will it cost to make this shape from perspex if the perspex costs $28.50 per m²?

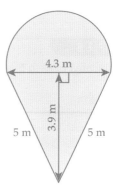

9 This postage stamp was issued by Semcirculonia to celebrate the king's birthday. Ignore the perforations, and assume an overall semicircular shape in your calculations.

a Find its area (correct to 1 decimal place).

b Forty stamps are printed on each sheet as shown.

i Find the total area of 40 stamps.

ii What is the area of the sheet of paper that they are printed on?

iii How much paper is wasted on each sheet (shaded)?

Trapezium

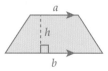

$A = \dfrac{1}{2} \times \text{height} \times (\text{sum of parallel sides})$

$A = \dfrac{1}{2} h(a + b)$

Kite and rhombus

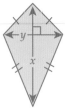 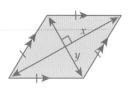

$A = \dfrac{1}{2} \times \text{diagonal 1} \times \text{diagonal 2}$

$A = \dfrac{1}{2} xy$

iStockphoto/olesiabilkei

EXAMPLE 5

Find the area of each quadrilateral.

a

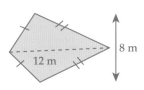

8 m

12 m

b 3.6 cm

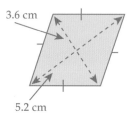

5.2 cm

c 18.4 m

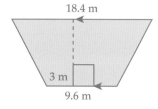

3 m

9.6 m

SOLUTION

a $A = \dfrac{1}{2} xy$

  kite

$= \dfrac{1}{2} \times 12 \times 8$

$= 48 \text{ m}^2$

b $A = \dfrac{1}{2} xy$

 rhombus

$= \dfrac{1}{2} \times 5.2 \times 3.6$

$= 9.36 \text{ cm}^2$

c $A = \dfrac{1}{2} h(a + b)$

 trapezium

$= \dfrac{1}{2} \times 3 \times (18.4 + 9.6)$

$= 42 \text{ m}^2$

ISBN 9780170351027

1   Find the area of a kite with diagonals 8 cm and 14 cm. Select the correct answer **A, B, C** or **D**.

   A 112 cm²   **B** 224 cm²   **C** 56 cm²   **D** 28 cm²

2   Find the area of a trapezium with parallel sides 4 m and 6 m and perpendicular height 11 m. Select **A, B, C** or **D**.

   A 55 m²   **B** 110 m²   **C** 34 m²   **D** 45 m²

3   Find the area of each of the figures below, to 2 decimal places if required.

a

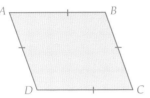

$AC = 20$ cm

$BD = 16$ cm

b

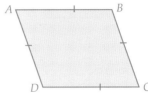

$AC = 2.4$ mm

$BD = 1.8$ mm

c

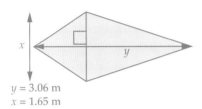

$y = 3.06$ m
$x = 1.65$ m

d

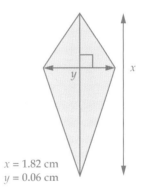

$x = 1.82$ cm
$y = 0.06$ cm

e

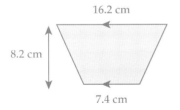

16.2 cm

8.2 cm

7.4 cm

f

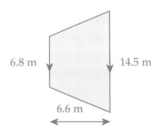

6.8 m   14.5 m

6.6 m

4 Find the area of each composite shape, to 2 decimal places if required. All dimensions are in centimetres.

a

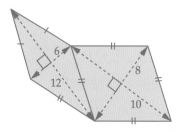

b

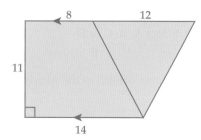

c

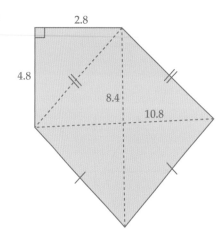

d

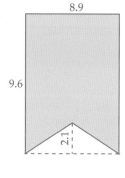

5 A garden is in the shape of a trapezium as shown and Elise wants to plant roses in it.
  a Find the area of Elise's garden.
  b If each rose plant needs 0.5 m² to grow, how many roses can Elise plant?
  c If each rose plant costs $4.60, find the cost of buying roses for the garden.

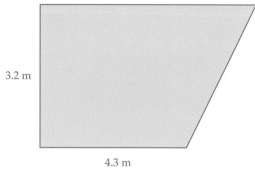

WORDBANK

surface area The sum of the areas of the faces of a solid shape.

net The faces of a solid shape laid out flat.

The surface area of a solid is easier to calculate by looking at its net. The number of faces and their shape can then be seen.

A cube and its net

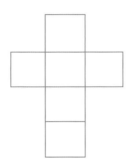

6 identical square faces

A rectangular prism and its net

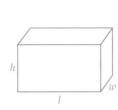
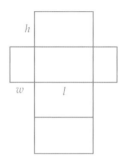

6 rectangular faces, with opposite faces identical: top and bottom, front and back, left and right

---

SURFACE AREA OF A CUBE

SA = 6 × (side length)^2
= 6s^2

SURFACE AREA OF A RECTANGULAR PRISM

SA = (2 × length × width) + (2 × length × height) + (2 × width × height)
= 2lw + 2lh + 2wh

Surface area is an area so it is measured in **square units**.

EXAMPLE 6

Find the surface area of each solid.

a

5 cm

b

9 m

5 m

11 m

SOLUTION

a SA = 6 × area of squares
        = 6 × 5^2
        = 150 cm^2

b SA = 2 × bottom + 2 × front + 2 × sides
        = 2 × 11 × 5 + 2 × 11 × 9 + 2 × 5 × 9
        = 398 m^2

1 What is the surface area of a cube with a side length of 4 cm? Select the correct answer **A, B, C** or **D**.

A 96 cm^2 **B** 64 cm^2 **C** 16 cm^2 **D** 32 cm^2

2 What is the surface area of a rectangular prism with length 8 m, width 6 m and height 9 m? Select **A, B, C** or **D**.

A 174 m^2 **B** 432 m^2 **C** 348 m^2 **D** 864 m^2

3 Find the area of each net.

a

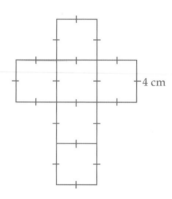

4 cm

b

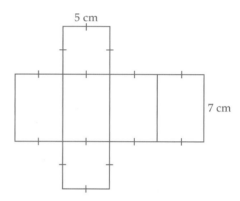

5 cm

7 cm

c

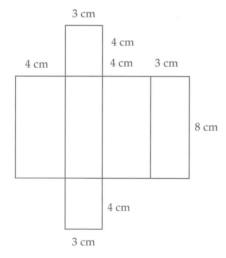

3 cm

4 cm

4 cm

4 cm   3 cm

8 cm

4 cm

3 cm

4   Draw the net of each solid, then find its surface area.

a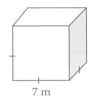

7 m

b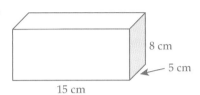

8 cm

5 cm

15 cm

5   Find the surface area of each prism, to 2 decimal places if required.

a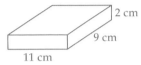

2 cm

9 cm

11 cm

b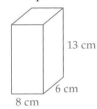

13 cm

6 cm

8 cm

c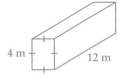

4 m

12 m

d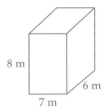

8 m

6 m

7 m

e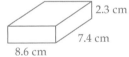

2.3 cm

7.4 cm

8.6 cm

f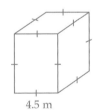

4.5 m

6   a   Find the surface area of a cube with side length 3.8 mm.

b   Would a rectangular prism with twice the length, the same width and half the height have the same surface area as this cube?

c   Check your answer by finding the surface area of the rectangular prism.

A triangular prism and its net

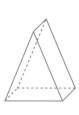
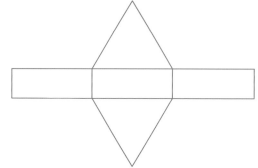

Find the surface area of the triangular prism with this net.

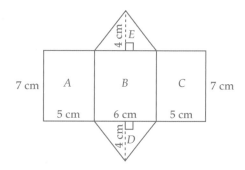

SOLUTION

Faces A and C are identical rectangles, B is another rectangle, D and E are identical triangles.

Surface area $= (2 \times 5 \times 7) + (6 \times 7) + (2 \times \dfrac{1}{2} \times 6 \times 4)$ ◄——— $2 \times$ area A + area B + $2 \times$ area E

$\qquad = 136$ cm^2

Find the surface area of this triangular prism.

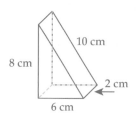

SOLUTION

5 faces: front and back are identical triangles, the other 3 faces are different rectangles.

Surface area $=$ Area of 2 triangles + area of base + area of LHS + area of RHS

$\qquad = (2 \times \dfrac{1}{2} \times 6 \times 8) + (6 \times 2) + (8 \times 2) + (10 \times 2)$ ◄——— $2 \times$ front + bottom + sides

$\qquad = 96$ cm^2

1 What are the shapes of the faces of a triangular prism? Select the correct answer **A**, **B**, **C** or **D**.

    A 3 triangles, 2 rectangles   **B** 1 triangle, 4 rectangles

    C 2 triangles, 2 rectangles   **D** 2 triangles, 3 rectangles

2 Each triangular face on a triangular prism has an area of 8 m² and each rectangular face has an area of 12 m². What is th total surface area of the prism? Select **A**, **B**, **C** or **D**.

    A 48 m²   **B** 56 m²   **C** 40 m²   **D** 52 m²

3 Draw the net of each triangular prism and find the surface area of each prism.

    a
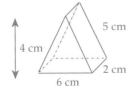

    b
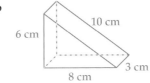

4 Find the surface area of each triangular prism.

    a

    b
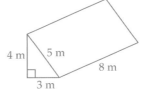

    c
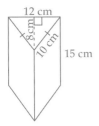

    d
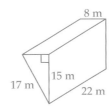

    e
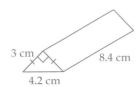

    f
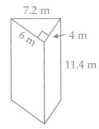

5 This chocolate bar has the shape of a triangular prism. How many cm² of paper is needed to wrap around the chocolate bar? (Assume that there is no overlap.)
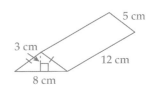

A cylinder and its net

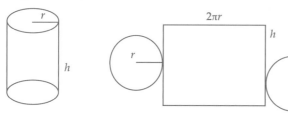

The net is made up of 2 identical circles and a rectangle. The rectangle is formed by cutting the cylinder along one edge (see dotted line) and opening it up. The length of this rectangle is $2\pi r$ because it is the circumference of the circle and the width of the rectangle is h, the height of the cylinder.

Area of both circles = $2 \times \pi r^2$

Area of rectangle = $2\pi r \times h$

SURFACE AREA OF A CYLINDER

SA = $2 \times \pi \times$ radius^2 + $2 \times \pi \times$ radius $\times$ height

  = $2\pi r^2 + 2\pi rh$

This formula is for a **closed** cylinder, where there is a top and a bottom.

An **open** cylinder has a base but no top so its **SA = $\pi r^2 + 2\pi rh$**, only one circle not two.

EXAMPLE 9

Find the surface area of each cylinder, correct to 1 decimal place.

a

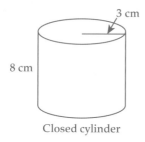

3 cm

8 cm

Closed cylinder

b

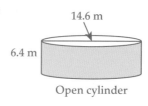

14.6 m

6.4 m

Open cylinder

SOLUTION

a Surface area = $2\pi r^2 + 2\pi rh$

    = $2 \times \pi \times 3^2 + 2 \times \pi \times 3 \times 8$

    = 207.345...

    ≈ 207.3 cm^2

b Surface area = $\pi r^2 + 2\pi rh$

    = $\pi \times 7.3^2 + 2 \times \pi \times 7.3 \times 6.4$

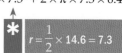

    ✱ $r = \dfrac{1}{2} \times 14.6 = 7.3$

    = 460.9658...

    ≈ 461.0 m^2

1   What are the shapes of the faces of a closed cylinder? Select the correct answer **A, B, C** or **D**.

   A 1 circle, 1 rectangle   **B** 2 circles, 1 rectangle

   C 2 circles, 2 rectangles   **D** 1 circle, 2 rectangles

2   Find the surface area of a cylinder with radius 4 cm and height 5 cm. Select **A, B, C** or **D**.

   A 175.9 cm^2 **B** 226.2 cm^2 **C** 351.9 cm^2 **D** 301.6 cm^2

3   **a** Draw the net of a cylinder with radius 6 cm and height 4 cm.

   b Calculate, correct to two decimal places, its surface area.

4   Find, correct to one decimal place, the surface area of each solid.

a
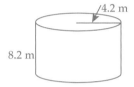
4.2 m

8.2 m

b

8.4 cm

11.5 cm

c

3 m

11 m

d
open

1.8 m

2.6 m

e
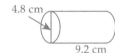
4.8 cm

9.2 cm

f
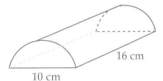
16 cm

10 cm

g
open
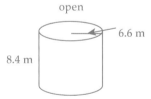
6.6 m

8.4 m

5   **a** The outside of a cylindrical water tank of height 12.6 m and diameter 8.4 m needs to be painted. The tank does not have a lid and the base cannot be painted. Find the surface area of the curved face of the tank, correct to two decimal places.

   b If one can of paint can cover 25 m^2 of the tank, how many full cans of paint are required?

   c If one can costs $64, what is the total cost of the paint required?

iStockphoto/jodie777

WORDBANK

volume The amount of space occupied by a solid shape, measured in cubic units such as cubic centimetres (cm^3) and cubic metres (m^3).

capacity The amount of liquid or material that a container can hold, measured in millilitres (mL), litres (L), kilolitres (kL) and megalitres (ML).

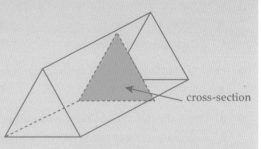

cross-section

cross-section A slice that is cut through a prism, parallel to the ends, that is the same shape all along the prism.

VOLUME AND CAPACITY

$1\ cm^3$ contains 1 mL
$1\ m^3$ contains 1 kL or 1000 L

1 mL

1 cm³   × 1 000 000 = $1\ m^3 = 1\ kL$

EXAMPLE 10

Convert:

a   3700 cm^3 to mL   **b** 15 000 cm^3 to L   **c** 2.4 m^3 to L

SOLUTION

a   3700 cm^3 = 3700 mL   ⟵——— $1\ cm^3 = 1\ mL$

b   15 000 cm^3 = 15 000 mL
                = (15 000 ÷ 1000) L   ⟵——— $1\ L = 1000\ mL$
                = 15 L

c   2.4 m^3 = 2.4 × 1000 L
            = 2400 L

VOLUME OF A PRISM

V = area of base or cross-section × height
$V = Ah$

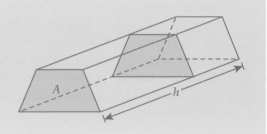

A h

EXAMPLE 11

Find the volume of each prism.

a

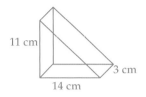

11 cm
3 cm
14 cm

b

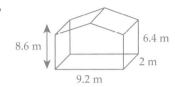

8.6 m
6.4 m
2 m
9.2 m

SOLUTION

Find A, the area of the base or cross-section first.

a $A = \dfrac{1}{2} \times 14 \times 11$ ⟵ Area of a triangle

$= 77$ cm²

$V = Ah$

$= 77 \times 3$ ⟵ $h = 3$

$= 231$ cm³

b $A = (\dfrac{1}{2} \times 9.2 \times 2.2) + (9.2 \times 6.4)$ ⟵ Area of a triangle + rectangle

✱ Triangle's height = 8.6 − 6.4 = 2.2 m

$= 69$ m²

$V = Ah$

$= 69 \times 2$ ⟵ $h = 2$

$= 138$ m³

EXERCISE 13-07

1 Convert 6.2 m³ to L. Select the correct answer **A**, **B**, **C** or **D**.

   A 620 000 **B** 62 000 **C** 6200 **D** 6 200 000

2 Convert 5 L to cm³. Select **A**, **B**, **C** or **D**.

   A 500 000 **B** 50 000 **C** 5000 **D** 500

3 Convert:

   a 20 cm³ to mL **b** 450 m³ to L **c** 72 500 cm³ to L

   d 655 mL to cm³ **e** 2560 L to m³ **f** 5.7 kL to m³

4 Draw a cross-section for each prism.

   a

   b

   c

5 Find the volume of each prism, to 1 decimal place if required.

a

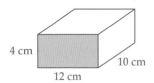

4 cm
10 cm
12 cm

b

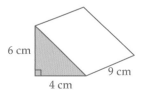

6 cm
4 cm
9 cm

c

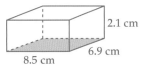

2.1 cm
6.9 cm
8.5 cm

d

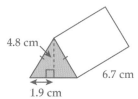

4.8 cm
6.7 cm
1.9 cm

e

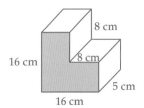

8 cm
16 cm
8 cm
5 cm
16 cm

f

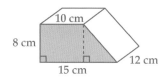

10 cm
8 cm
15 cm
12 cm

g

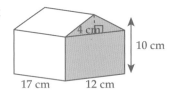

4 cm
10 cm
17 cm   12 cm

h

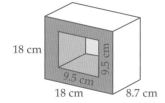

18 cm
9.5 cm
9.5 cm
18 cm   8.7 cm

6 A vase in the shape of a rectangular prism is 8 cm long, 5 cm wide and 16 cm high.
Find its capacity in:

a cubic centimetres

b millilitres

c litres.

7 A swimming pool has the shape of a rectangular prism that is 11 m long, 5 m wide and 1.8 m deep.

a Find the volume of the swimming pool.

b How many litres of water will it hold?

13-08 | Volume of a cylinder

VOLUME OF A CYLINDER

V = area of circular base × height
$V = \pi$ × radius^2 × height
$V = \pi r^2 h$

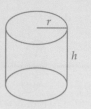

EXAMPLE 12

Find the volume of each cylinder correct to one decimal place.

a

4.5 cm

10.2 cm

b

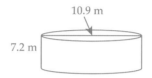

10.9 m

7.2 m

SOLUTION

a $V = \pi r^2 h$
$= \pi \times 4.5^2 \times 10.2$
$= 648.8959...$
≈ 648.9 cm^3

b $V = \pi r^2 h$
$= \pi \times 5.45^2 \times 7.2$ ⟵ $r = \dfrac{1}{2} \times 10.9 = 5.45$
$= 671.8547....$
≈ 671.9 m^3

EXERCISE 13-08

1 What is the volume of a cylinder with radius 4 m and height 12 m? Select the correct answer **A**, **B**, **C** or **D**.

    A 603.2 m^3 **B** 50.3 m^3 **C** 150.8 m^3 **D** 1206.4 m^3

2 What is the volume of a cylinder with diameter 14 cm and height 6.2 cm? Select **A**, **B**, **C** or **D**.

    A 3817.7 cm^3 **B** 954.4 cm^3 **C** 136.3 cm^3 **D** 1908.8 cm^3

3 Find the volume of each cylinder correct to 1 decimal place.

a

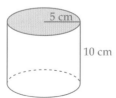

5 cm

10 cm

b

14 mm

radius = 2.5 mm

c

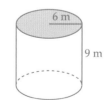

6 m

9 m

d

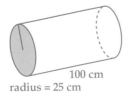

100 cm

radius = 25 cm

e

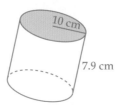

10 cm

7.9 cm

f

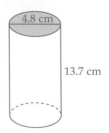

4.8 cm

13.7 cm

4   A can of soft drink has a diameter of 7 cm and a height of 8.5 cm.

    a   What is the volume of the can, correct to the nearest 0.1cm³?

    b   What is its capacity in mL?

5   A $1 coin has a radius of 12 mm and a thickness of 3 mm.
   What is its volume correct to the nearest cubic millimetre?

6   A water tank that is used to store rainwater has a diameter of 5 m
   and a height of 6 m.

    a   What volume (correct to two decimal places) of water can it hold
    in cubic metres?

    b   What is its volume, correct to the nearest litre?

7   A stool is supported by three legs in the shape of cylinders. Each leg has
   a radius of 3 cm and a height of 50 cm. What is the volume of the three
   legs? (Answer correct to the nearest cm³)

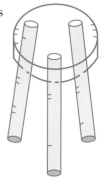

MIX AND MATCH

Match each diagram with its description below.

1

2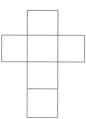

3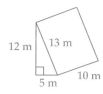

4

5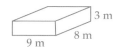

6

7

8

9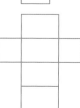

10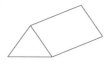

11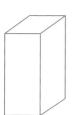

12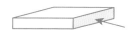

13

A Cube	**B** Square	**C** Triangular prism
E Net of a cube	**F** Rectangle	**H** Surface area is 360 m²
M Rectangular prism	**O** Triangle	**R** Face
S Net of a rectangular prism	**T** Edge	**U** Net of a triangular prism
Y Surface area is 246 m³		

Match each question number with an answer letter to decode the answer to this riddle:

What do surface area and a crowd have in common?

13-3-2-5   8-12-2   4-7-13-3   8   9-6-11   7-1   1-8-10-2-9

Part A General topics

Calculators are not allowed.

1 Evaluate $30 - 6 \times 4 + 20$

2 Simplify $4ab + ab^2 - 7ab^2 - 10ab$

3 Complete: $63 : 72 = 21 : \underline{\quad}$

4 Find y.

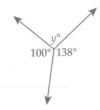

5 Convert $\dfrac{40}{6}$ to a mixed numeral.

6 Solve $3(3x - 7) = -12$

7 Does the point $(0, -4)$ lie on the x-axis or y-axis?

8 If $x = -7$, evaluate $8x - 9$.

9 Evaluate: $\dfrac{6}{7} - \dfrac{5}{6}$

10 If Ravi and Cindy share a prize of $2000 in the ratio $2 : 3$, what is Ravi's share?

Part B Area and volume

Calculators are allowed.

13-01 Area

11 Find the area of each shape.

a

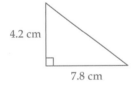

b

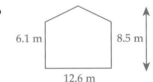

12 Find the area of a rectangle with length 9 cm and width 3.5 cm. Select the correct answer **A**, **B**, **C** or **D**.

    A 27 cm^2 **B** 31.5 cm^2 **C** 25 cm^2 **D** 12.5 cm^2

13-02 Area of a circle

13 Find the area of a circle with radius 4.2 m. Select **A**, **B**, **C** or **D**.

    A 55.4 m^2 **B** 26.4 m^2 **C** 13.2 m^2 **D** 221.7 m^2

14 Find the area of each shape, to 2 decimal places.

a

b

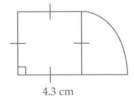

13-03 Areas of trapeziums, kites and rhombuses

15 Find the area of each shape.

a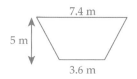

b

16 What is the area of a rhombus with diagonals 12.8 and 9.2 cm?

13-04 Surface area of a rectangular prism

17 Find the surface area of each prism.

a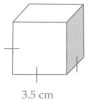

b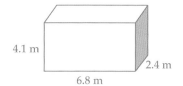

13-05 Surface area of a triangular prism

18 Draw the net of this triangular prism and then find its surface area.

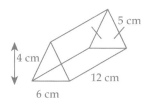

13-06 Surface area of a cylinder

19 Find the surface area of each cylinder correct to 1 decimal place.

a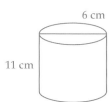

b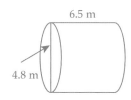

13-07 Volume of a prism

20 Find the volume of each prism.

a

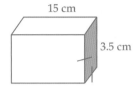

15 cm

3.5 cm

b

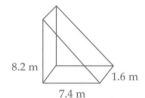

8.2 m

7.4 m

1.6 m

13-08 Volume of a cylinder

21 Find the capacity of a cylindrical swimming pool if its radius is 3.8 m and its height is 2.4 m.
Write your answer:

a   in m³, correct to three decimal places   b   correct to the nearest litre.

GRAPHING LINES

14

IN THIS CHAPTER YOU WILL:

- use a linear equation to complete a table of values
- identify points and quadrants on a number plane
- graph tables of values on the number plane
- graph linear equations on the number plane
- test if a point lies on a line
- find the equation of horizontal and vertical lines
- solve linear equations graphically

Shutterstock.com/iurii

EXAMPLE 1

Complete each table of values using the equation given.

a $y = x + 2$

x	0	1	2	3
y				

b $y = 2x - 1$

x	−1	1	4
y			

c $d = 3c + 5$

c	−2	1	4
d			

SOLUTION

Substitute the x-values from the table into each equation.

a $y = x + 2$
When $x = 0$, $y = 0 + 2 = 2$
When $x = 1$, $y = 1 + 2 = 3$, and so on.

x	0	1	2	3
y	2	3	4	5

b $y = 2x - 1$
When $x = -1$, $y = 2 \times (-1) - 1 = -3$
When $x = 1$, $y = 2 \times 1 - 1 = 1$, and so on.

x	−1	1	4
y	−3	1	7

c $d = 3c + 5$
When $c = -2$, $d = 3 \times (-2) + 5 = -1$
When $c = 1$, $d = 3 \times 1 + 5 = 8$, and so on.

c	−2	1	4
d	−1	8	17

Shutterstock.com/Sean Pavone

1 Find the value of y when $x = 2$ if $y = 2x - 3$. Select the correct answer **A**, **B**, **C** or **D**.

   A –7   **B** –1   **C** 1   **D** –5

2 Find the value of m when $n = -3$ if $m = 8 - 2n$. Select **A**, **B**, **C** or **D**.

   A 2   **B** 14   **C** 6   **D** 3

3 Copy and complete each table of values.

  a $y = 2x$

x	0	1	2	3
y				

  b $y = x \div 2$

x	10	8	6	4
y				

  c $y = x - 1$

x	4	3	2	1
y				

  d $y = x + 3$

x	0	1	2	3	4
y					

  e $y = 3x$

x	1	2	3	4
y				

  f $y = x - 3$

x	7	6	5	4	0	–1
y						

  g $y = 2 + x$

x	0	1	2	3	4
y					

  h $y = \dfrac{x}{3}$

x	12	9	6	3	0	–3	–6
y							

  i $y = 2x + 1$

x	1	2	3	4
y				

  j $y = 3x - 1$

x	4	3	2	1	0	–1	–2	–3
y								

4 Copy and complete each table of values.

  a $d = 5c + 1$

c	–1	0	1
d			

  b $h = 2g - 3$

g	1	2	3
h			

  c $q = 6 + 2p$

p	–1	0	1
q			

  d $t = 12 - 4s$

s	1	2	3
t			

  e $n = 3 - 2m$

m	–2	0	2
n			

  f $y = 5x - 6$

x	1	3	5
y			

A **number plane** is a grid for plotting points and drawing graphs.

It has an *x*-**axis** which is horizontal (goes across) and a *y*-**axis** which is vertical (goes up and down).

The **origin** is the centre of the number plane.

The number plane is divided into 4 quadrants (quarters).

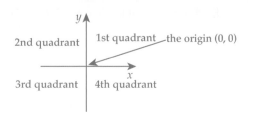

2nd quadrant | 1st quadrant —the origin (0, 0)

3rd quadrant | 4th quadrant

EXAMPLE 2

Plot each point on a number plane.

$A(1, 2)$, $B(–2, 3)$, $C(–3, –4)$, $D(2, –4)$, $E(0, 3)$, $F(–4, 0)$, $G(2, 0)$, $H(0, –1)$.

SOLUTION

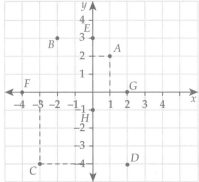

$A(1, 2)$ is 1 unit right and 2 units up from the origin.

$C(–3, –4)$ is 3 units left and 4 units down from the origin.

EXAMPLE 3

In which quadrant does each point lie?

$A (–1, 4)$ $B (2, –3)$ $C (–3, –3)$ $D (0, 2)$

SOLUTION

A is in the 2nd quadrant.

B is in the 4th quadrant.

C is in the 3rd quadrant.

D is on the *y*-axis, so it is not in any quadrant (between the 1st and 2nd quadrants).

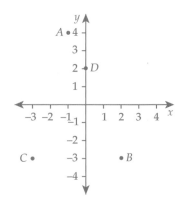

1 In which quadrant does the point (–2, –4) lie?

2 Where is the point (0, 0) positioned on the number plane? Select **A**, **B**, **C** or **D**.

  A Below the *x*-axis   **B** Below the *y*-axis

  C Where the *x* and *y*-axes meet   **D** In the 1st quadrant

3 **a** Write the coordinates of each point *A* to *F* shown.

  b State which quadrant or axis each point lies in.

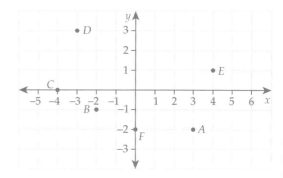

4 **a** On a number plane, plot the points below.

a $A(1, 2)$	**b** $B(-1, 2)$	**c** $C(1, -2)$	**d** $D(-1, -2)$	**e** $E(3, -5)$
f $F(-5, -2)$	**g** $G(4, -3)$	**h** $H(0, 4)$	**i** $I(2, 0)$	**j** $J(-2, -3)$
k $K(0, -3)$	**l** $L(-3, 0)$	**m** $M(-4, -1)$	**n** $N(2, -5)$	**o** $P(4, 5)$
p $Q(3, 3)$	**q** $R(2, 2)$	**r** $S(-2, -2)$	**s** $T(0, 0)$	**t** $V(0, -5)$

  b What type of figure is formed by the points:

  i *ABCD*?   **ii** *LFS*?   **iii** *ATBH*?

5 Picture puzzle: Draw a number plane with the *x*-axis from −10 to 10 and the *y*-axis from −4 to 6. Plot the points described below and join them as you go. Do not join points separated by a line. What familiar shape is formed?

(10, 4)	(−10, 4)	(−8, 0)	(−1, 0)	(−4, 3)
(8, 4)	(−8, 4)	(0, 0)	(−1, 3)	(−4, 6)
(8, −2)	(−8, −2)	(8, 0)	———	(−3, 3)
(10, −2)	(−10, −2)	———	(−2, 0)	(−2, 6)
(10, 4)	(−10, 4)	(6, 0)	(−2, 3)	(−1, 3)
———	———	(6, 1)	———	(0, 6)
(8, 2)	(−8, −1)	———	(−3, 0)	(1, 3)
(7, 3)	(−7, 0)	(5, 0)	(−3. 3)	(2, 6)
(6, 4)	(−6, 1)	(5, 2)	———	(3, 3)
(5, 5)	(−5, 2)	———	(−4, 0)	(4, 6)
(4, 6)	(−4, 3)	(4, 0)	(−4, 3)	(4, 3)
(3, 6)	(−3, 3)	(4, 3)	———	(6, 4)
(2, 6)	(−2, 3)	———	(−5, 0)	(5, 2)
(1, 6)	(−1, 3)	(3, 0)	(−5, 2)	(7, 3)
(0, 6)	(0, 3)	(3, 3)	———	(6, 1)
(−1, 6)	(1, 3)	———	(−6, 0)	(8, 2)
(−2, 6)	(2, 3)	(2, 0)	(−6, 1)	(7, 0)
(−3, 6)	(3, 3)	(2, 3)	———	———
(−4, 6)	(4, 3)	———	(−7, 0)	
(−5, 5)	(5, 2)	(1, 0)	(−8, 2)	
(−6, 4)	(6, 1)	(1, 3)	(−6, 1)	
(−7, 3)	(7, 0)	———	(−7, 3)	
(−8, 2)	(8, −1)	(0, 0)	(−5, 2)	
———	———	(0, 3)	(−6, 4)	
		———		

EXAMPLE 4

x	-2	-1	0	1	2
y	-1	0	1	2	3

Graph this table of values on a number plane.

SOLUTION

Reading the table of values in columns, we get the coordinates of the points.

x	-2	-1	0	1	2
y	-1	0	1	2	3

The points are: $(-2, -1)$ $(-1, 0)$ $(0, 1)$ $(1, 2)$ $(2, 3)$

Plotting these points on a number plane:

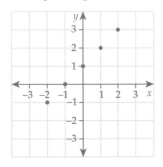

EXAMPLE 5

Graph this table of values after completing it.

$y = 2x - 3$

x	-1	2	0	3
y				

SOLUTION

x	-1	2	0	3
y	-5	1	-3	3

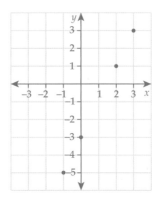

ISBN 9780170351027

1 If $y = 4 - x$, find y when $x = -2$. Select the correct answer **A**, **B**, **C** or **D**.

    A 2   **B** 6   **C** –2   **D** –6

2 Graph each table of values on a number plane.

a

x	-2	-1	0	1	2
y	0	1	2	3	4

b

x	-2	-1	0	1	2
y	-4	-2	0	2	4

c

x	-2	-1	0	1	2
y	-5	-4	-3	-2	-1

d

x	-2	-1	0	1	2
y	-1	$-\frac{1}{2}$	0	$\frac{1}{2}$	1

e

x	-2	-1	0	1	2
y	4	2	0	-2	-4

f

x	-3	-1	0	2	3
y	-5	-1	1	5	7

g

x	-4	-2	0	1	2
y	1	3	5	6	7

h

x	-2	-1	0	2	3
y	-6	-5	-4	-3	-2

i

x	-2	-1	0	1	2
y	-7	-4	-1	2	5

j

x	-3	-1	0	2	3
y	7	3	1	-3	-5

3 What do you notice about each set of points in question **2**?

4 Copy and complete each table and then graph the values on a number plane.

a $y = x + 3$

x	-1	0	1	2
y				

b $y = 6 - x$

x	-1	0	1	2
y				

c $y = 2x + 1$

x	-2	0	1	2
y				

d $y = 4x - 3$

x	-1	1	0	2
y				

e $y = -3 + x$

x	-1	0	1	2
y				

f $y = 12 - 5x$

x	1	0	3	2
y				

14-04 | Graphing linear equations

WORDBANK

linear equation An equation that connects two variables, usually x and y, whose graph is a straight line.

x-intercept The value where a line crosses the x-axis.

y-intercept The value where a line crosses the y-axis.

TO GRAPH A LINEAR EQUATION:
- complete a table of values using the equation
- plot the points from the table on a number plane
- join the points to form a straight line.

To be sure, it is best to find three points on the line using a table of values.
We can substitute any x-values into the linear equation, but $x = 0$, $x = 1$ or $x = 2$ are usually the easiest to use.

EXAMPLE 6

Graph each linear equation and state the x-intercept and y-intercept of each line.

a $y = 3x$ **b** $y = 2x - 1$

SOLUTION

a Complete a table of values.
$y = 3x$

x	0	1	2
y	0	3	6

Graph the table of values, rule the line and label it with the equation.

 Draw arrows on the ends of the line because a line has an infinite number of points and goes on endlessly in both directions.

The line crosses the x-axis at 0, so its x-intercept is 0.

The line crosses the y-axis at 0, so its y-intercept is also 0.

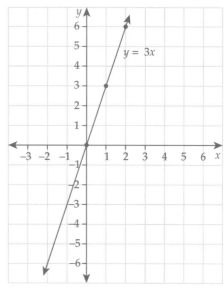

b $y = 2x - 1$

x	0	1	2
y	-1	1	3

The line crosses the x-axis at $\frac{1}{2}$, so its x-intercept is $\frac{1}{2}$.
The line crosses the y-axis at -1, so its y-intercept is -1.

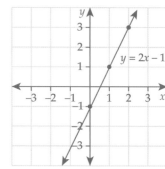

1 How many points are best for graphing a linear equation? Select the correct answer **A**, **B**, **C** or **D**.

   A 1   **B** 2   **C** 3   **D** 4

2 Copy and complete these sentences.

   To graph a linear equation, draw up a table of _____ and then plot each point from the _____ on the number plane. Join the _____ to form a straight _____.

3 Write down the x- and y-intercepts of each line below.

   a
   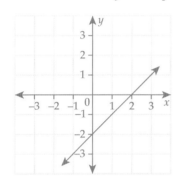

   b
   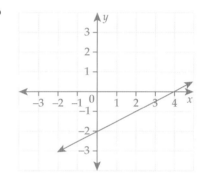

   c
   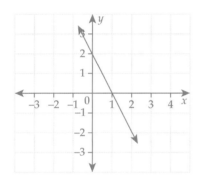

   d
   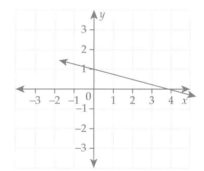

4 Graph each linear equation on a number plane after completing a table of values, and state the x-intercept and y-intercept of each line.

   a $y = 2x$

   b $y = x - 3$

   c $y = 5 - x$

   d $y = 3x - 1$

   e $y = 6 - 2x$

   f $y = 4 - 2x$

   g $y = -3x + 6$

   h $y = -4x - 1$

Shutterstock.com/Mincemeat

TO TEST IF A POINT LIES ON A LINE:
- substitute the *x*-value and the *y*-value into the linear equation
- if LHS = RHS, then the point lies on the line
- otherwise, the point does not lie on the line.

A point must satisfy the equation of the line to lie on the line.

EXAMPLE 7

Which of these points lie on the line with equation $y = 3x - 2$?

a (1, 1)   **b** (2, 5)   **c** (−1, −5)

SOLUTION

Substitute each point into the linear equation.

a For (1, 1), $x = 1, y = 1$
$y = 3x - 2$
$1 = 3 \times 1 - 2$
$1 = 1$ True
So (1, 1) lies on the line.

b For (2, 5), $x = 2, y = 5$
$y = 3x - 2$
$5 = 3 \times 2 - 2$
$5 = 4$ False
(2, 5) does not lie on the line.

c For (−1, −5), $x = −1, y = −5$
$y = 3x - 2$
$−5 = 3 \times (−1) - 2$
$−5 = −5$ True
So (−1, −5) lies on the line.

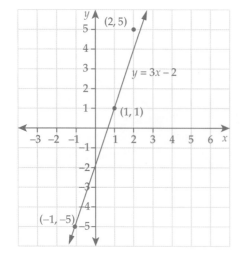

1   Which point lies on the line with equation $y = 2x + 3$? Select **A**, **B**, **C** or **D**.

   A $(1, 1)$ **B** $(-1, 5)$ **C** $(-2, -1)$ **D** $(2, 5)$

2   Which of these lines has the point $(-1, 3)$ on it? Select **A**, **B**, **C** or **D**.

   A $y = 2x - 1$ **B** $y = 2x + 5$ **C** $y = 2x + 1$ **D** $y = 2x - 3$

3   Copy and complete these sentences.

   To test if a point lies on a line, substitute the x-value and _____ into the linear equation.

   If the equation is true, then the point ____ on the line. If it is _____ the point does not ___ on the line.

4   Which of these points lie on the line $y = 4x - 1$?

   a $(1, 3)$ **b** $(-1, 4)$ **c** $(2, 6)$ **d** $(-2, -9)$ **e** $(0, -1)$

5   Graph $y = 4x - 1$ on a number plane and check your answers to question **4**.

6   Which of these points lie on the line $x + 2y = 3$?

   a $(1, 1)$ **b** $(-1, 2)$ **c** $(2, 3)$ **d** $(-2, -1)$ **e** $(0, 2)$

7   For the linear equations below, determine whether the points beside it lie on the line.

   a $y = 5x + 2$ $(2, -3)$ $(-1, -3)$

   b $y = 10 - x$ $(-2, 6)$ $(4, 6)$

   c $x - y = 8$ $(4, 6)$ $(3, -5)$

   d $2x - y = 8$ $(-1, 2)$ $(1, -2)$

   e $y = 5x - 10$ $(-4, 5)$ $(2, 0)$

   f $3x - 4y = 2$ $(2, 1)$ $(-3, 8)$

   g $y = 3x - 4$ $(0, -4)$ $(-1, 1)$

   h $x + 3y = 5$ $(1, 1)$ $(-2, 4)$

WORDBANK

horizontal A line that is flat, parallel to the horizon. ←——→

vertical A line that is straight up and down, at right angles to the horizon.

constant A number, not a variable.

A **horizontal line** has equation $y = c$, where c is a constant (number).
A **vertical line** has equation $x = c$, where c is a constant (number).

EXAMPLE 8

Graph each line.

a $y = 3$ **b** $y = -2$ **c** $x = 1$ **d** $x = -1$

SOLUTION

a $y = 3$ is a horizontal line going through 3 on the y-axis.

All points on $y = 3$ have a y-value of 3, such as $(-1, 3)$ and $(1, 3)$.

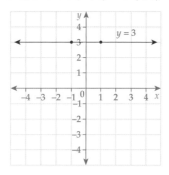

b $y = -2$ is a horizontal line going through -2 on the y-axis.

All points on $y = -2$ have a y-value of -2, such as $(-2, -2)$ and $(3, -2)$.

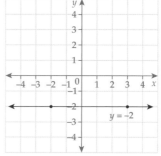

c $x = 1$ is a vertical line going through 1 on the x-axis.

All points on $x = 1$ have an x-value of 1, such as $(1, -1)$ and $(1, 2)$.

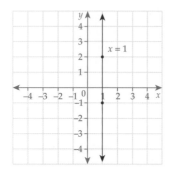

d $x = -1$ is a vertical line going through -1 on the x-axis.

All points on $x = -1$ have an x-value of -1, such as $(-1, 0)$ and $(-1, 4)$.

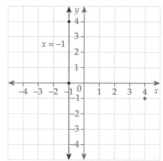

1 What types of lines are the graphs of $x = 1$ and $y = -2$? Select the correct answer **A**, **B**, **C** or **D**.

   A Both horizontal

   B Vertical and horizontal respectively

   C Both vertical

   D Horizontal and vertical respectively

2 What is the point of intersection between the lines with equations $x = 1$ and $y = -2$? Select **A**, **B**, **C** or **D**.

   A $(1, -2)$ **B** $(-2, 1)$ **C** $(-1, -2)$ **D** $(1, 2)$

3 Is each statement true or false?

   a $x = 4$ is a horizontal line. **b** $y = -4$ is a horizontal line.

   c $x = -1$ is a vertical line. **d** $y = 1$ is a vertical line.

4 Write down the equation of each line below.

   a 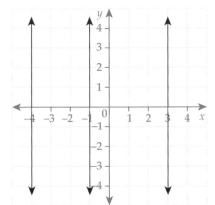 **b**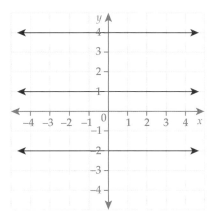

5 Graph each line on the same number plane.

   a $x = -2$ **b** $x = 0$ **c** $x = 2$ **d** $x = 3$

6 Graph each line on the same number plane.

   a $y = -3$ **b** $y = -1$ **c** $y = 0$ **d** $y = 2$

7 Write down the equation of the line that is:

   a horizontal with a y-intercept of 3

   b vertical with an x-intercept of -2

   c horizontal and passing through $(-1, 2)$

   d vertical and passing through $(2, -1)$.

8 Graph the lines $x = 3$ and $y = -1$ and write down their point of intersection.

In Chapter 9, Equations, we learnt how to solve linear equations such as $2x + 1 = 7$ algebraically. Linear equations can also be solved graphically.

EXAMPLE 9

Solve the linear equation $2x + 1 = 7$ graphically.

SOLUTION

Graph the lines $y = 2x + 1$ and $y = 7$ on the same number plane.

Graph $y = 2x + 1$ using a table of values.

x	0	1	2
y	1	3	5

The graph of $y = 7$ is a horizontal line going through 7 on the y-axis.

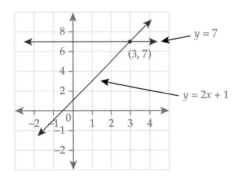

The two lines cross at (3, 7), where $x = 3$ and $y = 7$.

So the solution to $2x + 1 = 7$ is $x = 3$.

(Check: $2 \times \mathbf{3} + 1 = 7$)

Shutterstock.com/Ventura

ISBN 9780170351027

1   To solve $3x - 1 = 5$, which lines would you graph? Select the correct answer **A**, **B**, **C** or **D**.

   A $y = 3x - 5, y = 1$ **B** $y = 3x - 1, y = -5$

   C $y = 3x - 1, y = 5$ **D** $y = 3x - 5, y = -1$

2   To solve $2x + 3 = 6$, which lines would you graph? Select **A**, **B**, **C** or **D**.

   A $y = 2x + 6, y = 3$ **B** $y = 2x + 3, y = 6$

   C $y = 2x - 3, y = 6$ **D** $y = 2x + 3, y = -6$

3   Copy and complete.

   To solve the linear equation $4x - 2 = 6$ graphically, draw the graphs of _____ and $y = 6$.

   Find the point of intersection of the two _____. The solution to the linear equation will be the _____ of the point of intersection.

4   For each pair of lines shown, write:

   i   the point of intersection of the lines

   ii the linear equation whose solution can be read from the point of intersection

   iii the solution to this linear equation

   a 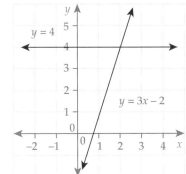   **b**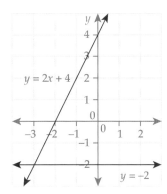

5   Graph $y = 2x - 3$ on a number plane and use this graph to solve each linear equation below.

   a $2x - 3 = 3$ **b** $2x - 3 = 7$ **c** $2x - 3 = -5$

6   Graph $y = 4x + 2$ on a number plane and use this graph to solve each linear equation below.

   a $4x + 2 = 6$ **b** $4x + 2 = 10$ **c** $4x + 2 = -6$

7   Graph $y = 8 - 2x$ on a number plane and use this graph to solve each linear equation below.

   a $8 - 2x = 10$ **b** $8 - 2x = 4$ **c** $8 - 2x = -2$

WORD CODE PUZZLE

Decode the rhyming pairs of words below by matching each ordered pair with a point and letter on the number plane.

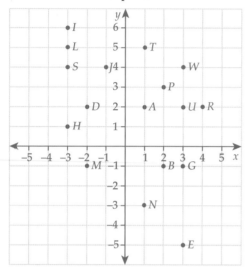

1 (2, 3) (–3, 5) (1, 2) (1, –3) (3, –5)   (2, –1) (4, 2) (1, 2) (–3, 6) (1, –3)

2 (1, –3) (3, 2) (–2, –1) (2, –1) (3, –5) (4, 2)   (–3, 4) (–3, 5) (3, 2) (–2, –1) (2, –1) (3, –5) (4, 2)

3 (2, 3) (0, 0) (–3, 6) (1, –3) (1, 5)   (–1, 4) (0, 0) (–3, 6) (1, –3) (1, 5)

4 (–2, 2) (0, 0) (1, 5)   (2, 3) (–3, 5) (0, 0) (1, 5)

5 (4, 2) (–3, 6) (3, –1) (–3, 1) (1, 5)   (–3, 1) (3, –5) (–3, 6) (3, –1) (–3, 1) (1, 5)

6 (–2, 2) (0, 0) (3, 4) (1, –3)   (1, 5) (0, 0) (3, 4) (1, –3)

ISBN 9780170351027

Part A General topics

Calculators are not allowed.

1 Find the sum of: $9.90, $1.40, $2.50 and $8.50.

2 Evaluate 0.0705×100.

3 Write 0.00951 in scientific notation.

4 Find the area of this triangle.

5 Evaluate $\dfrac{2}{3} - \dfrac{1}{4}$.

6 Decrease $180 by 15%.

7 Evaluate $15 \times 4 \times 7$.

8 Find the mean of the scores:

   5   10   6   11   8   16   6   7

9 Round 192.4598 to three decimal places.

10 If a die is rolled 90 times, how many times would you expect a 5 or a 6 to appear?

Part B Graphing lines

Calculators are allowed.

14–01 Tables of values

11 Complete each table of values.

  a $y = x - 5$

x	6	4	2
y			

  b $y = 3x - 4$

x	4	3	2
y			

12 Find the value of y when $x = -2$ if $y = 2x + 5$. Select the correct answer **A**, **B**, **C** or **D**.

  A 9   **B** 5   **C** 1   **D** 4

14–02 The number plane

13 In which quadrant does the point $(-1, 6)$ lie? Select **A**, **B**, **C** or **D**.

  A 1st   **B** 2nd   **C** 3rd   **D** 4th

14 Plot each point on a number plane.

  $A(1, 3)$ $B(-2, 5)$ $C(3, -5)$

  $D(-4, -3)$ $E(-4, 2)$

14–03 Graphing tables of values

15 Graph this table of values on a number plane.

x	-2	-1	0	1	2
y	-3	-1	1	3	5

14–04 Graphing linear equations

16 Write the x-intercept and y-intercept of each line.

a

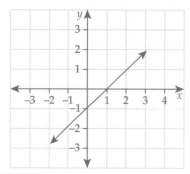

b

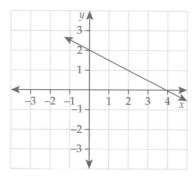

17 Graph each linear equation on a number plane.

    a $y = 3x - 1$ **b** $y = 6 - x$

14–05 Testing if a point lies on a line

18 Test whether each point lies on the line $y = 4x - 3$.

    a $(-1, 7)$ **b** $(2, -5)$ **c** $(3, 9)$

14–06 Horizontal and vertical lines

19 Sketch each linear equation.

    a $x = -1$ **b** $y = 4$ **c** $x = 2$ **d** $y = 0$

14–07 Solving linear equations graphically

20   **a**   Which two lines would you draw to solve $2x - 5 = 3$ graphically?

      b   Graph the two lines and use them to solve $2x - 5 = 3$.

PROBABILITY

15

IN THIS CHAPTER YOU WILL:

- calculate the probability of simple events
- calculate the probability of complementary events
- find the relative frequency of an event
- interpret and draw Venn diagrams and use them to solve probability problems
- read two-way tables and use them to solve probability problems

WORDBANK

probability The chance of an event occurring, expressed as a fraction from 0 to 1.

random Where each possible outcome cannot be predicted.

event An outcome or group of outcomes.

sample space The set of all possible outcomes in a situation.

This table shows some key words that are commonly used in probability.

Term	Example
Impossible: will not happen, probability of 0	Selecting a blue sock from a bag containing all red socks.
Unlikely: a low chance of happening	Winning a prize in a raffle.
Even chance: equally likely to happen or not happen, 50-50, probability of $\frac{1}{2}$	The next baby born being a girl.
Likely: a high chance of happening	Having a maths lesson on a school day.
Certain: will definitely happen, probability of 1	The sun will rise tomorrow morning.

These terms can be represented on a number line.

0   Impossible   Unlikely   Even Chance   Likely   Certain   1

EXAMPLE 1

Use a suitable probability term to describe the chance of each event.

a   Jen choosing a green marble from a bag containing only green marbles.

b   Jing choosing tails correctly when a coin is tossed.

c   Jai winning Powerball when she forgot to put her entry in.

d   Jacob being served in the next minute when all queues in the supermarket are long.

SOLUTION

a   Certain, it will definitely happen.

b   Even chance, the coin is equally likely to fall heads or tails.

c   Impossible, there is no chance of Jai winning if she hasn't put in an entry.

d   Unlikely, there is little chance of Jacob being served quickly.

The **probability of an event** has the abbreviation $P(E)$. If all outcomes are **equally likely** then:

$$P(E) = \frac{\text{number of outcomes in the event}}{\text{number of outcomes in the sample space}}$$

EXAMPLE 2

A die is rolled.

a List the sample space. Are all outcomes equally likely?

b Find the probability of rolling a number greater than 4.

SOLUTION

Use the formula in the box.

a Sample space = { 1, 2, 3, 4, 5, 6 }. Each number is equally likely to be rolled.

b A number greater than 4 is: 5 or 6.

$$P(\text{number greater than 4}) = \frac{2}{6}$$ 2 outcomes greater than 4: 5 and 6
6 outcomes in the sample space

$$= \frac{1}{3}$$

EXERCISE 15-01

1 What term describes the probability of rolling an even number on a die? Select the correct answer **A**, **B**, **C** or **D**.

   A impossible   **B** even chance   **C** likely   **D** certain

2 Which term describes the probability of rolling a number less than 7 on a die? Select **A**, **B**, **C** or **D**.

   A impossible   **B** even chance   **C** likely   **D** certain

3 Which term describes the probability of each event?

   a A traffic light shows red

   b Choosing a red marble from a bag of black marbles

   c A summer's day will be hot

   d Rolling an odd number on a die

   e Choosing a soft-centred chocolate from a bag containing all soft centres

   f Heads when a coin is tossed

   g It will be cold on a July day in Sydney

   h Choosing a green lolly from a bag containing only red and blue ones

   i Rolling 1 on a die

   j Winning Lotto when you have bought one ticket

4 List the sample space for each chance experiment.

   a Tossing a coin.   **b** The suit of a card in a standard pack of cards.

   c Traffic light colours.   **d** Even numbers between 13 and 23.

   e One note of each type in the Australian currency.   **f** Tossing two coins.

5 State whether the outcomes for the following are equally likely.

   a Tossing a coin.

   b Selecting the winner of the Melbourne Cup.

   c Choosing a card from a pack of cards.

   d Selecting the winner of a football match.

   e Rolling a die.

   f Choosing a marble from a bag with all identical marbles inside.

6 When rolling a die, find the probability of rolling:

   a a 4   **b** a 2

   c a number greater than 2   **d** a number less than 4

   e an even number   **f** an odd number

7 **a** A bag contains 3 black and 4 red marbles. If one is chosen from the bag, what is the probability that it is:

     i black?   **ii** red?

   b Add your probabilities in **i** and **ii** of part **a**. What is the sum of the probabilities?

8 A bag contains 2 black, 3 red and 4 white marbles. One is selected from the bag. What is the probability that it is:

   a black?   **b** red?   **c** white?

   d not black?   **e** not red?   **f** not white?

9 One card is selected from a pack of playing cards. What is the probability that it is:

   a a spade?   **b** a heart?   **c** an Ace?

   d a six?   **e** the six of hearts? **f** a King?

Shutterstock.com/Syda Productions

ISBN 9780170351027

The **complement** of an event is the event **not** taking place. For example, the complement of rolling 5 on a die is not rolling a 5, that is rolling 1, 2, 3, 4 or 6.

Probability is measured on a scale from 0 to 1, where 1 is certain, so if E is an event, then $P(E) + P(\text{not } E) = 1$. The complement of E is written as $\bar{E}$.

- $P(E) + P(\bar{E}) = 1$
- or $P(\bar{E}) = 1 - P(E)$
- or $P(\text{not } E) = 1 - P(E)$.

EXAMPLE 3

A bag contains 6 red, 4 blue and 5 yellow marbles. If a marble is selected at random from the bag, find the probability that it is:

a red　　　　　　**b** not red

iStockphoto/reginaaa

SOLUTION

a　Total marbles = 6 + 4 + 5

$$= 15$$

$$P(\text{red}) = \frac{6}{15}$$

$$= \frac{2}{5}$$

b　$P(\text{not red}) = 1 - P(\text{red})$

$$= 1 - \frac{2}{5}$$

$$= \frac{3}{5}$$

* This is also true because there are 9 marbles that are not red, out of 15 and $\frac{9}{15} = \frac{3}{5}$

1 If $P(\text{winning the match}) = \dfrac{3}{8}$ what is $P(\text{not winning the match})$? Select the correct answer **A, B, C** or **D**.

   A $\dfrac{3}{8}$ **B** $\dfrac{5}{8}$ **C** 1   **D** Not enough information

2 What would be the complementary event for 'coming first in a race'? Select **A, B, C** or **D**.

   A Coming second   **B** Coming in any place
   C Not coming first   **D** Coming last

3 Describe in words the complementary event for each event.

   a Tossing a die and getting a 3.
   b Choosing a red sock from a drawer containing red and yellow socks.
   c Throwing a coin and getting a head.
   d Coming second in a competition.
   e Choosing a heart from a pack of cards.
   f Picking the winner in a race.

4 Copy and complete this sentence.

   To find the probability of a complementary _____, find the probability of the event and then subtract it from ___.

5 If Nick rolls a die, what is the probability of rolling:

   a a 3?   **b** not a 3?
   c a number less than 3?   **d** a number not less than 3?

6 In a bag there are 7 red, 6 blue and 5 white balls. If Thanh selects one ball at random, what is the probability of choosing:

   a a blue ball?   **b** not a blue ball?
   c a white ball?   **d** not a white ball?
   e a red ball?   **f** not a red ball?

7 As Selina drives to work there is a 0.8 chance that one traffic light will be red. What is the probability that it won't be red?

8 Hannah has a 65% chance of passing her mathematics examination. What chance has she of not passing the examination?

9 The probability that Zoe wins first prize in a raffle is $\dfrac{1}{100}$.

   a What is the probability that Zoe does not win first prize?
   b If Zoe had bought 2 tickets, how many tickets were sold altogether?

WORDBANK

frequency The number of times an outcome occurs.

relative frequency The frequency of an event compared to the total number of trials in an experiment, used to estimate the probability of the event.

$$P(E) = \frac{\text{number of times the event occurred}}{\text{total number of trials}}$$
$$= \frac{\text{frequency of the event}}{\text{total frequency}}$$

EXAMPLE 4

Two coins are tossed 20 times and the number of tails were recorded.

Number of tails	Frequency
0	5
1	11
2	4

Total = 20

a Find the relative frequency of tossing two tails.

b What is the experimental probability of tossing 2 *heads*?

c If two coins are tossed 300 times, predict how many times 1 tail would occur.

SOLUTION

a Relative frequency of 2 tails $= \dfrac{4}{20}$ $\dfrac{\text{Frequency of the event}}{\text{Total frequency}}$

$$= \frac{1}{5}$$

b Experimental $P(2 \text{ heads}) = $ Experimental $P(0 \text{ tails})$

$$= \frac{5}{20}$$
$$= \frac{1}{4}$$

c Experimental $P(1 \text{ tail}) = \dfrac{11}{20}$

Expected number of '1 tail' $= \dfrac{11}{20} \times 300$

$$= 165$$

iStockphoto/ronstik

1   If a coin is tossed 30 times and 17 heads are thrown, what is the relative frequency of throwing a *tail*?

2   If a die is rolled 20 times and a 4 is rolled five times, what is the relative frequency of not rolling a 4? Select the correct answer **A**, **B**, **C** or **D**.

   A $\dfrac{3}{4}$ **B** $\dfrac{5}{20}$ **C** 1   **D** $\dfrac{5}{15}$

3   Copy and complete each sentence.
   The relative frequency of an event is the same as the experimental _____.
   It is found by dividing the frequency of the _____ by the total _____.
   The sum of the relative _____ is always ____.

4   Two coins are tossed 80 times and the results are recorded below.

Number of heads	Frequency	Relative frequency
0	16	
1	42	
2	22	
Total	80	

   a   Copy and complete the table.
   b   Find the experimental probability of throwing 1 head.
   c   Predict the frequency of 1 head when the coins are tossed 320 times.
   d   What is the relative frequency of throwing 2 tails?
   e   Find the sum of the relative frequencies.

5   A die is rolled 120 times and the results are as recorded below.

Score	Frequency	Relative frequency
1	18	
2	23	
3	16	
4	24	
5	22	
6	17	

   a   Copy and complete the table.
   b   Find the experimental probability of rolling a 4 on a die.
   c   Find the number of 4s you would expect when a die is tossed 600 times.
   d   Find the relative frequency of rolling a number less than 3.
   e   Find the sum of the relative frequencies.

A **Venn diagram** uses circles to group items into categories. Most Venn diagrams involve circles that **overlap**. For example, A could represent all swimmers and B could represent all netballers. Then the shaded region would represent **A and B**, which means people who are both swimmers and netballers.

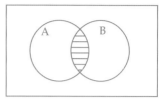

A or B means **A or B or both**

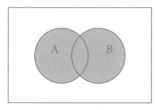

Swimmers or netballers or both

A only means **A but not B**

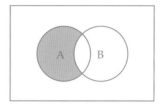

Swimmers only, not netballers

A Venn diagram may involve two groups that do not overlap.

These categories are **mutually exclusive** as it is not possible to be both a boy and a girl.

EXAMPLE 5

25 students were asked for their favourite ice-cream flavour. The results are shown in the Venn diagram.

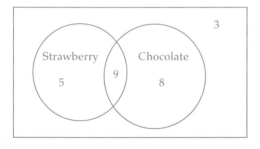

Dreamstime/Designsstock

a   How many students preferred strawberry ice-cream?

b   How many students preferred both strawberry and chocolate?

c   What is the probability of choosing a student who prefers chocolate only?

d   What is the probability that a student prefers strawberry or chocolate?

e   Describe the three students who are represented outside the circles in the Venn diagram.

SOLUTION

a 14 students preferred strawberry ice cream. $\longleftarrow$ 5 + 9 = 14

b 9 students preferred both strawberry and chocolate. $\longleftarrow$ The overlap section

c $\dfrac{8}{25}$ $\longleftarrow$ 8 students prefer chocolate but not strawberry

d $\dfrac{22}{25}$ $\longleftarrow$ 5 + 9 + 8 = 22 students prefer strawberry or chocolate or both

e These students do not prefer strawberry or chocolate.

EXAMPLE 6

A group of 72 students at an international school were surveyed on whether they spoke English or Japanese fluently. The results were English: 52 students, Japanese: 18 students, English and Japanese: 5 students. Represent this information on a Venn diagram.

SOLUTION

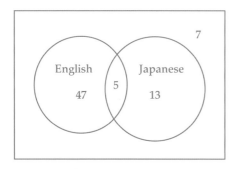

✱ | Fill in the centre overlap section first

- 5 students speak both English and Japanese so write 5 in the overlap
- This leaves 52 – 5 = 47 students who speak only English
- This leaves 18 – 5 = 13 students who speak only Japanese.
- 47 + 5 + 13 = 65, but 72 students were surveyed, so 72 – 65 = 7 students speak neither English nor Japanese, so write 7 outside the circles.

1 If 30 students are surveyed on whether they prefer hockey or basketball, and 12 like hockey and 25 like basketball, how many like both? Select the correct answer **A**, **B**, **C** or **D**.

   A 4   **B** 5   **C** 6   **D** 7

2 In question **1**, how many students like hockey only? Select **A**, **B**, **C** or **D**.

   A 4   **B** 5   **C** 6   **D** 7

Shutterstock.com/mooinblack

3 This Venn diagram shows how students get to school each day.

   a How many students were surveyed?

   b How many students:

      i came by bus only?

      ii walked and caught a bus?

      iii walked?

      iv neither walked nor caught a bus?

      v did not catch a bus?

   c Are walking and catching a bus mutually exclusive?

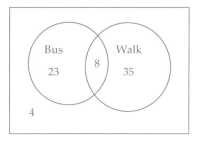

4 This Venn diagram shows the number of students learning musical instruments.

   a How many students were surveyed?

   b How many students learn:

      i the piano?

      ii guitar?

      iii both?

      iv neither piano nor guitar?

      v either piano or guitar?

   c Are learning piano and guitar mutually exclusive?

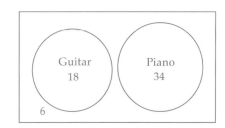

5   A sample of 35 families were asked whether they had a pet cat or dog. The results were Cats: 15 families, Dogs: 22 families, Cats and dogs: 8 families. Draw a Venn diagram to represent this data, then count how many families had:

iStockphoto/GlobalP

a   cats but no dogs   b   dogs and no cats

c   neither cats nor dogs

6   This Venn diagram shows the type of food liked by a group of Year 9 students.

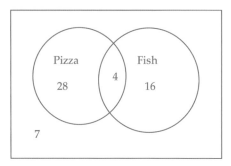

What is the probability that a student chosen at random likes:

a   only fish?   b   neither pizza nor fish?

c   pizza or fish but not both?   d   at least one type of food?

7   a   Draw a Venn diagram to represent these results for preferred football teams. Knights 36, Eels 28, Knights and Eels 6, Neither 4.

   b   What is the probability that a person selected at random supports

      i   only the Eels?   ii   both teams?

      iii only 1 team?   iv only the Knights?

EXAMPLE 7

Fifty people were surveyed for left- and right-handedness, with the results shown in the two-way table below. Copy and complete the table.

	Male	Female	Total
Right-handed	20	21	41
Left-handed	6	3	
TOTAL		24	

a   How many people were left-handed?
b   How many males were right-handed?
c   What is the probability that a person selected at random will be female and left-handed?
d   What is the probability of selecting a male?

SOLUTION

	Male	Female	Total
Right-handed	20	21	41
Left-handed	6	3	9
Total	26	24	50

a   Total left-handed = 9   b   Number of right-handed males = 20

c $P(\text{left-handed female}) = \dfrac{3}{50}$ d $P(\text{male}) = \dfrac{26}{50} = \dfrac{13}{25}$

EXAMPLE 8

This Venn diagram shows whether 45 students preferred sweet or savoury food. Show this information on a two-way table.

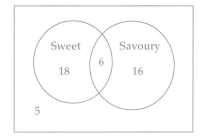

SOLUTION

✱ | Start at the centre overlap section first

* 6 students like both sweet and savoury
* 18 students like sweet, but not savoury
* 16 students like savoury, but not sweet
* 5 students like neither

Complete the totals column and row.

	Sweet	Not Sweet	Total
Savoury	6	16	22
Not Savoury	18	5	23
Total	24	21	45

1   If there are 26 students studying Geography and 28 students studying History and 46 students altogether, how many students study both subjects? Select **A**, **B**, **C** or **D**.

    A 6   **B** 8   **C** 10   **D** 54

2   In question **1**, how many students study Geography only? Select **A**, **B**, **C** or **D**.

    A 18   **B** 20   **C** 26   **D** 8

3   Copy and complete this two-way table.

	Female	Male	Total
Blue eyes	8	10	
Not blue eyes	18	12	
Total			

    a   How many people had blue eyes?

    b   How many males were not blue-eyed?

    c   What is the probability that a person selected at random will be female and have blue eyes?

    d   What is the probability of selecting a male?

4   Copy and complete this table showing people who like or don't like football.

	Girls	Boys	Total
Like football	12	18	
Don't like football	11	9	
Total			

    a   How many people liked football?

    b   How many boys did not like football?

    c   What is the probability that a student selected at random will be a girl and like football?

    d   What is the probability of selecting a boy who likes football?

Shutterstock.com/Neale Cousland

5 a This Venn diagram shows whether 50 preschoolers preferred cake or lollies. Copy and complete the two-way table.

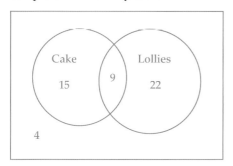

	Cake	Not cake	Total
Lollies			
Not lollies			
Total			

b What is the probability that a preschooler selected at random will like:

i both cake and lollies?   ii cake or lollies but not both?

6 a Copy and complete the two-way table.

	Swimming	Not swimming	Total
Running	32	28	
Not running	14	16	
Total			

b What is the probability that a person selected at random will be:

i swimming but not running?   ii running and swimming?

EXAMPLE 9

Ben selected a card at random from a pack of cards, recorded its suit and then returned it to the pack. He did this 100 times. The results were: hearts 28, diamonds 23, spades 21 and clubs 28.

a   Find as a decimal the experimental and theoretical probabilities for selecting a hearts card.

b   Explain why the experimental and theoretical probabilities are different.

c   What could Ben do to make the experimental and theoretical probabilities more similar?

SOLUTION

a   Experimental P(hearts card) $= \dfrac{28}{100} = 0.28$

Theoretical P(hearts card) $= \dfrac{13}{52} = \dfrac{1}{4} = 0.25$ ⟵——— 52 cards in a deck, 13 are hearts

b   The experimental and theoretical probabilities are different as the experimental one is a result from an experiment and the theoretical one is calculated from the number of cards in a deck.

c   Increase the number of trials of the experiment, for example, 200 times. With more trials, the experimental probability should get closer to the theoretical probability.

EXAMPLE 10

Consider this claim: For the spinner below, the probability of spinning a 4 is $\dfrac{1}{5}$.
Is this statement correct or incorrect? Justify your answer.

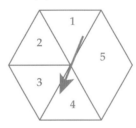

SOLUTION

The statement is incorrect because while there are 5 numbers, there are actually 6 equal sections, with 5 taking up two of the sections.

1   If there are 6 red, 5 blue and 3 green marbles in a bag, what is the probability of choosing a red marble? Select the correct answer **A**, **B**, **C** or **D**.

   A $\dfrac{5}{14}$ **B** $\dfrac{6}{11}$ **C** $\dfrac{3}{7}$ **D** $\dfrac{8}{14}$

2   For the bag of marbles in question **1**, what is the probability of choosing a marble that is not blue? Select **A**, **B**, **C** or **D**.

   A $\dfrac{5}{14}$ **B** $\dfrac{11}{14}$ **C** $\dfrac{3}{7}$ **D** $\dfrac{9}{14}$

3   The letters of the alphabet from A to J are written on separate cards and shuffled. Mila chooses one at random. What is the probability that this letter is:

   a   a vowel?   **b**   a consonant ?   **c**   part of the word BIG?

   d   part of the word HOLIDAY?   **e**   part of the word MATHEMATICS?

4   The chocolates in a box have different centres: 7 toffees, 4 cherries, 6 peppermints and 13 nuts. If Rhys takes one chocolate at random, what is the probability that it is:

   a   a cherry centre?   **b**   a peppermint centre?

   c   a nut centre?   **d**   not a toffee centre?

5   Emily selected a card at random from a pack of cards, recorded its suit and then returned it to the pack. She did this 200 times. The results were hearts 58, diamonds 43, spades 36 and clubs 63.

   a   Find the experimental probability of selecting a spades card:

       i   as a fraction   **ii**   as a decimal

   b   Find the theoretical probability of selecting a spades card:

       i   as a fraction   **ii**   as a decimal

   c   Are the experimental and theoretical probabilities from parts **a** and **b** similar?

   d   What should happen to the experimental probability if Emily selects a card 500 times?

6   For the spinner below, the probability of spinning a 3 is $\dfrac{1}{6}$.

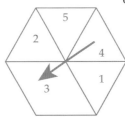

   a   Is this statement correct or incorrect? Justify your answer.

   b   What is the probability of spinning a 3?

7   The Venn diagram shows whether a group of students liked chips or chicken.

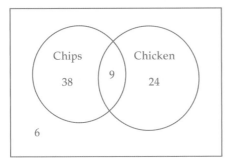

a   How many students liked:

   i   chicken only?   **ii**   only one type of food?

b   What is the probability that a student prefers:

   i   only chips?   **ii**   neither chips nor chicken?

8   Copy and complete this table.

	Girls	Boys	Total
Dancing	23	12	
Not dancing	8	18	
Total			

What is the probability that a student selected at random is:

a   a girl who likes dancing?

b   a boy who likes dancing?

Shutterstock.com/YanLev

CODE PUZZLE

Each letter of the alphabet is represented by a number below.

A = 1	G = 7	M = 13	S = 19	Y = 25
B = 2	H = 8	N = 14	T = 20	Z = 26
C = 3	I = 9	O = 15	U = 21	
D = 4	J = 10	P = 16	V = 22	
E = 5	K = 11	Q = 17	W = 23	
F = 6	L = 12	R = 18	X = 24	

Use the code to uncover the following probability words and phrases.

1   3-8-1-14-3-5

2   3-5-18-20-1-9-14

3   9-13-16-15-19-19-9-2-12-5

4   21-14-12-9-11-5-12-25

5   5-22-5-14-20

6   15-21-20-3-15-13-5

7   16-18-15-2-1-2-9-12-9-20-25

8   20-8-5-15-18-5-20-9-3-1-12

9   5-17-21-1-12-12-25   12-9-11-5-12-25

10 19-1-13-16-12-5   19-16-1-3-5

Part A General topics

Calculators are not allowed.

1 Find the sales price of a $50 pair of sunglasses after a 15% discount.

2 Complete: $2 : 3 = 14 : \underline{\ \ \ }$.

3 Evaluate $-1 \times 4 + (7 - 9)$.

4 Find the area of this parallelogram.

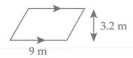

9 m   3.2 m

5 Factorise $16xy^2 - 8ky$.

6 Write the recurring decimal 2.1464646... using dot notation.

7 What type of line has a gradient of 0?

8 Simplify: $\dfrac{14p^4 y}{56p^2 y^3}$

9 Find the mean of these scores:

12   11   22   8   9

10 What is the time 14 hours and 20 minutes after 3:35 p.m.?

Part B Probability

Calculators are allowed.

15-01 Probability

11 What term describes the probability of throwing a number greater than 6 when a die is tossed? Select the correct answer **A**, **B**, **C** or **D**.

A impossible   C likely

B even chance   D certain

12 A card is selected at random from a pack of playing cards. What is the probability of choosing a diamond? Select **A**, **B**, **C** or **D**.

A $\dfrac{1}{2}$ B $\dfrac{1}{4}$ C $\dfrac{1}{13}$ D $\dfrac{1}{12}$

15-02 Complementary events

13 What would be the complementary event for 'winning 2nd prize in a raffle'? Select **A**, **B**, **C** or **D**.

A winning 1st prize   B not winning a prize

C winning 3rd prize   D not winning 2nd prize

14 If Jake tosses a die, what is the probability of throwing:

a a number less than 2?

b a number not less than 2?

15-03 Relative frequency

15 Two coins are tossed 40 times. The results are recorded below.

Number of heads	Frequency	Relative frequency
0	8	
1	23	
2	9	
Total	40	

a   Copy and complete the table.

b   What is the experimental probability of throwing a head?

c   Based on the experimental probability, if a coin is tossed 100 times, how many times would you expect to get 1 head?

d   What is the relative frequency of throwing 2 tails?

15-04 Venn diagrams

16 The Venn diagram below shows how many students own smartphones and tablets.

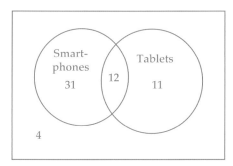

a   How many students were surveyed?

b   How many students had tablets only?

c   How many students had only 1 device?

d   What is the probability that a student has only a smartphone?

17 Draw a Venn diagram illustrating the information below for a region of farms.

Farms with cows 45   Farms with sheep 54

Farms with sheep and cows 28   Farms with neither 16

15-05 Two-way tables

18 Copy this table and complete it using the information from question **16**.

	Smartphone	No smartphone	Total
Tablet			
No tablet			
Total			

19 Fifty people were surveyed and were asked if they had a driver's licence. Copy and complete the table.

	Male	Female	Total
Licence	24	18	
No Licence	5	3	
Total			

a How many people had a licence?

b How many males did not have a licence?

c What is the probability that a person selected at random will be a female with a licence?

15-06 Probability problems

20 Lachlan selected a card at random from a pack of cards, recorded its suit and then returned it to the pack. He did this 100 times. The results were: hearts 18, diamonds 26, spades 17 and clubs 39.

a Find, as a decimal, the experimental probability of selecting a diamonds card.

b Find, as a decimal, the theoretical probability of selecting a diamonds card.

c Are the experimental and theoretical probabilities from parts **a** and **b** similar?

RATIOS AND RATES

16

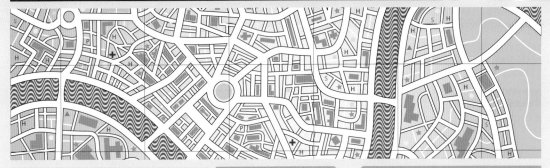

IN THIS CHAPTER YOU WILL:

- find equivalent ratios and simplify ratios
- solve ratio problems
- understand how to read a scale diagram
- interpret scale maps and diagrams
- divide a quantity into a given ratio
- write and simplify rates
- solve rate problems, including those related to speed

Shutterstock.com/Robert Adrian Hillman

A **ratio** compares quantities of the same kind, consisting of two or more numbers that represent parts or shares. For example, when mixing ingredients for a cake we could have a ratio of flour to milk of 3 : 2. This means that there are 3 parts of flour to 2 parts of milk. The **order** is important. The ratio of milk to flour would be 2 : 3.

Operations with ratios are similar to operations with fractions.

To find an equivalent ratio, multiply or divide each term by the same number.

EXAMPLE 1

Complete each pair of equivalent ratios.

a 4 : 3 = 16 : ___   b 10 : 25 = ___ : 5

c 2 : 3 : 7 = 6 : ___ : ___

SOLUTION

Examine the term that has been multiplied or divided from LHS to RHS.

a 4 : 3 = 16 : ___   ◄─────── $4 \times 4 = 16$
   4 : 3 = 16 : 12   ◄─────── Do the same to 3 to complete the ratio: $3 \times 4 = 12$

b 10 : 25 = ___ : 5   ◄─────── $25 \div 5 = 5$
   10 : 25 = 2 : 5   ◄─────── Do the same to 10 to complete the ratio: $10 \div 5 = 2$

c 2 : 3 : 7 = 6 : ___ : ___ ◄─── $2 \times 3 = 6$
   2 : 3 : 7 = 6 : 9 : 21   ◄─────── $3 \times 3 = 9, 7 \times 3 = 21$

To simplify a ratio, divide each number in the ratio by the highest common factor (HCF).

EXAMPLE 2

Express each ratio in simplest form.

a 18 : 27   b $2 : 25c   c 0.4 : 0.08   d $\dfrac{1}{3} : \dfrac{2}{5}$

SOLUTION

a $18 : 27 = \dfrac{18}{9} : \dfrac{27}{9}$ ◄─────── Dividing both terms by the HCF, 9

   $= 2 : 3$

 Simplifying ratios is similar to simplifying fractions.

b $2 : 25c = 200c : 25c   ◄─────── Convert both to cents

   $= \dfrac{200}{25} : \dfrac{25}{25}$ ◄─────── Dividing both terms by the HCF, 25

   $= 8 : 1$

c $0.4 : 0.08 = 0.4 \times 100 : 0.08 \times 100$ ◄─── Multiplying both terms by 100 to make them whole
   $= 40 : 8$
   $= 5 : 1$ ◄─────── Dividing both terms by the HCF, 8

d $\dfrac{1}{3} : \dfrac{2}{5} = \dfrac{1}{3} \times 15 : \dfrac{2}{5} \times 15$ ◄─────── Multiplying both terms by the LCD, 15, to make them whole

   $= 5 : 6$

1 Which ratio is equivalent to 3 : 5? Select the correct answer **A**, **B**, **C** or **D**.

    A 6 : 15   **B** 9 : 15   **C** 12 : 25   **D** 24 : 35

2 Simplify 75 : 45. Select **A**, **B**, **C** or **D**.

    A 25 : 9   **B** 50 : 30   **C** 15 : 9   **D** 5 : 3

3 Copy and complete each equivalent ratio.

    a 3 : 5 = 9 : ___   **b** 8 : 3 = 32 : ___

    c 4 : 7 = ___ : 21   **d** 12 : 5 = ___ : 30

    e 21 : 60 = 7 : ___   **f** 48 : 80 = ___ : 10

    g 22 : 55 = ___ : 5   **h** 28 : 56 = 4 : ___

    i 16 : 36 = ___ : 9   **j** 108 : 72 = ___ : 6

    k 2 : 5 : 8 = 8 : ___ : ___   **l** 25 : 45 : 75 = 5 : ___ : ___

4 Write each ratio in simplest form.

    a 10 : 20   **b** 3 : 9

    c 4 : 12   **d** 25 : 5

    e 2 : 8   **f** 5 : 15

    g 7 : 28   **h** 18 : 3

    i 36 : 9   **j** 4 : 32

    k $\frac{1}{2}$: 3   **l** $\frac{1}{4}$: $\frac{3}{4}$

    m $\frac{3}{4}$: 2   **n** $\frac{1}{2}$: $\frac{3}{4}$

    o $\frac{1}{3}$: $\frac{3}{8}$ **p** 0.3 : 0.4

    q 0.9 : 2   **r** 0.04 : 5

    s 0.2 : 0.03   **t** 1.2 : 0.08

5 Simplify each ratio.

    a 5 mm : 1 cm   **b** 10 mg : 1 g

    c 4 cm : 1 m   **d** 250 m : 1 km

    e $1.20 : $1.40   **f** 500 mL : 2 L

    g 10 min : 1 h   **h** 3c : $1

    i 6 litres : 600 mL   **j** $2.45 : $3.00

    k 4 h : 1 day   **l** 5 years : 1 decade

    m 26 weeks : 1 year   **n** 35 s : 2 min

    o 400 g : 3 kg   **p** 25 years : 1 century

    q 5 months : 5 years   **r** 500 kg : 5 tonnes

    s $2\frac{1}{2}$ h : 100 min   **t** $2\frac{1}{2}$ kg : 200 g

EXAMPLE 3

In a class of 30 students there are 16 boys. Find the ratio, in simplest form, of:

a   girls : boys   b   boys : girls   c   girls : whole class

SOLUTION

a   Number of girls = 30 – 16 = 14
       Girls : boys = 14 : 16
                   = 7 : 8

b   Boys : girls = 16 : 14   c   Girls : whole class = 14 : 30
                   = 8 : 7   = 7 : 15

EXAMPLE 4

Jonathon is making paste by mixing cornflour and boiling water in the ratio 2 : 3.

a   How much cornflour should be mixed with 15 cups of water?

b   If 5 cups of cornflour are used, how much water is needed?

SOLUTION

Cornflour : water = 2 : 3

a   3 parts (water) = 15 cups
           1 part = 15 ÷ 3   ◄——————— Using the unitary method to find one part
                  = 5 cups
       2 parts (cornflour) = 2 × 5
                          = 10 cups
   10 cups of cornflour are required.
       OR: Cornflour : water = 2 : 3 = ___ : 15   ◄——————— Using equivalent ratios
                       2 : 3 = **10** : 15
   10 cups of cornflour are required.

b   2 parts (cornflour) = 5 cups
       1 part = 5 ÷ 2   ◄——————— Using the unitary method to find one part
           $= 2\frac{1}{2}$ cups
       3 parts (water) $= 3 \times 2\frac{1}{2}$

                       $= 7\frac{1}{2}$ cups

   $7\frac{1}{2}$ cups of water are required.
       OR: Cornflour : water = 2 : 3 = 5 : ___   ◄——————— Using equivalent ratios
                       $2 : 3 = 5 : 7\frac{1}{2}$
   $7\frac{1}{2}$ cups of cornflour are required.

ISBN 9780170351027

1 If there are 5200 people in a crowd and 1800 are males, what is the ratio of females to males? Select the correct answer **A**, **B**, **C** or **D**.

   A 17 : 9   **B** 52 : 18   **C** 9 : 17   **D** 18 : 52

2 In question **1**, what is the ratio of females to the crowd? Select **A**, **B**, **C** or **D**.

   A 34 : 18   **B** 17 : 26   **C** 52 : 34   **D** 26 : 17

3 A class of 18 students has 12 brunettes, 4 blondes and 2 redheads. Find the ratio of:

   a blondes : brunettes   **b** redheads : brunettes   **c** brunettes : blondes

4 In a pack of animals there are 25 cows, 12 horses and 6 sheep. Find the ratio of:

   a horses : cows   **b** sheep : cows   **c** cows : sheep

5 Allan and Effie share the rent of a house in the ratio 5 : 4. If Allan pays $205, find Effie's share of the rent. Also find the whole rent.

6 A car dealer finds that the ratio of sedans to hatchbacks sold was 18 : 5. If 162 sedans were sold, how many hatchbacks were sold?

7 The ratio of the Dragons' wins to losses was 5 : 3. If the team lost 21 games, how many games did it win?

8 Ray and Sandy own a shop in the ratio 5 : 7. If Sandy's share is $59 500, what is Ray's share?

9 Sand and cement are mixed in the ratio 4 : 1 to make concrete. What mass of cement is needed to mix with 120 kg of cement?

10 The ratio of sushi to salad rolls sold at a school canteen was 7 : 6. If 91 sushi rolls were sold, how many salad rolls were sold?

11 The ratio of adults to children at the cinema was 3 : 4. If there were 78 adults, how many children were there?

12 Lauren and Harley share the profits of a garage sale in the ratio 7 : 6. If Harley receives $462, how much does Lauren get?

13 A recipe for making paste involves mixing cornflour and boiling water in the ratio 2 : 3. How much boiling water must be added to 10 cups of cornflour to make paste?

14 A survey of car buyers found that they purchased cars with colours in this ratio:
   White : red : other colours = 12 : 7 : 5

   a If a dealer ordered 20 cars in colours other than red or white, how many red cars should he order?

   b What fraction of the cars surveyed were white?

WORDBANK

scale diagram A miniature or enlarged drawing of an actual object or building, in which lengths and distances are in the same ratio as the actual lengths and distances.

scale The ratio on a scale diagram that compares lengths on the diagram to actual lengths.

scaled length A length on a scale diagram that represents an actual length.

Scale maps and diagrams are an important use of ratios. A scale of 1: 20 means the actual lengths on the object are 20 times larger than on the scale diagram, whereas a scale of 20 : 1 means that the actual lengths are 20 times smaller than on the scale diagram.

> The scale on a scale diagram is written as the ratio **scaled length : actual length**
> - The first term of the ratio is usually smaller, meaning that the diagram is a miniature version.
> - If the first term of the ratio is larger, then the diagram is an enlarged version.

EXAMPLE 5

Measure and find the actual length that each interval represents if a scale of 1 : 50 has been used.

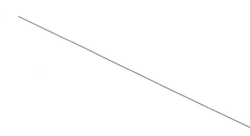

a ————————

b

c

SOLUTION

Scale is 1 : 50

a   Scaled length = 4 cm
    Actual length = 4 cm × 50 ⟵——— The actual length is 50 times larger
                  = 200 cm
                  = 2 m

b   Scaled length = 6 cm
    Actual length = 6 cm × 50
                  = 300 cm
                  = 3 m

c   Scaled length = 7.5 cm
    Actual length = 7.5 cm × 50
                  = 375 cm
                  = 3.75 m

EXAMPLE 6

Given the scale in each diagram, measure and calculate the actual length of each object.

a

Scale 8 : 1

b

Scale 1 : 10 000

SOLUTION

a Scaled length = 4 cm
   Ant's length = 4 cm ÷ 8 ◄———— Divide as the scale diagram is an enlargement
              = 0.5 cm

b Scaled length = 7 cm
   Bridge's length = 7 cm × 10 000 ◄———— The actual bridge is 10 000 times larger
                 = 70 000 cm
                 = 700 m

EXERCISE 16–03

1 What is the actual length of an interval if the scale length is 5 cm and the scale is 1 : 20? Select the correct answer **A**, **B**, **C** or **D**.

   A 20 cm   **B** 100 cm   **C** 4 cm   **D** 25 cm

2 What is the length of a dragonfly if its scale length is 2 cm and the scale is 8 : 1? Select the correct answer **A**, **B**, **C** or **D**.

   A 4 cm   **B** 16 cm   **C** 10 cm   **D** 0.25 cm

3 Find the actual length that each interval represents if a scale of 1 : 40 has been used.

   a ├———————————┤

   b ├———————————————————┤

   c ├———————————————————————┤

4   Find the actual length of each object.

a

Scale 1 : 3

b

6 cm

Scale 1 : 20

c

Scale 1 : 1.5

d

Scale 1 : 6200

5   A spider is drawn to a scale of 5 : 1.

   a   How long are its legs if they are drawn 15 cm long?

   b   How long is its body if it is drawn 22.5 cm long?

6   The area set aside for a tennis court is 120 m by 65 m.

   a   Draw a scale diagram of this tennis court using a scale of 1 : 1000.

   b   If the boundary lines for doubles are drawn 1 m in from each side edge, what are the dimensions of the singles court on the scale diagram?

   c   What are the actual dimensions of the singles court?

7   Convert the scales below to the same units and express them as whole numbers.

   a   1 cm : 5 m   b   1 mm : 40 cm   c   2 cm : 5 km   d   3 mm : 12 cm

8   The scale used to draw this plan of a house is 1 cm to represent 2.5 m.

   a   What is the actual length of the house?

   b   What is the actual width?

9   This tree is drawn to a scale of 1 mm to represent 20 cm.
   What is the actual height of the tree in metres?

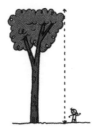

10 Use the map of Taree find the required distances rounded to the nearest 0.1 km.

a From Taree High School to the Riverview Motor Inn

b From Taree High School to the Youth Club

c From Taree Hospital to Edinburgh Park

d From Marco Polo Motor Inn to the RSL Club

e From the Youth Club to the Rowing Club

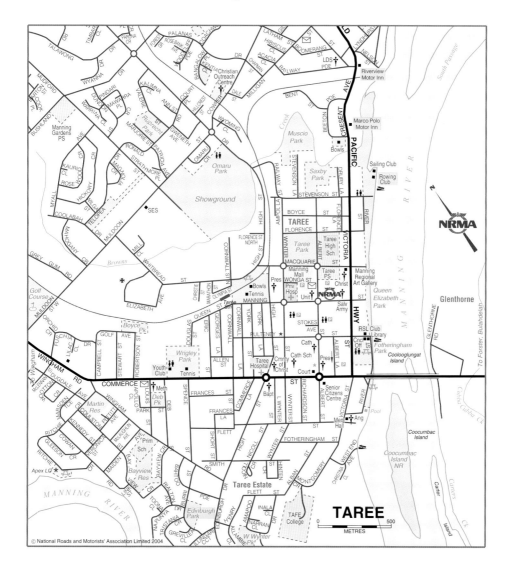

EXAMPLE 7

Ed and Manal won $5000 and share it in the ratio 3 : 5. How much will each person receive?

SOLUTION

Total number of parts = 3 + 5 = 8

One part = $5000 ÷ 8   ← Using the unitary method to find one part

       = $625

Ed's share = 3 × $625   ← 3 parts

       = $1875

Manal's share = 5 × $625   ← 5 parts

       = $3125

Check: $1875 + $3125 = $5000 ← The two shares add up to the whole prize.

EXAMPLE 8

Jenny, Jacqui and Jules share 36 chocolates in the ratio 4 : 3 : 2. How many chocolates does each person receive?

SOLUTION

Total number of parts = 4 + 3 + 2 = 9

     One part = 36 ÷ 9

        = 4

     Jenny's share = 4 × 4   ← 4 parts

        = 16

     Jacqui's share = 3 × 4   ← 3 parts

        = 12

     Jules' share = 2 × 4   ← 2 parts

        = 8

Check: 16 + 12 + 8 = 36   ← The shares add up to the whole.

Getty Images/KidStock

ISBN 9780170351027

1 Find the size of one part if $360 is divided in the ratio 7 : 5. Select the correct answer
   A, **B**, **C** or **D**.

   A $180   **B** $30   **C** $72   **D** $90

2 What does Bill receive if $3500 is divided between Bill and Ben in the ratio 3 : 2?
   Select **A**, **B**, **C** or **D**.

   A $700   **B** $1400   **C** $2800   **D** $2100

3 Copy and complete this table.

Ratio	Total number of parts	Total amount	One part	Ratio
3 : 1	3 + 1 = 4	64	64 ÷ 4 = 16	48 : 16
2 : 3		75		
1 : 5		600		
3 : 5		720		
4 : 5		1800		
7 : 6		3900		

4 Divide $25 between Asha and Juno in the ratio 2 : 3.

5 Divide $7200 between Nathan and Tom in the ratio 3 : 5.

6 If $880 is to be shared among three people in the ratio 2 : 3 : 6,

   a how much is the largest share?

   b how much is the smallest share?

7 Three people share a prize of $96 000 in a lottery. How much does each one receive, if it is
   to be shared in the ratio 3 : 4 : 5?

8 Tara invests $15 000 and Gemma invests $25 000 in a business. If the profit at the end of
   the year is $150 000, how much should each receive if the profits are shared in the same
   ratio as their investments?

9 A soccer team played 15 matches in a season. If the ratio of wins to losses was 4 : 1, how
   many matches did the team lose?

10 Gina is 12 years old and Liam is 8 years old. They were given $568 to be shared in the
   ratio of their ages. How much should Liam get?

11 Divide 1260 in the ratio 2 : 3 : 4.

12 In a spelling test of 50 words, Amber spelt 45 words correctly. Is it true or false to say that
   the ratio of words spelt correctly to words spelt incorrectly was 9 : 1?

A **rate** compares two quantities of different kinds or different units of measure. For example, a heartbeat is measured in beats/minute and the cost of petrol is measured in cents/litre. 120 beats **per** minute means 120 beats in 1 minute and is written as 120 beats/minute.

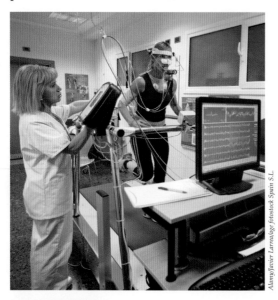

Alamy/Javier Larrea/age fotostock Spain S.L.

AAP Image/Julian Smith

EXAMPLE 9

Write each measurement as a simplified rate.

a   270 km in 3 hours   **b** $240 for 8 hours work   **c** $6.60 for a 4-minute call

SOLUTION

a   270 km in 3 hours = 270 ÷ 3 km/h
                    = 90 km/h

b $240 for 8 hours = 240 ÷ 8 $/h
                    = $30/h

c $6.60 for 4 minutes = $6.60 ÷ 4
                    = $1.65/min

Alamy/Maurice Savage

Shutterstock.com/Air Images

Shutterstock.com/Diego Cervo

1   Simplify the rate $280 for 14 hours work. Select the correct answer **A**, **B**, **C** or **D**.

    A $28/h   **B** $7/h   **C** $14/h   **D** $20/h

2   Simplify the rate 180 words in 3.6 minutes. Select **A**, **B**, **C** or **D**.

    A 50 words/min   **B** 60 words/min   **C** 30 words/min   **D** 20 words/min

3   Copy and complete each sentence.

    a $3.60 for 12 apples is at a rate of ☐ c/apple.

    b   250 km in 5 hours is at a rate of ☐ km/h.

    c $18.40 for 10 kg butter is at a rate of $☐/kg

    d   1200 m for 120 s is at a rate of ☐ m/s

    e $48.40 for 4 hours work is at a rate of $☐/h

    f $75 000 for 250 ha is at a rate of $☐/ha

    g   96 teachers for 1920 students is at a rate of ☐ students/teacher

    h   7000 revolutions in 5 min is at a rate of ☐ revs/min

    i $696 for 3 days at a hotel is at a rate of $☐/day

    j   5 wickets for 115 runs is at a rate of ☐ runs/wicket

4   A car uses 5 litres of petrol to travel 40 km. At what rate is the car using petrol, in km/L?

5   **a**   The cost of staying at a certain motel is $830 for 5 days. What is the daily rate, in $/day?

    b   Another motel in the city centre charges $420 for 3 days. What is the daily rate?

    c   Which motel is the best value if you are trying to save money?

6   Travis works 5 hours and earns $115. What is his hourly rate of pay?

7   A plumber took 3 hours to complete a job. If he charged $198, work out the rate that he charged per hour.

8   While an elephant was browsing peacefully for 5 minutes in the grasslands, its heart beat 140 times. What was the elephant's heart rate in beats/min?

9   Jorja's car travelled 195 km in 3.5 hours. What was her average speed in km/h, correct to 1 decimal place?

10 Which rate is faster, 320 words in 5 mins or 680 words in 8 mins?

WORDBANK

fuel consumption The amount of fuel used by a car compared to distance travelled, measured in litres/100 km.

To solve a rate problem, write the units in the rate as a fraction $\dfrac{x}{y}$
- To find x (the numerator amount), **multiply** by the rate.
- To find y (the denominator amount), **divide** by the rate.

EXAMPLE 10

In the supermarket, apples cost $3.20/kg.

a How much does it cost to buy 3.5 kg of apples?

b How many kilograms of apples can be bought for $30.40?

Shutterstock.com/racorn

SOLUTION

The units of the rate are $\dfrac{\$}{\text{kg}}$ ←——— Writing the units as a fraction

a To find $, multiply by the rate.
Cost of 3.5 kg of apples = 3.5 × $3.20 ←——— 1 kg costs $3.20 so multiply by 3.5
= $11.20

b To find kg, divide by the rate.
Number of kilograms = $30.40 ÷ $3.20 ←——— 1 kg costs $3.20 so divide by $3.20
= 9.5
We can buy 9.5 kg of apples for $30.40.

EXAMPLE 11

Tyler took his new car for a drive to test its fuel consumption. He travelled 520 km on 65 litres of petrol. What is his fuel consumption in L/100 km?

SOLUTION

Fuel consumption $= \dfrac{65\,\text{L}}{520\,\text{km}}$ ←——— Writing the rate as a fraction in L/km first

$= \dfrac{65\,\text{L}}{5.2 \times 100\,\text{km}}$ ←——— Write 520 as a multiple of 100

$= \dfrac{65}{5.2}\,\text{L}/100\,\text{km}$

$= 12.5\,\text{L}/100\,\text{km}$

1 Write the rate 250 beats in 4 minutes in beats/min as a simplified fraction. Select the correct answer **A**, **B**, **C** or **D**.

   A $\dfrac{250}{4}$ **B** $\dfrac{4}{250}$ **C** $\dfrac{2}{125}$ **D** $\dfrac{125}{2}$

2 Write the fuel consumption 72 L per 450 km in L/100 km. Select **A**, **B**, **C** or **D**.

   A 16 L/100 km   **B** 6.25 L/100 km   **C** 12 L/100 km   **D** 9 L/100 km

3 Emily earns $22/h working in a boutique.

   a Write the units in the rate as a fraction.

   b How much will Emily earn if she works 25 hours?

   c How long will it take Emily to earn $286?

4 David types 480 words in 5 minutes.

   a How many words/min does he type?

   b Write the units in the rate as a fraction.

   c How long will it take him to type 4032 words?

   d How many words will David type in half an hour?

5 Dianne paid income tax at the rate of 33c per dollar of income earned.

   a How much tax does she pay on her income of $15 800?

   b What is her annual income if she pays $9240 tax p.a.?

6 Handmade chocolates cost $24.50/kg.

   a How much will 350 g of chocolates cost? (answer correct to the nearest cent)

   b How many grams of chocolate could you buy for $20? (answer correct to the nearest gram)

7 Mr Joucevski pays council rates at the rate of 3.327c in the dollar on his unimproved property value. If his property is valued at $35 000, how much does he have to pay in council rates?

8 Kayla can type at an average rate of 48 words per minute. How long should it take for her to type a 5000-word essay? (Answer correct to the nearest minute.)

9 Chris took his new car for a drive to test its fuel consumption. He travelled 400 km on 45 litres of petrol. What is his fuel consumption in L/100 km?

10 Eric's car travels 680 km on 81.6 L of petrol. What is his car's fuel consumption in L/100 km?

11 Taylor washes 25 cars in 9 hours and 10 minutes.

   a How many minutes does she take to wash 25 cars?

   b Find Taylor's car washing rate in minutes per car.

   c How long will it take for Taylor to wash 20 cars (in hours and minutes)?

WORDBANK

speed A rate that compares distance travelled with time taken. The units are kilometres per hour (km/h) or metres per second (m/s).

THE SPEED FORMULA

$$\text{Speed} = \frac{\text{Distance travelled}}{\text{Time taken}} \qquad S = \frac{D}{T}$$

You can memorise this triangle to help you remember this rule.

Cover S with your finger. You are left with $\frac{D}{T}$.

If you want to find the distance, cover D and you are left with $S \times T$.

$$D = S \times T$$

In the same way, if you want to find time, cover T and you are left with $\frac{D}{S}$.

$$T = \frac{D}{S}$$

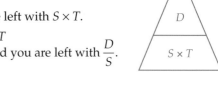

EXAMPLE 12

Renee drove from Sydney to Newcastle, a distance of 164 km, in 2 hours. What was her average speed?

SOLUTION

$$S = \frac{D}{T}$$

$$S = \frac{164 \text{ km}}{2 \text{ h}}$$

$$= 82 \text{ km/h}$$

EXAMPLE 13

a   If Jarrod jogged at an average speed of 5 m/s, how long would he take to jog 1 km?

b   Jarrod slowed down to 4 m/s and jogged for half an hour. How far did he go in this time?

SOLUTION

a $T = \dfrac{D}{S}$ ←——— Rearrange the formula $S = \dfrac{D}{T}$ to find time, T.

     $= \dfrac{1000}{5}$ ←——— $S = 5$ m/s, $D = 1$ km $= 1000$ m

     $= 200$ s

     $= (200 \div 60)$ min   ←——— 1 min $= 60$ s

     $= 3.333\ldots$ min

     $= 3$ min 20 s   ←——— Enter [° "] or [2ndF] [—] on a calculator

                        or calculate $0.3333\ldots \times 60$ for seconds

b $D = ST$ ←——— Rearrange the formula $S = \dfrac{D}{T}$ to find distance, D.

     $= 4 \times 1800$ ←——— $S = 4$ m/s, $T = 30$ min $= 30 \times 60$ s $= 1800$ s

     $= 7200$ m

     $= 7.2$ km

1   If it takes 3 hours to travel 243 kilometres, what is the average speed? Select **A**, **B**, **C** or **D**.

   A 80 km/h   **B** 81 km/h   **C** 82 km/h   **D** 240 km/h

2   Travelling at 110 km/h, how long will it take to travel 935 km? Select **A**, **B**, **C** or **D**.

   A 8.5 h   **B** 9 h   **C** 9.5 h   **D** 7.5 h

3   Copy and complete this table to find the average speed.

Distance	Time	Speed
120 km	3h	
650 km	5h	
1250 km	25 h	
680 m	10 s	
4860 m	80 s	

4   Jacob drives 110 km in 2 hours. Find his average speed.

5   A racing pigeon flies 420 km in 6 hours. Calculate its average speed.

6   Chantal takes 1 hour to cycle 10 km and then walks for another hour, travelling a further 6 km.

   a   What is the total distance travelled?

   b   How much time has she taken for the whole distance?

   c   Find her average speed.

7   Copy and complete the table.

Distance	Time	Speed
450 km		50 km/h
	80 s	20 m/s
472.5 km		105 km/h

8   A car travels 225 km in 3 hours.

   a   What is its average speed, in km/h?

   b   How far would this car travel in 4 hours?

   c   How long would this car take to travel 412.5 km?

9   A racing car driver does one lap of a 5.5 km race track in 2 minutes.

   a   How far would the driver travel in 60 minutes?

   b   What is the speed in km/h?

   c   How many metres would the driver travel in 60 seconds?

   d   Find his speed in m/s.

10 A car is travelling at 60 km/h.

   a   How far does it travel in one minute?   **b**   How many metres is this?

   c   Find the car's speed in m/s.

11 Nina has a practice on a race track. She does 5 laps of the 12 km track in 15 minutes.

   a   How far would she travel in 60 minutes?   **b**   What is her speed in km/h?

   c   How many metres did she travel in 1 minute?   **d**   Find her speed in m/s.

COMPOUND WORDS

The words in the list below are called **compound words** as they are made up of two smaller words. For each compound word, identify the two smaller words, count the number of letters in each word, and write the pair of numbers as a simplified ratio.

For example,   NEWCASTLE = NEW + CASTLE, with ratio 3 : 6 = 1 : 2.

1 ROUNDABOUT
2 TIMETABLE
3 PIECEWORK
4 BREAKAGE
5 OVERALL
6 ANYWHERE
7 SUNSET
8 MOONLIGHT
9 RATTLESNAKE
10 SEASHORE
11 TEXTBOOK
12 HONEYSUCKLE
13 MATCHBOX
14 SUNGLASSES
15 TYPEWRITER
16 CHEESEBURGER
17 WAVELENGTH

Shutterstock.com/sabza

Shutterstock.com/Rusty Dobson

Shutterstock.com/harmizi rizali

Part A General topics

Calculators are not allowed.

1. Convert 8:25 p.m. to 24-hour time.

2. Simplify 45 : 60.

3. Evaluate 8^{-2}.

4. Write Pythagoras' theorem for this triangle.

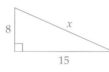

5. Factorise: $20x^2y + 15xy^2$

6. Write down the prime numbers between 10 and 20.

7. Find $\frac{2}{3}$ of $33.

8. Expand $-2(3a + 9)$

9. Write $\frac{4}{5}$, 0.85, 0.803 and 86% in ascending order.

10. What is the probability of rolling a number less than 3 on a die?

Part B Ratio and rates

Calculators are allowed.

16–01 Ratios

11. Simplify 120 : 36. Select the correct answer **A**, **B**, **C** or **D**.

    A 20 : 6   **B** 40 : 12   **C** 12 : 3.6   **D** 10 : 3

12. Which ratio is equivalent to 1 : 3 : 4? Select **A**, **B**, **C** or **D**.

    A 3 : 9 : 8   **B** 5 : 15 : 20   **C** 4 : 12 : 20   **D** 7 : 21 : 35

16-02 Ratio problems

13. In a room full of 54 adults, there are 28 males. What is the ratio of females : adults? Select **A**, **B**, **C** or **D**.

    A 26 : 28   **B** 13 : 27   **C** 28 : 54   **D** 54 : 26

14. Matt is mixing juice using 2 parts apple, 3 parts orange and 1 part watermelon.

    a If he uses 780 mL of orange juice, how much apple juice does he use?

    b How many litres of juice are made?

16-03 Scale maps and diagrams

15 Find the actual length of each drawing.

a

Scale 1 : 180

b

Scale 1 : 12

16-04 Dividing a quantity in a given ratio

16 Divide a prize of $25 000 between Lachlan and Courtney in the ratio 1 : 4.

17 Ruben invests $6000, John $4500 and Tim $8000 in a new business. If their profit in the first year was $111 000, work out each person's share if it is in the same ratio as their investments.

16-05 Rates

18 Simplify each rate.

   a $180 for 4.5 hours work

   b $7.20 for 12 apples

   c 434 words in 7 minutes

16-06 Rate problems

19 In the supermarket, lamb costs $18.60/kg.

   a How much will it cost to buy 350 g of lamb?

   b How many grams of lamb could I buy for $15.80?

16-07 Speed

20 a A ship travels at 36 km/h for $5\frac{1}{2}$ hours. How far did it travel?

   b How long will it take to travel 480 km at this rate? Answer in hours and minutes.

CHAPTER 1

Exercise **1-01**

1 C
2 a 9   b 49   c 121   d 196
   e 73.96   f 81   g 146.41   h 3.61
   i 16   j 262.44   k 64   l 92.16
3 a 4   b 6   c 8
   d 11   e 14
4 B
5 a 4.24   b 5.66   c 9.06
   d 12.25   e 14.83
6 D
7 a True   b False   c False   d True
   e False
8 25 m^2
9 7 m
10 Side = 8 m

Exercise **1-02**

1 A   2 C
3 a $p^2 = q^2 + r^2$ b $i^2 = m^2 + n^2$
   c $a^2 = b^2 + c^2$ d $e^2 = d^2 + f^2$
   e $x^2 = y^2 + z^2$ f $r^2 = s^2 + t^2$
4 a $13^2 = 5^2 + 12^2$, True   b $16^2 = 9^2 + 12^2$, False
   c $17^2 = 8^2 + 15^2$, True
5 a and c

Exercise **1-03**

1

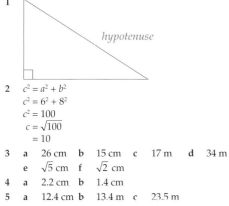

2 $c^2 = a^2 + b^2$
   $c^2 = 6^2 + 8^2$
   $c^2 = 100$
   $c = \sqrt{100}$
     $= 10$
3 a 26 cm   b 15 cm   c 17 m   d 34 m
   e $\sqrt{5}$ cm   f $\sqrt{2}$ cm
4 a 2.2 cm   b 1.4 cm
5 a 12.4 cm   b 13.4 m   c 23.5 m

Exercise **1-04**

1 A   2 B
3 $20^2 = b^2 + 12^2$
   $b^2 + 12^2 = 20^2$
   $b^2 = 20^2 - 12^2$
     $= 400 - 144 = 256$
   $b = 16$
4 a 24 cm   b 24 m   c 5 m   d 7 cm
5 a 13.4 m   b 8.5 m   c 8.3 m   d 11.8 m

Exercise **1-05**

1 A   2 B
3 a 8 cm   b $\sqrt{288}$ m
   c $\sqrt{337}$ m   d $\sqrt{1475}$ m
   e 41 m   f $\sqrt{2380}$ cm
4 a 9.7 cm   b 10.4 m   c 19.1 m
5 hypotenuse

Exercise **1-06**

1 Substitute given values into Pythagoras' theorem, and check if the left-hand side is the same as the value of the right-hand side.
2 B   3 C
4 $17^2 = 15^2 + 8^2$
   $289 = 225 + 64$
   289 = 289 Yes, so the triangle is right-angled.
5 a $41^2 \ne 10^2 + 40^2$, No   b $25^2 = 20^2 + 15^2$, Yes
   c $22^2 \ne 15^2 + 16^2$, No   d $2^2 = 1.6^2 + 1.2^2$, Yes
   e $6.1^2 = 1.1^2 + 6^2$, Yes   f $74^2 \ne 8^2 + 22^2$, No
6 b $\angle R$ d $\angle Q$ e $\angle Q$

Exercise **1-07**

1 three Pythagoras'
2 C   3 B
4 $13^2 = 5^2 + 12^2$
   $169 = 25 + 144$
   169 = 169, So {5, 12, 13} is a Pythagorean triad.
5 a $41^2 = 9^2 + 40^2$, Yes   b $25^2 \ne 7^2 + 20^2$, No
   c $20^2 = 12^2 + 16^2$, Yes   d $52^2 \ne 11^2 + 50^2$, No
   e $7^2 \ne 5^2 + 6^2$, No   f $25^2 = 7^2 + 24^2$, Yes
   g $17^2 = 8^2 + 15^2$, Yes   h $12^2 \ne 4^2 + 8^2$, No
6 {10, 24, 26} and {15, 36, 39}
7 a $34^2 = 16^2 + 30^2$ b {8, 15, 17}
8 {14, 48, 50} and {21, 72, 75}
9 a True   b True   c False
10 Teacher to check, Yes.

Exercise **1-08**

1 C   2 6.5 m   3 17 cm
4 5 cm   5 2.5 m
6 a

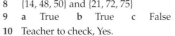

           b 61 m
7 2.6 m   8 23 m
9 32 − 23 = 9 m   10 13.3 km

Practice test **1**

1 9:08 p.m.   2 608 mm   3 220
4 44 m   5 $\dfrac{1}{14}$ 6 $-2x + 8$
7 $24   8 $\bar{x} = 4$ 9 5   10 $16.55
11 a 5.83   b 6.71   c 7.21
12 $\sqrt{82}, \sqrt{28}, \sqrt{75}, \sqrt{68}$
13 $k, k^2 = h^2 + j^2$

14 a $c = \sqrt{61}$ b $r = \sqrt{250}$
15 a 5.74 cm b 10.32 m
16 a 10.7 m b 175.3 m c 23.4 km
17 a $34^2 = 16^2 + 30^2$, Yes b $2.2^2 \neq 1.6^2 + 0.4^2$, No
18 a $3^2 = 2.4^2 + 1.8^2$, Yes b $26^2 \neq 7^2 + 24^2$, No
19 75.6 cm
20 56 cm

CHAPTER 2

Exercise 2-01

1 B
2 a 19   b 39   c 37   d 148
   e 39   f 44   g 59   h 77
   i 108 j 59   k 122 l 108
3 b and e, g and j, i and l
4 a 82   b 82   c 148 d 189
5 a They are the same.
   b Just add 41 to c's answer
6 a T   b F   c F   d T
7 a 2   b 3   c 2   d 300
8 C
9 a 856 b 1146 c 1125 d 1545
   e 658 f 982 g 1859 h 3641
10 a 120 b 90   c 120 d 80
   e 316 f 942 g 674 h 683
11 a

3	6	6
8	5	2
4	4	7

   b

4	7	7
9	6	3
5	5	8

Exercise 2-02

1 A
2 a 11   b 16   c 14   d 28
   e 6   f 52   g 16   h 59
   i 224 j 436 k 663 l 838
3 a $84 - 9 = 84 - 10 + 1 = 74 + 1 = 75$
   b $358 - 41 = 358 - 40 - 1 = 318 - 1 = 317$
4 D
5 a 63   b 43   c 47   d 89
   e 51   f 87   g 102 h 123
   i 408 j 501 k 468 l 492
6 $6 + 30 + 3 = 39$
7 a 25   b 71   c 38   d 38
   e 47   f 37   g 35   h 57
8 a F   b T   c T   d F
9 5070

Exercise 2-03

1 C   2 B
3 a F   b T   c T
   d T   e F
4 a 3   b 7   c 13   d 21
   e 25   f 7   g 16   h 121
   i 264 j 18
5 a 0.4 b 1.7 c 0.9 d 1.7
   e 3.2 f 0.7 g 2.2 h 3.4
   i 12.5 j 2.0
6 a 0.22   b 1.42 c 5.52 d 11.32
   e 126.22 f 0.43 g 3.27 h 0.92
   i 7.23   j 1.90
7 a 0.264 b 1.731 c 5.844
   d 11.321 e 126.232
8 a $5   b $52   c $27   d 45
   e $29   f $10   g $232 h $501
   i $227 j $402
9 a $5.71   b $12.43 c $32.57 d $4.23
   e $218.93 f $110.43 g $1.68   h $502.92
   i $22.30 j $43.90
10 a $490 b $487 c $486.80 d $486.85

Exercise 2-04

1 D   2 A
3 a 0.03 b 0.26 c 0.53 d 3.36
4 a 0.2   b 2.2   c 2.9   d 2.8
5 a 15.98   b 51.05   c 133.368
   d 132.58 e 22.14   f 338.9
   g 28.814 h 147.63 i 28.3
   j 9.48   k 307.037 l 420.3
6 a 0.3   b 4.1   c 0.9   d 1.2
   e 0.06 f 2.22 g 0.3   h 1.86
   i 3.8   j 11.6 k 2.1   l 11.79
7 a $23.50 b $17.30 c $7.75
   d $57.40 e $55.22 f $31.01
   g $20.65 h 51.18
8 a $17   b Yes   c $17.40 d $2.60
9 $14.39   10 $4885

Exercise 2-05

1 B   2 C
3 a 30   b 24   c 28   d 18
   e 40   f 24   g 63   h 36
   I 42   j 30   k 21   l 72
   m 48   n 27   o 35   p 16
   q 45   r 18
4 a T   b F   c T
   d T   e T   f F
5 a 140 b 506 c 567 d 328
   e 350 f 416 g 150 h 1164
   i 340 j 570 k 2960 l 6120
   m 8100 n 28800 o 41000 p 364000

ANSWERS

<table>
<tr><td>6</td><td>a</td><td>5, 10, 330</td><td></td><td>b</td><td>25, 100, 4200</td><td></td></tr>
<tr><td>7</td><td>a</td><td>270</td><td>b 1500</td><td>c 2600</td><td>d 1400</td></tr>
<tr><td></td><td>e</td><td>540</td><td>f 5800</td><td>g 3600</td><td>h 82 000</td></tr>
</table>

Exercise 2-06

1 B   2 D

<table>
<tr><td>3</td><td>a</td><td>260</td><td>b</td><td>54</td><td>c</td><td>6800</td><td>d</td><td>4.5</td></tr>
<tr><td></td><td>e</td><td>22650</td><td>f</td><td>154.2</td><td>g</td><td>38</td><td>h</td><td>204.15</td></tr>
<tr><td>4</td><td>a</td><td>F</td><td>b</td><td>T</td><td>c</td><td>F</td><td></td><td></td></tr>
<tr><td>5</td><td>a</td><td>1</td><td>b</td><td>1</td><td>c</td><td>2</td><td>d</td><td>3</td></tr>
<tr><td>6</td><td>a</td><td>4.5</td><td>b</td><td>16.8</td><td>c</td><td>0.14</td><td>d</td><td>0.012</td></tr>
<tr><td>7</td><td>a</td><td>22.4</td><td>b</td><td>8.4</td><td>c</td><td>171</td><td>d</td><td>259.2</td></tr>
<tr><td></td><td>e</td><td>8.52</td><td>f</td><td>18.825</td><td>g</td><td>0.368</td><td>h</td><td>0.8722</td></tr>
<tr><td></td><td>i</td><td>108.18</td><td>j</td><td>0.0096</td><td>k</td><td>10.155</td><td>l</td><td>51.212</td></tr>
<tr><td>8</td><td>a</td><td>16.64</td><td>b</td><td>1.664</td><td>c</td><td>1.664</td><td>d</td><td>0.1664</td></tr>
</table>

9 Estimate = $20 × 6 = $120, Answer = $130.80
10 Estimate = 9 × 1 = 9 m², Answer = 7.74 m²

Exercise 2-07

1 B   2 D

<table>
<tr><td>3</td><td>a</td><td>5</td><td>b</td><td>4</td><td>c</td><td>4</td><td>d</td><td>5</td></tr>
<tr><td></td><td>e</td><td>6</td><td>f</td><td>7</td><td>g</td><td>5</td><td>h</td><td>5</td></tr>
<tr><td></td><td>i</td><td>11</td><td>j</td><td>9</td><td>k</td><td>7</td><td>l</td><td>8</td></tr>
<tr><td></td><td>m</td><td>5</td><td>n</td><td>9</td><td>o</td><td>5</td><td>p</td><td>9</td></tr>
<tr><td></td><td>q</td><td>7</td><td>r</td><td>4</td><td></td><td></td><td></td><td></td></tr>
<tr><td>4</td><td>a</td><td>F</td><td>b</td><td>T</td><td>c</td><td>T</td><td>d</td><td>T</td></tr>
<tr><td></td><td>e</td><td>F</td><td>f</td><td>F</td><td>g</td><td>T</td><td></td><td></td></tr>
<tr><td>5</td><td>a</td><td>43</td><td>b</td><td>39</td><td>c</td><td>32</td><td>d</td><td>107</td></tr>
<tr><td></td><td>e</td><td>23</td><td>f</td><td>153</td><td>g</td><td>16</td><td>h</td><td>14</td></tr>
<tr><td></td><td>i</td><td>80</td><td>j</td><td>31</td><td>k</td><td>52</td><td>l</td><td>17</td></tr>
<tr><td>6</td><td>a</td><td>12</td><td>b</td><td>54</td><td>c</td><td>33</td><td>d</td><td>51</td></tr>
<tr><td></td><td>e</td><td>24.5</td><td>f</td><td>2.18</td><td>g</td><td>2.9</td><td>h</td><td>3.21</td></tr>
<tr><td></td><td>i</td><td>0.43</td><td>j</td><td>0.62</td><td>k</td><td>0.048</td><td>l</td><td>0.036</td></tr>
<tr><td>7</td><td>a</td><td>594</td><td>b</td><td>650</td><td>c</td><td>316</td><td>d</td><td>478</td></tr>
<tr><td>8</td><td>a</td><td>4</td><td></td><td></td><td></td><td></td><td></td><td></td></tr>
</table>

    b Tinsel $1.05 Baubles $0.65 Gift wrap $1.70
      Cards $0.90

Exercise 2-08

1 D   2 C

<table>
<tr><td>3</td><td>a</td><td>0.038</td><td>b</td><td>2.56</td><td>c</td><td>0.0724</td><td>d</td><td>78</td></tr>
<tr><td></td><td>e</td><td>0.1283</td><td>f</td><td>0.1278</td><td>g</td><td>0.095</td><td>h</td><td>0.025</td></tr>
<tr><td>4</td><td>12.9</td><td></td><td></td><td></td><td></td><td></td><td></td><td></td></tr>
<tr><td>5</td><td>a</td><td>7.6</td><td>b</td><td>10.81</td><td>c</td><td>30.9</td><td>d</td><td>22.5</td></tr>
<tr><td></td><td>e</td><td>31.6</td><td>f</td><td>103.02</td><td>g</td><td>14.4</td><td>h</td><td>164.25</td></tr>
</table>

6    $\overset{228.2}{684.6, 3)684.6}$

<table>
<tr><td>7</td><td>a</td><td>4</td><td>b</td><td>6.2</td><td>c</td><td>12.06</td><td>d</td><td>42.6</td></tr>
<tr><td></td><td>e</td><td>249</td><td>f</td><td>8.6</td><td>g</td><td>1184.6</td><td>h</td><td>1830</td></tr>
<tr><td></td><td>i</td><td>1.15</td><td>j</td><td>415</td><td>k</td><td>260</td><td>l</td><td>649</td></tr>
<tr><td>8</td><td>4.641</td><td></td><td>9</td><td>1.19 m</td><td></td><td>10</td><td>$6.28</td><td></td></tr>
</table>

Exercise 2-09

1 D   2 C
3 a T b F c T

<table>
<tr><td>4</td><td>a</td><td>0.7̇</td><td>b</td><td>0.6̇2̇</td><td>c</td><td>2.1̇</td><td>d</td><td>3.46̇</td></tr>
<tr><td></td><td>e</td><td>5.2̇3̇</td><td>f</td><td>0.52̇8̇</td><td>g</td><td>7.2̇3̇</td><td>h</td><td>12.42̇5̇</td></tr>
<tr><td>5</td><td>a</td><td>3.1875</td><td>b</td><td>6.52</td><td>c</td><td>95.1̇3̇</td><td></td><td></td></tr>
<tr><td></td><td>d</td><td>2.825</td><td>e</td><td>30.4̇3̇</td><td>f</td><td>8.7̇1̇</td><td></td><td></td></tr>
<tr><td></td><td>g</td><td>11.35</td><td>h</td><td>9.40571428</td><td></td><td></td><td></td><td></td></tr>
</table>

6 a, b, d, g terminating; c, e, f, h recurring
7 a Divide 3 by 8   b 0.375
   c terminating

<table>
<tr><td>8</td><td>a</td><td>0.25</td><td>b</td><td>0.3̇</td><td>c</td><td>0.625</td></tr>
<tr><td></td><td>d</td><td>0.6</td><td>e</td><td>0.16̇</td><td>f</td><td>0.2̇85714̇</td></tr>
<tr><td></td><td>g</td><td>0.416̇</td><td>h</td><td>0.875</td><td></td><td></td></tr>
</table>

9 a $\frac{2}{3}$ b 0.6̇   c $600

Language activity
DECIMAL DILEMMA

Practice test 2

<table>
<tr><td>1</td><td>720</td><td>2</td><td>12000</td><td>3</td><td>1, 2, 3, 6</td><td></td></tr>
<tr><td>4</td><td>x</td><td>5</td><td>7</td><td>6</td><td>9</td><td></td></tr>
<tr><td>7</td><td>$12</td><td>8</td><td>d = 14</td><td>9</td><td>36rv</td><td>10 16</td></tr>
<tr><td>11</td><td>a 38</td><td>b</td><td>81</td><td>c</td><td>471</td><td></td></tr>
<tr><td>12</td><td>B</td><td></td><td></td><td></td><td></td><td></td></tr>
<tr><td>13</td><td>a 49</td><td>b</td><td>97</td><td></td><td></td><td></td></tr>
<tr><td>14</td><td>a 2.68</td><td>b</td><td>0.925</td><td></td><td></td><td></td></tr>
<tr><td>15</td><td>a 16.25</td><td>b</td><td>19.71</td><td></td><td></td><td></td></tr>
<tr><td>16</td><td>a 960</td><td>b</td><td>5300</td><td></td><td></td><td></td></tr>
<tr><td>17</td><td>a 4760</td><td>b</td><td>0.028</td><td>c</td><td>0.948</td><td></td></tr>
<tr><td>18</td><td>T</td><td></td><td></td><td></td><td></td><td></td></tr>
<tr><td>19</td><td>a 44</td><td>b</td><td>339</td><td>c</td><td>586</td><td></td></tr>
<tr><td>20</td><td>a 1.642</td><td>b</td><td>17.15</td><td>c</td><td>457.5</td><td></td></tr>
<tr><td>21</td><td>F</td><td></td><td></td><td></td><td></td><td></td></tr>
<tr><td>22</td><td>a 6.1̇2̇</td><td>b</td><td>0.4̇8̇</td><td>c</td><td>12.4̇12̇</td><td></td></tr>
<tr><td>23</td><td>0.2̇</td><td></td><td></td><td></td><td></td><td></td></tr>
</table>

CHAPTER 3

Exercise 3-01

1 B   2 C

<table>
<tr><td>3</td><td>a</td><td>6</td><td>b</td><td>7 left</td><td>c</td><td>8 left</td><td>d</td><td>4, 2 left</td></tr>
<tr><td>4</td><td>a</td><td>5</td><td>b</td><td>−1</td><td>c</td><td>2</td><td>d</td><td>2</td></tr>
<tr><td></td><td>e</td><td>0</td><td>f</td><td>−1</td><td>g</td><td>−1</td><td>h</td><td>3</td></tr>
<tr><td></td><td>i</td><td>3</td><td>j</td><td>0</td><td>k</td><td>7</td><td>l</td><td>9</td></tr>
<tr><td>5</td><td>a</td><td>3</td><td>b</td><td>5 + (−3)</td><td>c</td><td>−5 + (−3)</td><td></td><td></td></tr>
<tr><td>6</td><td>a</td><td>3</td><td>b</td><td>7</td><td>c</td><td>−9</td><td>d</td><td>−4</td></tr>
<tr><td></td><td>e</td><td>8</td><td>f</td><td>−7</td><td>g</td><td>−7</td><td>h</td><td>9</td></tr>
<tr><td></td><td>i</td><td>1</td><td>j</td><td>16</td><td>k</td><td>6</td><td>l</td><td>21</td></tr>
<tr><td>7</td><td>a</td><td>6</td><td>b</td><td>23</td><td>c</td><td>6</td><td>d</td><td>12</td></tr>
<tr><td></td><td>e</td><td>38</td><td>f</td><td>−24</td><td>g</td><td>35</td><td>h</td><td>−11</td></tr>
<tr><td>8</td><td>$51</td><td></td><td>9</td><td>13th floor</td><td></td><td>10</td><td>22⁰</td><td></td></tr>
</table>

Exercise 3-02

1 C   2 D
3 a F b T c F d T

4

×	−2	5	−6	8	−4
3	−6	15	−18	24	−12
−7	14	−35	42	−56	28
−9	18	−45	54	−72	36
10	−20	50	−60	80	−40

5   a −12   b −30   c 21   d −16
    e 36   f 27   g −42   h −25
    i −72   j 30   k 16   l 42
    m 36   n −168   o −100   p −135

6   a −8   b −2   c 9   d −8
    e 7   f −4   g −8   h −7
    i −9   j 30   k −40   l −7
    m 8   n −2   o 16   p 20

7   a 32   b −18   c 12   d 2

8 $1500

Exercise 3-03

1 B   **2** C

3   a ×   b ÷   c ÷
    d ÷   e ×   f ÷
    g ×   h ×   i ×

4   a 27   b 44   c 129
    d 37   e −42   f 85
    g −45   h 78   i 80

5   a −18   b −25   c 142
    d 17   e −12   f 32

6   a 5   b 9   c 122   d 286

7   a −36   b 15.6   c −5
    d 95.1   e −31   f 27.9
    g −236   h 2   i −4.05

8 $1685

Exercise 3-04

1 A   **2** D   **3** $\dfrac{5}{4}, \dfrac{12}{5}, \dfrac{9}{8}, \dfrac{8}{5}$

4   a $1\dfrac{1}{4}$ b $3\dfrac{1}{2}$ c $1\dfrac{3}{5}$ d $1\dfrac{5}{8}$
    e $2\dfrac{1}{3}$ f $2\dfrac{1}{7}$ g $2\dfrac{1}{2}$ h $1\dfrac{2}{9}$

5   a $\dfrac{3}{2}$ b $\dfrac{10}{3}$ c $\dfrac{22}{5}$ d $\dfrac{15}{8}$
    e $\dfrac{13}{5}$ f $\dfrac{19}{6}$ g $\dfrac{17}{9}$ h $\dfrac{14}{5}$

6 23 pieces

7   a T   b F   c T
    d T   e F

8   a $\dfrac{2}{5}$ b $\dfrac{2}{5}$ c $\dfrac{4}{5}$ d $\dfrac{3}{4}$
    e $\dfrac{5}{12}$ f $\dfrac{2}{5}$ g $\dfrac{21}{32}$ h $\dfrac{18}{25}$
    i $\dfrac{1}{2}$ j $\dfrac{3}{4}$

9   a 8 pieces   b $\dfrac{1}{3}$ c $\dfrac{2}{9}$

Exercise 3-05

1 C   **2** B

3   a T   b F   c T   d F

4   a 40   b 8   c 60   d 400
    e 10 mins   f 10 m   g $80   h 12 L
    i 80 cm   j 600g   k 15 h   l $450

5   a $640   b $\dfrac{1}{3}$

6   a $350   b $\dfrac{7}{12}$

7   a $9000   b $15 000   c $\dfrac{1}{3}$

8   a 3 ha   b 4 ha   c 5 ha

Exercise 3-06

1 C   **2** B

3   a T   b T   c F   d F
    e T   f F   g T   h T

4 $\dfrac{11}{70}, \dfrac{1}{5}, \dfrac{7}{10}, \dfrac{6}{7}$ **5** $\dfrac{5}{6}, \dfrac{3}{4}, \dfrac{3}{8}, \dfrac{1}{12}$

6   a <   b >   c =   d <
    e >   f <   g <   h =

7 Convert fractions to 12ths:
$\dfrac{1}{2} = \dfrac{6}{12}, \dfrac{3}{4} = \dfrac{9}{12}, \dfrac{1}{3} = \dfrac{4}{12}, \dfrac{5}{6} = \dfrac{10}{12}, \dfrac{7}{12}$

Number line from 0 to 1 marked: $\dfrac{1}{3}$, $\dfrac{1}{2}$, $\dfrac{7}{12}$, $\dfrac{3}{4}$, $\dfrac{5}{6}$

8 Her daughter.

Exercise 3-07

1 D   **2** B

3   a 3   b 8   c $\dfrac{3}{12} + \dfrac{8}{12} = \dfrac{11}{12}$
    d 9   e 4   f $\dfrac{9}{12} - \dfrac{4}{12} = \dfrac{5}{12}$

4   a $\dfrac{5}{6}$ b $\dfrac{9}{20}$ c $\dfrac{11}{15}$ d $\dfrac{5}{8}$
    e $\dfrac{49}{40} = 1\dfrac{9}{40}$ f $\dfrac{29}{28} = 1\dfrac{1}{28}$
    g $3\dfrac{4}{5}$ h $\dfrac{13}{8} = 1\dfrac{5}{8}$ i $5\dfrac{2}{3}$ j $1\dfrac{7}{12}$

5   a $\dfrac{1}{4}$ b $\dfrac{7}{15}$ c $\dfrac{3}{8}$ d $\dfrac{4}{9}$
    e $\dfrac{1}{6}$ f $\dfrac{5}{24}$ g $3\dfrac{2}{3}$ h $\dfrac{1}{12}$
    i $2\dfrac{1}{4}$ j $3\dfrac{1}{12}$

6   a $5\dfrac{13}{20}$ b $2\dfrac{1}{24}$ c $8\dfrac{9}{40}$
    d $3\dfrac{3}{8}$ e $2\dfrac{3}{4}$

7   a $\dfrac{2}{3}$ b $\dfrac{1}{3}$ c 2

ANSWERS

Exercise 3-08

1 C   2 D   3 $\dfrac{2\times1}{3\times5}=\dfrac{2}{15}$

4 a $\dfrac{1}{12}$ b $\dfrac{1}{2}$ c $\dfrac{3}{40}$ d $\dfrac{7}{32}$

   e $\dfrac{1}{2}$ f $\dfrac{10}{27}$ g 8 h $\dfrac{1}{6}$

   i $7\dfrac{1}{5}$ j $\dfrac{1}{6}$

5 $\dfrac{2}{3}\times\dfrac{5}{1}=\dfrac{10}{3}=3\dfrac{1}{3}$

6 a $\dfrac{1}{2}$ b $\dfrac{1}{2}$ c $1\dfrac{7}{8}$ d $\dfrac{1}{3}$

   e $\dfrac{1}{4}$ f $1\dfrac{7}{8}$ g $\dfrac{1}{18}$ h $1\dfrac{1}{2}$

   i $\dfrac{1}{20}$ j $\dfrac{7}{15}$

7 a $6\dfrac{4}{5}$ b $8\dfrac{7}{16}$ c $2\dfrac{11}{16}$

   d $1\dfrac{19}{22}$ e $9\dfrac{9}{32}$

8 a $\dfrac{1}{8}$ b $\$2100$ c $\$1050$

Language activity
FRACTION FRENZY

Practice test 3

1 2.19   2 $\dfrac{4}{5}$ 3 $15x+2y$

4 24cm^2 5 $4\dfrac{1}{2}$ 6 4.15

7 $\$6$ 8 $3x^2$

9 7 6 3 0 –2 –5

10 12.7

11 a –2   b –11   c –14

12 A

13 a 28   b –7   c –108

14 B

15 a 6   b 60   c –31

16 a $1\dfrac{1}{4}$ b $1\dfrac{5}{8}$ c $2\dfrac{7}{10}$

17 a $\dfrac{2}{3}$ b $\dfrac{4}{5}$ c $\dfrac{1}{3}$

18 a 30   b $\$90$ c 150 cm

19 a T   b F   c T

20 a $1\dfrac{4}{15}$ b $\dfrac{11}{24}$ c $2\dfrac{3}{4}$ d $4\dfrac{1}{8}$

21 a $\dfrac{2}{15}$ b $2\dfrac{1}{7}$ c $9\dfrac{3}{8}$ d $2\dfrac{2}{3}$

22 a 8   b $\dfrac{7}{16}$

CHAPTER 4

Exercise 4-01

1 C   2 A

3 a $a+b$ b $a-b$ c ab d $\dfrac{a}{b}$

4 a T   b F   c F   d T

5 a $3a-8$ b $12-2b$ c $2(m+n)$
   d $3(x-y)$ e $2n+7$ f w^2-5
   g $3y+w$ h $2(x+y)$ i $(mn)^2$
   j $24-g^2$ k $120+3d$ l m^2+2n
   m $12-3b$ n $h+2j$

6 a $5a$ cents   b $\$6q$ c $\$15d$
   d $\$\dfrac{r}{3}$ e $\$kw$ f $\$\dfrac{d}{12}$

7 a $4g$ b $m+2n$ c $2d+2w$

8 a g^2 b $\dfrac{1}{2}mh$ c dw

9 a $2m$ b $m-4$ c $2m+10$ d $m+8$

10 a $4x$ b $2y$ c $4x+2y$ d $x+y$

Exercise 4-02

1 A   2 D

3 a F   b F   c F   d T

4 a –20   b 19   c 50   d –24
   e –3   f 24   g –62   h 150

5

	$3m-n$	$2m^2$	$\dfrac{4m}{n}$	$3(2m+n)$	$6mn^2$
$m=2,$ $n=4$	2	8	2	24	192
$m=-1,$ $n=3$	–6	2	$-\dfrac{4}{3}$	3	–54
$m=5,$ $n=-2$	17	50	–10	24	120
$m=-1,$ $n=-4$	1	2	1	–18	–96
$m=3,$ $n=-2$	11	18	–6	12	72

6 a 12.6   b 48   c 45.92
   d 90.3   e 13   f 2.4

7 a $y=3x-1$

x	–1	0	1	2
y	–4	–1	2	5

   b $y=12-x$

x	–1	0	1	2
y	13	12	11	10

ISBN 9780170351027

c $y = 2x^2$

x	–2	0	2	4
y	8	0	8	32

d $y = \dfrac{2x}{3} + 1$

x	–3	0	3	6
y	–1	1	3	5

8 311.6 cm³

Exercise 4-03

1 B   **2** D

3 **a** $3x, 2x$ **b** $4a. a$ **c** $5w, –2w$
    d $2ab, 3ba$ **e** $2ab, 2ba$ **f** $–6m, 18m$

4 **a** F **b** T **c** F

5 **a** $3a$ **b** $6x$ **c** $6w – 3$
    d $7m$ **e** $2y$ **f** $–2n +7$
    g $–2a$ **h** $15m$ **i** $–3t$

6 **a** $4m – 2n – 6m + n = 4m – 6m – 2n + n$
                     $= – 2m – n$

    b $8x + 5y – 2y – x = 8x – x + 5y – 2y$
                       $= 7x + 3y$

7 **a** $7x – y$ **b** $6x – 6y$ **c** $11m – 3n$
    d $10m – n$ **e** $3x + 4y$ **f** $9a – 5b$
    g $15r – 8s$ **h** $–2ab + 6$ **i** $–xy – 7y + 7$

8 **a** $9x + 1$ **b** $24x + 12$ **c** $6x – 6$

9 All the terms are unlike and cannot be added or subtracted.

Exercise 4-04

1 D   **2** B

3 **a** F **b** T **c** T **d** T

4 **a** $5a \times 3b = 5 \times 3 \times a \times b$
                $= 15ab$

    b $–6x \times 8y = –6 \times 8 \times x \times y$
                 $= –48 xy$

    c $5a^2 \times 2a = 5 \times 2 \times a^2 \times a$
                $= 10\,a^3$

    d $–4mn \times (–7m^2) = –4 \times (–7) \times mn \times m^2$
                         $= 28\ m^3n$

5 **a** $16m$ **b** $12xy$ **c** $–18mn$ **d** $–20ab$
    e $–12cd$ **f** $10wv$ **g** $27ab$ **h** $6x^2y$
    i $60mn$ **j** $24a^2b^2$ **k** $12a^3$ **l** $15m^4$
    m $6w^3$ **n** $–20xy^3$ **o** $–24ab^3$ **p** $18x^3y^2$
    q $–32r^4s^2$ **r** $–72w^3$ **s** $21m^4$ **t** $24a^2b^2$

6 **a** $9x^2$ **b** $12bh$ **c** $27vw$

Exercise 4-05

1 A   **2** D

3 **a** F **b** T **c** T **d** F

4 **a** $4a$ **b** $–3b$ **c** $2d$ **d** $–3ab$
    e $–3bc$ **f** $–2xy$ **g** $–4m$ **h** $\dfrac{9a}{c}$

i $–\dfrac{4v}{x}$ **j** $–2r$ **k** $–\dfrac{6x}{y}$ **l** $–\dfrac{12w}{y}$

5 **a** $\dfrac{4x^2}{y}$ **b** $\dfrac{5a^3}{b^2}$ **c** $\dfrac{4m^3}{5n^2}$

6 **a** $2ab^4$ **b** $4x^2y$ **c** $–\dfrac{5m^2}{n^3}$

7 **a** $3x \times 4$ **b** $14a \div 7$ **c** $–24ab \div (–6b)$

8 **a** $12 – 12x$ **b** $6 + 2a$ **c** $8a^2$

9 **a** $18 + 12x$ **b** $13m$ **c** $6a$ **d** $–18y^2$
    e $6m^2$ **f** $24 + 4y$ **g** $18m$ **h** 0

Exercise 4-06

1 B   **2** C

3 **a** $4(3x – 2) = 4 \times 3x + 4 \times (–2)$
                $= 12x + (–8)$
                $= 12x – 8$

    b $–6(2a – 1) = –6 \times 2a + (–6) \times (–1)$
                  $= –12a + 6$

    c $a(3a + 2) = a \times 3a + a \times 2$
                 $= 3a^2 + 2a$

    d $–2x(3x – 1) = –2x \times 3x + (–2x) \times (–1)$
                   $= –6x^2 + 2x$

4 **a** $5x + 10$ **b** $3a – 12$
    c $8a – 4$ **d** $24x + 16$
    e $24 – 12a$ **f** $2a^2 – 3a$
    g $3x^2 + 4x$ **h** $2m^2 – 12m$
    i $6w^2 + 15w$ **j** $12a^2 – 24a$
    k $–8a – 6$ **l** $–40 + 15m$
    m $–2x + 8$ **n** $–3a – 9$
    o $–12w + 48$

5 **a** Both sides equal 45.
    b Both sides should still be equal.

6 **a** $3(2y – 1) + 5y = 6y – 3 + 5y$
                     $= 11y – 3$

    b $12 + 4(2m – 1) = 12 + 8m – 4$
                     $= 8m + 8$

    c $2(3x – 4) – 3(x + 4) = 6x – 8 – 3x – 12$
                     $= 3x – 20$

    d $5(2a – 2) + 7(3a + 1) = 10a – 10 + 21a + 7$
                     $= 31a – 3$

7 **a** $5x – 23$ **b** $3x + 9$ **c** $a + 34$
    d $3x + 10$ **e** $5x – 9$ **f** $4a + 54$
    g $–2x + 32$ **h** $x – 37$ **i** $–a + 42$

8 $14w – 28v – 6v – 15w = – w – 34v$

Exercise 4-07

1 C   **2** B

3 **a** The factors of 8 are: 1, 2, 4, 8.
      The factors of 20 are: 1, 2, 4, 5, 10, 20.
      The HCF of 8 and 20 is 4.

    b The factors of 16 are: 1, 2, 4, 8, 16.
      The factors of 40 are: 1, 2, 4, 5, 8, 10, 20, 40.
      The HCF of 16 and 40 is 8.

4 **a** 2 **b** 3 **c** 4 **d** 15
    e 2 **f** 6 **g** 4 **h** 9

5 Any 4 of or combination of:

a $1, 3, 9, a, b, ab$ b $1, 2, 7, 14, m, n, mn$

c $1, 2, 4, 8, x, x^2$ d $1, 2, 3, 4, 6, 12, b, c, bc$

e $1, 2, 4, m, m^2$ f $1, 2, 4, 8, 16, b, b^2$

g $1, 2, 4, 5, 10, 20, a, a^2, a^3$

h $1, 2, 3, 6, m, m^2, m^3, m^4$

6 $16, xy, 16 \times xy = 16xy$

7 a $4a$ b 2 c 2 d 3

e $8m$ f $4a$ g $2w$ h $2m$

i bc j mn k $3s$ l $6uv$

8 $4bc$

Exercise **4–08**

1 D   **2** B

3 a $3a + 3b = 3(a + b)$ b $5x − 10y = 5(x − 2y)$

c $6m + 18n = 6(m + 3n)$ d $4b^2 − 4 = 4(b^2 − 1)$

e $9a + 3b = 3(3a + b)$ f $15x − 12y = 3(5x − 4y)$

g $7u + 14v = 7(u + 2v)$ h $8n^2 − 2 = 2(4n^2 − 1)$

4 a $2(a + 6)$ b $6(p − 3)$

c $5(n + 4)$ d $8(y − 8)$

e $12(a + 5)$ f $16(x − 3)$

g $18(a + 3)$ h $14(c − 2)$

i $15(a + 5)$ j $5(3x − y)$

k $a(a + b)$ l $y(x + z)$

m $n(m + n)$ n $9(p − 3q)$

o $xy(z − 3)$

5 a $3a^2 + 6ab = 3a(a + 2b)$

b $15xy − 10y^2 = 5y (3x − 2y)$

c $16mno + 24noq = 8no(2m + 3q)$

d $48b^2 − 16bc = 16b(3b − c)$

e $27ab + 18bc = 9b(3a + 2c)$

f $25x^2 − 15xy = 5x(5x − 3y)$

6 a $4p(a − 3q)$ b $4s(4r + 3t)$

c $7y(4x − 6y)$ d $32bc(2a − d)$

e $8m(m + 6n)$ f $11b(2b − a)$

g $18g(2f + g)$ h $x(yz − 7uv)$

i $7b(2a^2 − 3bc)$ j $9w(6w + uv)$

k $8t(3r^2 − 2v^2)$ l $25mn(3n + 2m)$

7 $7bc(6a^2 − 2b + 4a)$

Exercise **4–09**

1 C   **2** B

3 a $−5$ b $−4$ c $−6$ d $−5$

e $−5$ f $−4x$ g $−6n$ h $−5v$

4 a $−4a − 4b = −4(a + b)$

b $−5x − 10y = −5(x + 2y)$

c $−16m − 8n = −8(2m + n)$

d $−4b^2 − 4 = −4(b^2 + 1)$

e $−9a + 3b = −3(3a − b)$

f $−35x − 7y = −7(5x + y)$

g $−7u + 14v = −7(u − 2v)$

h $−8n^2 + 2 = −2(4n^2 − 1)$

5 a $−2(a + 9)$ b $−6(p + 4)$

c $−4(n + 5)$ d $−7(y + 7)$

e $−12(a + 5)$ f $−16(x − 2)$

g $−18(a − 2)$ h $−4(c + 7)$

i $−15(a − 5)$ j $−25(x + 2y)$

k $−a(a + c)$ l $−y(x − z)$

m $−n(m − n)$ n $−3(p + 9q)$

o $−xy(z − 5)$

6 a $−3a^2 − 6ab = −3a(a + 2b)$

b $−5xy − 20y^2 = −5y(x + 4y)$

c $−8mno + 24noq = −8no(m − 3q)$

d $−32b^2 − 16bc = −16b(2b + c)$

e $−9ab + 18bc = −9b(a − 2c)$

f $−50x^2 − 25xy = −25x(2x + y)$

7 a $−3p(a + 4q)$ b $−4s(2r + 3t)$

c $−7y(x + 6y)$ d $−16bc(a − 2d)$

e $−6m(m − 8n)$ f $−11b(3b + a)$

g $−6g(f − 3g)$ h $−x(yz + 7uv)$

i $−7b(a^2 + 3bc)$ j $−3w(2w − 3uv)$

k $−8t(r^2 + 2v^2)$ l $−15mn(n − 3m)$

8 $−12bc(2a^2 + b − 3a)$

Language activity

1 C   **2** S   **3** G   **4** R

5 A   **6** L   **7** B   **8** I

9 T   **10** E   **11** N

ABSTRACT ALGEBRA ANTICS

Practice test **4**

1 150   **2** 180^0 **3** $3b$ **4** 96 cm^3

5 \$30   **6** \$35.15   **7** 5   **8** 2

9 9.9   9.909   9.91   9.95   **10** 27, 243

11 a $2a − b$ b $12xy$

c $3(m + n)$ d $\dfrac{7a}{3b}$

12 D

13 a $−2.4$ b 29.2 c 53.8

14 a $2v − 11w$ b $13ab − 4b$

c $16mn − m^2$

15 a $6ab$ b $−24x^3y$ c $72uv^2w$

16 $21xy$

17 a $−4a$ b $6x$ c $−\dfrac{b}{2a}$

18 a F   b T   c F

19 a $5m − 14$ b $−3a^2 − 7a$

20 a $9m$ b $4x$ c $16mn$

22 a $6(a − 2b)$ b $2x(7 − 8y)$

c $2u(3u − 14v)$ d $7ab(4b − a)$

23 a $−6(m + 4n)$ b $−8x(3 − 2y)$

c $−15u(u + 3vw)$ d $−8b(6bc + a^2)$

CHAPTER 5

Exercise **5–01**

1 C   **2** B

3 a P   b Q   c r d p

4 a

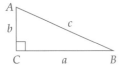

b

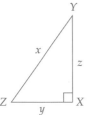

c

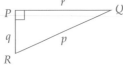

5 a c **b** x **c** p

6 a hypotenuse = 13, opposite = 12, adjacent = 5
b hypotenuse = 10, opposite = 8, adjacent = 6
c hypotenuse = 17, opposite = 15, adjacent = 8
d hypotenuse = q, opposite = r, adjacent = p
e hypotenuse = y, opposite = x, adjacent = z
f hypotenuse = u, opposite = v, adjacent = w

7

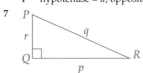

Exercise **5-02**

1 C   **2** B

3 $\sin A = \dfrac{\text{opposite}}{\text{hypotenuse}}$ $\cos A = \dfrac{\text{adjacent}}{\text{hypotenuse}}$

$\tan A = \dfrac{\text{opposite}}{\text{adjacent}}$

4 a $\dfrac{8}{10}, \dfrac{6}{10}$ and $\dfrac{8}{6}$

b $\dfrac{6}{10}, \dfrac{8}{10}$ and $\dfrac{6}{8}$

5 a i $\dfrac{b}{c}$ **ii** $\dfrac{r}{q}$

b i $\dfrac{a}{c}$ **ii** $\dfrac{p}{q}$

c i $\dfrac{b}{a}$ **ii** $\dfrac{r}{p}$

6 a $\sin F = \dfrac{9}{15}$ **b** $\cos F = \dfrac{12}{15}$

c $\tan F = \dfrac{9}{12}$ **d** $\sin D = \dfrac{12}{15}$

e $\cos D = \dfrac{9}{15}$ **f** $\tan D = \dfrac{12}{9}$

7

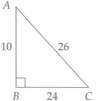

Exercise **5-03**

1 $\tan 60° \approx 1.73$ **2** $\tan 23° \approx 0.42$
3 $\tan 71° \approx 2.90$ **4** $\tan 45° = 1$
5 $\tan 38° \approx 0.78$

Exercise **5-04**

1 C   **2** B
3 a 1.111   **b** 0.743   **c** 0.669
4 a 42° The 3 angles of a triangle add up to 180°.
b 0.900, 0.669, 0.743
5 a 0.866   **b** 1.732   **c** 0.707   **d** 0.643
e 0.990   **f** 1.036   **g** 0.891   **h** 0.934
i 3.732   **j** 0.970
6 a 0.6157   **b** 2.0503   **c** 0.1045   **d** 0.4663
e 0.1219   **f** 0.3249   **g** 0.7193   **h** 1.0000
i 1.0000   **j** 1.6003
7 a 0.8   **b** 0.9   **c** 5.7   **d** 0.4
e 0.6   **f** 0.9   **g** 3.5   **h** 0.5
i 0.3   **j** 0.0
8 a 39°   **b** 49°   **c** 83°   **d** 73°
e 11°   **f** 39°   **g** 44°   **h** 68°
i 55°
9 a 43°   **b** 58°   **c** 26°   **d** 60°
e 58°   **f** 15°   **g** 75°   **h** 27°
i 37°   **j** 90°   **k** 44°   **l** 90°
m 37°   **n** 0°   **o** 37°   **p** 30°
q 45°   **r** 60°
10 42°

Exercise **5-05**

1 C   **2** C
3 $\tan 28° = \dfrac{m}{5}$
$5 \times \tan 28° = m$
$m = 2.658547158$
$m = 2.66$
4 a $c = 11.52$ **b** $q = 7.55$
c $x = 5.46$
5 11.62
6 a $a = 3.7$ **b** $r = 15.8$ **c** $y = 19.7$
7 13.86 m   **8** 156.9 m

Exercise **5-06**

1 A   **2** D
3 a F   **b** T   **c** T   **d** F
4 a 33°   **b** 71°   **c** 37°
5 a 62°   **b** 28°   **c** 80°

6 a

$$\tan P = \frac{5}{12}, P = 23°$$

7 67°   **8** 64°

Exercise 5-07

1 B   **2** A

3 $\tan 35° = \dfrac{x}{15.8}$

   $15.8 \tan 35° = x$

   $x = 11.0632791$

   $x = 11.1$ m   (1 DP)

4 a 11.49 **b** 2.07 **c** 9.37 **d** 15.65

  e 3.40 **f** 0.84 **g** 3.56 **h** 7.59

5 2.11 m   **6** 63.377 m

Exercise 5-08

1 A   **2** C   **3** $c = \dfrac{8}{\cos 47°}$

4 a sin 46° **b** cos 32° **c** tan 16°

5 a 11.12 m **b** 5.54 m **c** 19.53 cm

6 $\tan 53° = \dfrac{8.3}{d}$

   $d = \dfrac{8.3}{\tan 53°}$

   $d = 6.254\,498\,616$

   $d \approx 6.3$ cm

7 a 12.66 m **b** 11.12 m

8 No

Exercise 5-09

1 C   **2** B

3 $\sin R = \dfrac{7}{11}$

   $R = 39.52119636$

   $R = 40°$ (nearest degree)

4 a 32° **b** 58° **c** 52°

5 a 41° **b** 60° **c** 51°

6 a

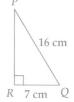

  $\cos Q = \dfrac{7}{16}$ $Q = 64°$

  b 26°

7 21°   **8** 58°

Language activity

triangles trigonometry label vertices small
cosine adjacent opposite hypotenuse CAH
calculator degrees

Practice test 5

1 $102   **2** $-y - 3$ **3** $x = 72$ **4** 60 m

5 $1\dfrac{1}{6}$ **6** 4   **7** 13.5   **8** yes

9 $4x^2 - 6x$ **10** $\dfrac{1}{2}$

11 hypotenuse = 10, opposite = 6, adjacent = 8

12 B

13 a 0.391 **b** 6.314 **c** 0.588

14 a 81° **b** 51° **c** 69°

15 26.69 cm **16** 14.8 cm **17** 27°

18 a 16.1 cm **b** 23.2 m

19 a 8.5 m **b** 4.3 m

20 a 65° **b** 75° **c** 26°

21 67°

CHAPTER 6

Exercise 6-01

1 D   **2** B

3 a $\dfrac{1}{100}$ **b** $\dfrac{3}{100}$

  c $\dfrac{10}{100} = \dfrac{1}{10}$ **d** $\dfrac{15}{100} = \dfrac{3}{20}$

  e $\dfrac{20}{100} = \dfrac{1}{5}$ **f** $\dfrac{50}{100} = \dfrac{1}{2}$

  g $\dfrac{62}{100} = \dfrac{31}{50}$ **h** $\dfrac{34}{100} = \dfrac{17}{50}$

  i $\dfrac{60}{100} = \dfrac{3}{5}$ **j** $\dfrac{40}{100} = \dfrac{2}{5}$

  k $\dfrac{12}{100} = \dfrac{3}{25}$ **l** $\dfrac{24}{100} = \dfrac{6}{25}$

  m $\dfrac{75}{100} = \dfrac{3}{4}$ **n** $\dfrac{82}{100} = \dfrac{41}{50}$

  o $\dfrac{68}{100} = \dfrac{17}{25}$

4 a 0.7 **b** 0.9 **c** 1.4 **d** 0.18

  e 0.35 **f** 0.06 **g** 0.24 **h** 0.96

  i 1.08 **j** 2.43 **k** 0.08 **l** 0.075

  m 2.2 **n** 0.04 **o** 0.44 **p** 0.12

  q 0.125 **r** 0.145 **s** 0.125 **t** 0.025

5 a 40% **b** 30% **c** 87.5% **d** 25%

  e 5% **f** 80% **g** 7.5% **h** 31.25%

  i 48% **j** 42.5% **k** 162.5% **l** 525%

  m 135% **n** 91.7% **o** 112.5%

6 a 2% **b** 6% **c** 5% **d** 9%

  e 8% **f** 17% **g** 13% **h** 31%

  i 45% **j** 47% **k** 92% **l** 30%

  m 70% **n** 450% **o** 112.5%

7 Fraction: $\dfrac{1}{10}, \dfrac{1}{4}, \dfrac{3}{10}, \dfrac{2}{5}, \dfrac{1}{2}, \dfrac{3}{5}, \dfrac{3}{4}, \dfrac{4}{5}, \dfrac{9}{10}, 1$

Decimal: 0.1, 0.25, 0.3, 0.4, 0.5, 0.6, 0.75, 0.8, 0.9, 1.0

Percentage: 10%, 25%, 30%, 40%, 50%, 60%, 75%, 80%, 90%, 100%

8 **a** $\dfrac{3}{5}$, 0.655, 68.5% **b** $\dfrac{3}{8}$, 0.38, 38.2%

  c 0.805, 82.5%, $\dfrac{5}{6}$

Exercise 6-02

1 C **2** B

3 **a** $\dfrac{1}{4}$ **b** $\dfrac{1}{8}$ **c** $\dfrac{4}{5}$ **d** $\dfrac{1}{2}$

  e $\dfrac{1}{3}$ **f** 1 **g** $\dfrac{1}{10}$ **h** $\dfrac{2}{3}$

  i $\dfrac{3}{4}$ **j** $\dfrac{2}{5}$

4 **a** $10\% \text{ of } \$200 = \dfrac{1}{10} \times 200$
$$= \$20$$

  b $75\% \text{ of } 80 \text{ cm} = \dfrac{3}{4} \times 80$
$$= 60 \text{ cm}$$

5 **a** 70c **b** 6 mm **c** 21c **d** $1.80
  e $32 **f** 25 km **g** $105 **h** $33
  i 63 km **j** $60 **k** 60c **l** $30
  m 30c **n** $28 **o** $2.80 **p** $720
  q 6 years **r** 36 km

6 **a** $12.50 **b** 1.62 m
  c 6 m **d** 0.2 year
  e 64.8 cm **f** 5.85 yr
  g 4.88 L **h** 3.6 h
  i 840 m **j** $652.80
  k 2.375 days **l** $2400

7 **a** 25 200 **b** 30 800

8 225 L **9** $1408

Exercise 6-03

1 D **2** C

3 **a** $\dfrac{12}{30} = \dfrac{2}{5}$ **b** $\dfrac{75}{120} = \dfrac{5}{8}$

4 **a** $\dfrac{2}{5}$ **b** $\dfrac{1}{4}$ **c** $\dfrac{9}{50}$

  d $\dfrac{1}{5}$ **e** $\dfrac{11}{250}$ **f** $\dfrac{1}{8}$

  g $\dfrac{1}{3}$ **h** $\dfrac{3}{500}$ **i** $\dfrac{1}{8}$

5 **a** 40% **b** 25% **c** 18%
  d 20% **e** 4.4% **f** 12.5%

  g $33\dfrac{1}{3}\%$ **h** 0.6% **i** 12.5%

6 **a** 20, 75 **b** $27, 33\dfrac{1}{3}$

7 **a** 25% **b** 12.5% **c** 12.5% **d** 20%
  e 5% **f** 5% **g** 20% **h** 2.5%
  i 8.3% **j** 25% **k** 25% **l** 20%

8 **a** 68% **b** 90% **c** 80%, Gina

9 35%

10 **a** 48% **b** $\dfrac{1}{5}$ **c** 32%

Exercise 6-04

1 D **2** A **3** add, subtract

4 **a** $55 **b** 910 m **c** $6250
  d $34 000 **e** 4125 kg **f** 90 t

5 **a** $170 **b** 2000 m **c** $48 000
  d $250 000 **e** 240 ha **f** 4000 kg

6 **a** $338 **b** $552.50 **c** $832

7 $700 **8** $721.60.

9 $16 675 **10** $115.20, No

Exercise 6-05

1 D **2** B

3 **a** T **b** F **c** F

4 **a** 15% of an amount = $450
  1% of the amount = $450 ÷ 15
$$= \$30$$
  100% of the amount = $30 × 100
$$= \$3000$$

  b 60% of an amount = $960
  1% of the amount = $960 ÷ 60
$$= \$16$$
  100% of the amount = $16 × 100
$$= \$1600$$

5 **a** $1600 **b** $400 **c** $800
  d 2000 m **e** 2000L **f** 30 m
  g 8 h **h** $20 000 **i** 3000 kg

6 440 **7** $6000 **8** 3500

9 **a** $1148.95
  b Find 28.6% of it. It should be the tax paid.

10 **a** $160 **b** $96

Exercise 6-06

1 A **2** B

3

Cost price	Selling price	Profit or loss
$12.00	$15.00	$3.00 profit
$25.00	$30.00	$5.00 profit
$60.00	$20.00	$40.00 loss
$250.00	$200.00	$50.00 loss
$735.00	$780.00	$45.00 profit
$1590.90	$1470.10	$120.80 loss
$424.75	$450.50	$25.75 profit
$6550.80	$6250.35	$300.45 loss

4 a $5 b $33\frac{1}{3}$% c 25%

5 a $50 500 b 9%

6 a 37.5% b 27.27%

7 $66.70 8 $535

9 a $15.80, $173.80 b $4.50, $49.50
   c $8.40, $92.40 d $12, $132
   e $3.60, 39.60 f $2.80, $30.80
   g $5.20, $57.20 h $2.60, $28.60

10 a $104.28 b $29.70 c $55.44 d $79.20
    e $23.76 f $18.48 g 34.32 h $17.16

Exercise **6-07**

1 D 2 A

3 a $250 b $2500

4 a $P = \$450, R = 5\% = 0.05$ as a decimal,
      $N = 2$ years
      $I = PRN$
      $= 450 \times 0.05 \times 2$
      $= \$ 45$

5 a $60 b $180 c $162.50
   d $502.50 e $558.90 f $599.69

6 $P = \$2500, R = 9\% = 0.09$ as a decimal
   p.a. $= 0.0075$ as a decimal per month, $N = 8$ months
   $I = PRN$
   $= 2500 \times 0.0075 \times 8$
   $= \$ 150$

7 a $45 b $180 c $4080
   d $200 e $3000 f $882

8 a $910 b $2310

9 4 years 10 1.1%

Practice test **6**

1 $6 2 $6xy$ 3 168 4 8

5 $4\frac{1}{2}$ 6 7, 6, 3, 0, −2, −5

7 4.15 8 12

9 $-12x - 30$ 10 5h 45 min

11 a $\frac{3}{25}$, 0.12 b $\frac{7}{10}$, 0.7 c $\frac{13}{20}$, 0.65

12 C 13 D

14 a 6m b 3 h c 48L

15 a $\frac{13}{15}$ b $\frac{1}{5}$ c $\frac{1}{400}$

16 a $86\frac{2}{3}$% b 20% c 0.25%

17 $300 18 $750 19 $1009.73

20 a $300 b 31.6%

21 $28 146.80

22 a $90 b $238

23 5%

CHAPTER 7

Exercise **7-01**

1 D 2 C

3 a $5 \times 5 \times 5$ b 5 c 3

4 a 3^4 b 2^5 c 7^2 d 8^3
   e x^5 f $5^2 \times a^3$ g $3^2 \times 6^3$ h wv^3
   i $5^3 \times a^2$ j m^2n^3 k 4^2u^2v l $15mn$

5 a 3^7 b 5^8 c 2^{11} d 6^3
   e m^7 f x^5 g p^7 h n^{11}

6 a $3 \times 4 \times m^3 \times m^2$ b $5 \times (-3) \, w^4 \times w^7$
      $= 12m^5$ $= -15w^{11}$

7 a $10n^7$ b $24a^7$ c $60n^7$ d $48w^8$
   e $-12v^7$ f $-63n^7$ g $45a^7$ h $-64a^7$
   i $-72c^{10}$ j $48a^7b^4$ k $-32m^7n^6$ l $24a^7c^5$

8 a F b T c F d T

9 No, 1296

10 a 576 b 324 c 1125 d 2048
    e 1600 f 1764 g −2000 h 288

Exercise **7-02**

1 C 2 A

3 a $2^{8-3} = 2^5$ b $x^{9-3} = x^6$ c $12^{6-3} = 12^3$

4 a 2^3 b 4^5 c 3^7 d 3^5
   e d^5 f m g x^4 h m^3

5 a $3n^4$ b $4a^8$ c $3n$ d $2w^{10}$
   e $-3v^4$ f $-7n^5$ g $5a^3$ h $-4a^7$
   i $-4c$ j $8a^4b^3$ k $-4m^3n$ l $8a^4c^3$

6 a F b F c T d T

7 No, $\frac{1024}{27}$

8 a 202.27 b 157464 c 38.04 d 32768

Exercise **7-03**

1 C 2 B

3 a $5^{3 \times 2} = 5^6$ b $x^{4 \times 3} = x^{12}$

4 a 2^{15} b 6^8 c 3^{12} d 5^{18}
   e 7^{40} f x^{12} g n^{18} h m^{16}
   i w^{20} j a^{18} k 9^{15} l x^8
   m 5^{24} n q^{24} o p^{56}

5 a F b T c T

6 a $4x^6$ b $27a^{12}$ c $25n^{12}$ d $64m^{24}$
   e $343c^{15}$ f $32w^{40}$ g $216b^{12}$ h $81t^8$
   i $-32a^{30}$ j $64w^{24}$ k $-27c^{21}$ l $64q^{54}$

7 a T b F c F

8 a $\frac{a^{20}}{32}$ b $\frac{m^4}{9}$ c $\frac{9}{w^{12}}$ d $\frac{25}{m^{12}}$
   e $\frac{x^{24}}{64}$ f $\frac{w^{21}}{-8}$ g $\frac{64}{c^{24}}$ h $\frac{81}{n^{24}}$

9 a T b T

10 a $\frac{81x^{24}}{16}$ b $\frac{m^{24}}{n^{12}}$ c $\frac{729}{8a^{15}}$ d $\frac{a^{40}}{b^{60}}$

Exercise 7-04

1 B 2 C

3 a T b F c F d T

4 a 1 b 1 c 1 d 1
  e 1 f 1 g 1 h 2
  i 1 j 6 k 1 l 6

5 a $\dfrac{1}{4^3}$ b $\dfrac{1}{w^5}$ c $\dfrac{4}{n^3}$

6 a $\dfrac{1}{3^2}$ b $\dfrac{1}{2^4}$ c $\dfrac{1}{4}$

  d $\dfrac{1}{5^2}$ e $\dfrac{1}{3^4}$ f $\dfrac{1}{7^2}$

7 a 1 b $\dfrac{1}{243}$ c $\dfrac{1}{16}$

  d $\dfrac{1}{6}$ e 1 f $\dfrac{1}{128}$

8 a F b T c F
  d F e F f T

9 a $\dfrac{3}{x^4}$ b $\dfrac{6}{a^5}$ c $\dfrac{4}{m^7}$ d $\dfrac{2}{w^3}$

  e $\dfrac{1}{3x^2}$ f $\dfrac{1}{4n^3}$ g $\dfrac{2}{3x^4}$ h $\dfrac{3}{5a^6}$

  i $\dfrac{3}{2u^5}$ j $\dfrac{5}{w^6}$

10 a $1\dfrac{1}{2}$ b $\dfrac{8}{9}$ c $1\dfrac{3}{4}$

  d $-\dfrac{7}{10}$ e $\dfrac{1}{2}$

Exercise 7-05

1 B 2 C

3 a F b T c F d T

4 a 4^2 b 3^3 c 7^{-5} d 5^{-4}
  e w f c^3 g x^{-7} h a^{-5}

5 a 6^6 b 7^{-2} c 5^6 d 2^{-1}
  e x^{-2} f c^9 g w^{-1} h a^{-3}

6 a 2^8 b 3^6 c 4^9 d 5^8
  e 3^{-4} f 2^{-12} g 4^{-6} h 1
  i x^9 j a^{12} k w^{48} l p^{24}
  m a^{-6} n x^{-14} o 1 p x^{12}

7 a $6x^9$ b $4a^2$ c $16m^{12}$ d $12a^3$
  e $5m^8$ f $16a^{10}$ g $81x^{-12}$ h $2w^{-4}$
  i $-32a^6b^5$ j $-3m^{11}n$ k $-40w^{12}$ l $-7xy^6$

8 a T b T c F d F
  e T f F g T h F

Exercise 7-06

1 D 2 B

3 a 23 600 b 587 000
  c 81 600 d 3570
  e 4 230 000 f 45 600
  g 32 500 000 h 428 000
  i 408 000 j 12 100 000
  k 60 300 000 l 2 100 000

4 a 0.035 b 0.0038 c 0.83 d 2.4
  e 0.0031 f 0.072 g 0.000 91 h 0.70
  i 4.8 j 0.051 k 56 l 0.71

5 a 70 000 b 69 000 c 68 700

6 0.099

7 a 57 464 b 57 460

8 0.36

Exercise 7-07

1 D 2 A

3 a 3.3×10^4 b 6×10^5
  c 2.8×10^3 d 8.24×10^5
  e 9.02×10^7 f 2.6×10^8

4 a 1.2×10^4 b 4.56×10^5
  c 1.2×10^8 d 4.5×10^7
  e 6.803×10^6 f 3.45×10^7
  g 8×10^{10} h 5.2×10^8

5 a 72 000 b 824 000
  c 410 000 000 d 603
  e 1 980 000 f 3728

6 a 50 000 b 3800
  c 1 680 000 d 24 500 000
  e 4000 f 92
  g 300 400 000 h 1 090 000 000

7 a 7×10^4 2.8×10^6 6.11×10^6 2.006×10^7
  b 2.006×10^7 6.11×10^6 2.8×10^6 7×10^4

8 a $4.568\ 21 \times 10^5$ b 6.8×10^4
  c 2.29×10^7 d 1.52×10^8

9 a 1.23×10^3 b 1.44×10^4
  c 7.392×10^3 d 9.49×10^6
  e 2.52×10^4

10 1×10^6 or $1.0485\ 76 \times 10^6$

Exercise 7-08

1 C 2 A

3 a 5×10^{-3} b 1.2×10^{-3}
  c 6.83×10^{-4} d 7×10^{-5}
  e 4.2×10^{-5} f 1.8×10^{-7}

4 a 9×10^{-3} b 1.2×10^{-4}
  c 9.72×10^{-5} d 4×10^{-7}
  e 5×10^{-2} f 7.508×10^{-4}
  g 7.214×10^{-4} h 8×10^{-4}

5 a 0.008 b 0.000 624
  c 0.005 01 d 0.000 000 72
  e 0.942 f 0.000 028 34

6 a 0.012 b 0.0035
  c 0.000 06 d 0.000 482
  e 0.9 f 0.000 007 6
  g 0.000 010 82 h 0.000 000 004 06

7 a 6.1×10^{-4} 5×10^{-5} 1.8×10^{-5} 3.129×10^{-7}
  b 3.129×10^{-7} 1.8×10^{-5} 5×10^{-5} 6.1×10^{-4}

8 a 8×10^{-5} b 7×10^{-3} c 4.8×10^{-6}
  d 1.5×10^{-3} e 1×10^{-12}

9 1 micrometre = 1×10^{-6} metre (one-millionth)
1 nanometre = 1×10^{-9} metre (one-billionth)

Exercise **7-09**

1 B   **2** C

3 a 1800   **b** 2 400 000
c 7 000 000 000   **d** 2 00 560 000
e 0.42   **f** 8.9
g 50 000 000   **h** 0.030 09

4 a 3.2×10^{13} **b** 2.44×10^{15}
c 6.83×10^2 or 683   **d** 5.17×10^6
e 5.25×10^5 **f** 7.54×10^{-6}
g 3.07×10^{10} **h** 1.38×10^{-3} or 0.001 38
i 3.59×10^{-7} **j** 1.11×10^{19}

5 a 2.2×10^4 **b** 4.5×10^9
c 1.3×10^{15} **d** 6.3×10^{21}
e 2.0×10^{20} **f** 2.7×10^{-12}
g 1.9×10^{14} **h** 3.3×10^{-15}

6 a 4.2×10^4 **b** 5.6×10^4
c 8×10^3 **d** 8.6×10^4
e 5.7×10^5 **f** 8.9×10^3
g 7.3×10^{-2} **h** 3.5×10^{-1}

7 a 1.156×10^9 **b** 3.144×10^{17}
c 2.073×10^3 **d** 1.817×10^3

8 a Venus by 1476 times
b Jupiter by 3 times
c Neptune by 17.25 times
d Uranus by 1.3 times
e Mercury, Mars, Uranus, Venus, Earth, Neptune, Saturn, Jupiter

Language activity

An impatient index

Practice test **7**

1 $51   **2** 14   **3** −1
4

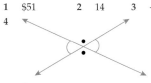

5 $\dfrac{1}{2}$ **6** 30%   **7** 7:25 pm

8 $3pq(9p - 5q)$ **9** 82.69   **10** $\dfrac{1}{4}$

11 a a^{13} **b** 3^{11} **c** $12x^9$ **d** $15n^7$
12 a $5n^8$ **b** $6x^{12}$ **c** $4x^4$
13 a 2^{20} **b** m^{18} **c** $81x^{12}$ **d** $\dfrac{2^{15}}{n^5}$
14 a 1   **b** 1   **c** 5
15 a $\dfrac{1}{7^2}$ **b** $\dfrac{4}{a^3}$ **c** $\dfrac{1}{2x^4}$
16 a $-10a^3b^3$ **b** $4m^3n^3$ **c** $36a^6$
17 a 522 000   **b** 6 040 000
c 0.00209

18 a 3.872×10^5 **b** 6.72×10^6
c 1.9×10^{10}
19 a 1.86×10^{-2} **b** 4.05×10^{-4}
c 4×10^{-9}
20 8×10^{-4} 3.8×10^6 6.2×10^9
21 a 1.8942×10^5 **b** 9.39×10^{19}

CHAPTER 8

Exercise **8-01**

1 D   **2** C

3 a $\angle BAC$ **b** $\angle YXZ$ **c** $\angle PQR$
d $\angle DEF$ **e** $\angle ABC$ **f** $\angle EDF$

4 a 33°   **b** 91°   **c** 230°
d 90°   **e** 152°   **f** 300°

5 a acute   **b** obtuse   **c** reflex
d right   **e** obtuse   **f** reflex

6 a

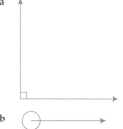

b

c

d

e

f

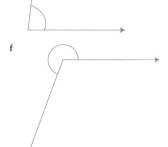

7 **a**

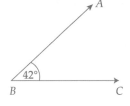

b

c

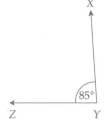

d

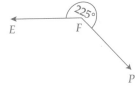

e

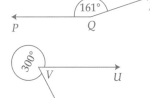

f

8 **a** acute **b** reflex **c** obtuse **d** right
   e reflex **f** obtuse **g** acute **h** straight
   i reflex **j** revolution

9 Sample answers
   a angle between 2 walls
   b path around a roundabout
   c angle when a door is wide open
   d angle when a page is open in a book
   e angle between your fingers
   f angle from north to east anticlockwise

Exercise 8-02

1 C   **2** A
3 **a** 50°   **b** 15°   **c** 45°
     d 56°   **e** 37°
4 **a** 120°   **b** 145°   **c** 35°
     d 74°   **e** 124°
5 **a**

   b

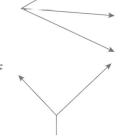

   c

   d

   e

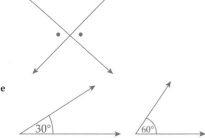

6 **a** $x + 58 = 90$ (complementary angles)
            $x = 90 - 58$
            $x = 32$
     b $n = 42$ (vertically opposite angles)
         $m + 42 = 180$ (supplementary angles)
            $m = 180 - 42$
            $m = 138$
7 **a** $a = 70$ complementary angles
     b $b = 52$ complementary angles
     c $x = 130$ supplementary angles
     d $y = 48$ supplementary angles
     e $m = 30$ angles at a point
     f $n = 123$ angles at a point
     g $s = 78$ vertically opposite angles
     h $t = 26$ vertically opposite angles
     i $x = 134$ supplementary angles

j $y = 38$ angles at a point

k $a = 171$ vertically opposite angles

l $b = 57$ complementary angles

8   Other answers possible.

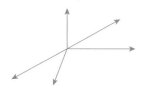

Exercise 8-03

1   C   **2**   D

3

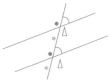

4

5

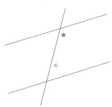

The other pair of co-interior angles are on the other side of the transversal and in between the parallel lines.

6   **a**   alternate angles

     b   corresponding angles

     c   co-interior angles

7   **a** $a = 70$ alternate angles

     b $b = 110$ corresponding angles

     c $c = 50$ co-interior angles

     d $x = 115$ corresponding angles

     e $y = 112$ co-interior angles

     f $d = 121$ alternate angles

     g $e = 125$ alternate angles

     h $x = 82$ alternate angles

     i $y = 115$ corresponding angles

     j $d = 47$ co-interior angles

     k $x = 120$ corresponding angles, $y = 60$ supplementary angles, $z = 60$ corresponding angles

     l $a = 88$ alternate angles, $b = 92$ supplementary angles, $c = 92$ alternate angles

8   **a**   No alternate angles are not equal.

     b   Yes, corresponding angles are equal.

     c   No, co-interior angles are not supplementary.

Exercise 8-04

1   C   **2**   A

3   **a**

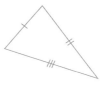

     b

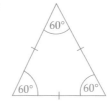

     c

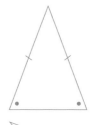

     d

     e

     f

4   **a**   equilateral   **b**   isosceles

     c   scalene   **d**   isosceles

     e   isosceles

5   **a**   acute-angled   **b**   acute-angled

     c   right-angled   **d**   acute-angled

     e   obtuse-angled

ANSWERS

6   a   right-angled   b   acute-angled

     c   obtuse-angled   d   obtuse-angled

     e   right-angled

7   a   scalene   b   scalene   c   scalene

     d   scalene   e   isosceles

8

Type of triangle	Number of equal sides	Number of equal angles
Scalene	0	0
Isosceles	2	2
Equilateral	3	3

9   a   Yes   b   Yes   c   Yes

10   a

     b

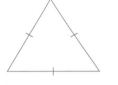

     c

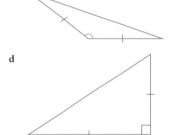

     d

Exercise 8-05

1   D   2   B

3   a   isosceles, acute, x and y

     b   equilateral, acute, m, n and o

     c   isosceles, right-angled, v and w

4   a   i g ii d and f

     b   i q ii p and s

5   a $52 + 65 + x = 180$ (angle sum of triangle)

             $x = 180 - 52 - 65$

             $x = 63$

     b $n = 72 + 46$ (exterior angle of triangle)

         $n = 118$

6   a $a = 80$, angle sum of triangle

     b $y = 22$, angle sum of triangle

     c $e = 60$, angle sum of triangle

     d $b = 35$, angle sum of triangle

     e $d = 105$, exterior angle of a triangle

     f $f = 134$, angle sum of triangle

     g $c = 112$, exterior angle of a triangle

     h $p = 145$, angle sum of triangle

     i $g = 65$, exterior angle of a triangle

     j $x = 81$, angle sum of triangle

     k $q = 50$, exterior angle of a triangle

     l $h = 45$, angle sum of triangle

Exercise 8-06

1   B, C

2   a

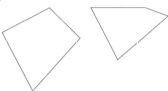

     b

3   a

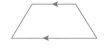

     b

     c

     d

4   Teacher to check.   5   Teacher to check.

6   a   kite   b   trapezium

     c   rhombus   d   rectangle

     e   square   f   parallelogram

     g   trapezium   h   rhombus

     i   rectangle   j   kite

Exercise **8-07**

1 B   2 A

3

Quadrilateral	Sides	Angles	Diagonals
Parallelogram	Opposite sides equal and parallel	Opposite angles equal	Diagonals bisect each other
Trapezium	One pair of parallel sides	All angles different	Diagonals different
Rhombus	All sides equal Opposite sides parallel	Opposite angles equal	Diagonals bisect each other at right angles Diagonals bisect the vertex angles
Rectangle	Opposite sides equal and parallel	All angles are 90°	Diagonals are equal and bisect each other
Square	All sides equal Opposite sides are parallel	All angles are 90°	Diagonals are equal and bisect each other at right angles Diagonals bisect the vertex angles

4    $x + 125 = 180$ (co-interior angles on parallel lines)
   $x = 180 - 125$
   $x = 55$

5    $n + 126 + 90 + 78 = 360$ (angle sum of quadrilateral)
   $n = 360 - 126 - 90 - 78$
   $n = 66$

6   a $x = 28$ angle sum of quadrilateral
   b $p = 90$ angle of a rectangle
   c $n = 133$ angle sum of quadrilateral
   d $x = 124$ cointerior angles on parallel lines
   e $v = 115$ opposite angles of parallelogram
   f $n = 90$ angle sum of quadrilateral
   g $e = 45$ diagonals bisect the angles of a square
   h $n = 96$ co-interior angles on parallel lines
   i $x = 72$ corresponding angles on parallel lines

Exercise **8-08**

1    $\angle R$ 2   C

3   a   translation   b   rotation
   c   reflection   d   translation
4   a   Yes   b   Yes
5   a   No   b   The size of the angles.
6   Matching sides: AB and RP, AC and RQ, BC and PQ
    Matching angles: $\angle A$ and $\angle R$, $\angle B$ and $\angle P$,
    $\angle C$ and $\angle Q$
7   Congruent shapes are men, rotation.

Exercise **8-09**

1   A   2   D
3   a $LM = RS, LN = RT, MN = ST$
   b RST c   SSS
4   A and C
5   a $\triangle ABC \equiv \triangle LMN$ b   AAS
6   B, C and D
7    $\triangle ABC \equiv \triangle IHG$ RHS, $\triangle PQR \equiv \triangle EFD$ SSS
    $\triangle WXY \equiv \triangle MON$ SAS
8   Teacher to check.

Language activity

WHEN FINDING ANGLES
IN QUADRILATERALS AND TRIANGLES
REMEMBER ANGLE SUM
AND THE ANSWER WILL COME.

Practice test **8**

1    $168   2 $\dfrac{k}{2}$ 3 $20   4   64 cm^3

5   10   6 $\dfrac{7}{25}$ 7   5 : 2   8 $14x + 4$

9   38   10   6h 40min
11   A   12   C
13   a $x = 58$ vertically opposite angles
   b $n = 18$ complementary angles
14

15    $x = 48$
16   a   scalene   b   isosceles
17   a   obtuse-angled   b   right-angled
18   a $x = 28$, isosceles triangle
   b $n = 71$, exterior angle of a triangle
19   a   parallelogram   b   trapezium
20   a $x = 45$, angle sum of a quadrilateral
   b $m = 74$, opposite angles of a parallelogram
21   EDF 22   AAS

CHAPTER 9

Exercise 9-01

1 B   **2** D

3
- **a** $x = 3$
- **b** $m = 1$
- **c** $p = 3$
- **d** $a = 5$
- **e** $q = 1$
- **f** $x = 7$
- **g** $y = 15$
- **h** $m = 11$
- **i** $p = 11$
- **j** $q = 15$
- **k** $x = 5$
- **l** $y = 4$
- **m** $a = 4$
- **n** $y = 2$
- **o** $m = 2$
- **p** $x = 20$
- **q** $a = 12$
- **r** $m = 12$
- **s** $y = 14$
- **t** $a = 24$

4
- **a** 4
- **b** add
- **c** divide 5
- **d** multiply 4

5
- **a** $x = 2$
- **b** $y = 7$
- **c** $x = 10$
- **d** $y = 16$
- **e** $m = 2$
- **f** $x = 13$
- **g** $a = 6$
- **h** $b = 9$
- **i** $y = 9$
- **j** $x = 8$
- **k** $p = 3$
- **l** $n = 3$
- **m** $x = 11$
- **n** $b = 3$
- **o** $t = 4$
- **p** $x = 27$
- **q** $t = 22$
- **r** $n = 10$
- **s** $p = 16$
- **t** $q = 32$

6
- **a** $m = -4$
- **b** $x = -5$
- **c** $e = 2\frac{4}{7}$
- **d** $v = -20$
- **e** $w = -1\frac{1}{8}$
- **f** $x = -21$
- **g** $s = -80$
- **h** $n = -21$

Exercise 9-02

1 D   **2** C

3 Various answers such as:
- **a** $x + 6 = 10$
- **b** $3x = 12$
- **c** $2x - 5 = 3$

4
- **a**
$$5a - 3 + 3 = 17 + 3$$
$$5a = 20$$
$$\frac{5a}{5} = \frac{20}{5}$$
$$a = 4$$
- **b**
$$2x + 6 - 6 = 14 - 6$$
$$2x = 8$$
$$\frac{2x}{2} = \frac{8}{2}$$
$$x = 4$$

5
- **a** $x = 3$
- **b** $a = 2$
- **c** $b = 4$
- **d** $m = 4$
- **e** $b = -2$
- **f** $a = 4$
- **g** $x = 5$
- **h** $a = 7$
- **i** $b = 8$
- **j** $y = 4$
- **k** $m = 5$
- **l** $n = 9$

6 Teacher to check.

7
- **a** $x = 2\frac{2}{3}$
- **b** $n = 3\frac{4}{5}$
- **c** $a = -2$
- **d** $x = 2\frac{1}{3}$
- **e** $s = -1\frac{1}{4}$
- **f** $y = -1\frac{1}{3}$
- **g** $r = 8$
- **h** $w = -5\frac{4}{5}$
- **i** $x = 29$
- **j** $n = -33$
- **k** $u = -26$
- **l** $x = -54$

8
- **a** T
- **b** F
- **c** T
- **d** F

Exercise 9-03

1 C   **2** A

3
- **a** $3x - 4$
- **b** $5a + 6$
- **c** $8 - 2x$

4
- **a** $2x + 9$
- **b** $3a - 4$
- **c** $x + 5$

5
- **a** $x = 13$
- **b** $a = -5$
- **c** $x = 1$

6
- **a**
$$2w + 16 = w - 8$$
$$2w + 16 - w = w - 8 - w$$
$$w + 16 = -8$$
$$w + 16 - 16 = -8 - 16$$
$$w = -24$$

- **b**
$$4x - 15 = 2x + 17$$
$$4x - 15 - 2x = 2x + 17 - 2x$$
$$2x - 15 = 17$$
$$2x - 15 + 15 = 17 + 15$$
$$2x = 32$$
$$x = 16$$

7
- **a** $a = 12$
- **b** $w = 9$
- **c** $b = -8$
- **d** $m = -10$
- **e** $a = 6$
- **f** $x = 13$
- **g** $a = 3$
- **h** $b = -7$
- **i** $x = 6$
- **j** $w = 2$
- **k** $x = 10$
- **l** $r = -5$
- **m** $a = 2$
- **n** $b = 7$
- **o** $a = -3$

8
- **a** $x = 2$
- **b** $n = 1\frac{1}{4}$
- **c** $x = -30$
- **d** $d = 2\frac{2}{3}$
- **e** $m = 6\frac{1}{7}$
- **f** $w = -3$
- **g** $x = -\frac{1}{7}$
- **h** $n = 6\frac{1}{3}$
- **i** $x = -18$

Exercise 9-04

1 B   **2** D

3
- **a** F
- **b** F
- **c** T
- **d** T

4
- **a**
$$3(w - 4) = 15$$
$$3w - 12 = 15$$
$$3w - 12 + 12 = 15 + 12$$
$$3w = 27$$
$$w = 9$$
- **b**
$$-2(a + 4) = 12$$
$$-2a - 8 = 12$$
$$-2a - 8 + 8 = 12 + 8$$
$$-2a = 20$$
$$a = -10$$

5
- **a** $x = 7$
- **b** $a = -2$
- **c** $b = 4$
- **d** $x = 1$
- **e** $a = 2$
- **f** $b = 12$
- **g** $m = 1$
- **h** $a = 10$
- **i** $v = 11$
- **j** $z = 3$
- **k** $a = -15$
- **l** $w = 12$

6 Teacher to check.

7
- **a** $x = 2$
- **b** $a = 2$
- **c** $b = 1\frac{2}{3}$
- **d** $y = 5$
- **e** $a = 2$
- **f** $b = -1$
- **g** $x = 2$
- **h** $w = 2$
- **i** $a = -3\frac{1}{3}$
- **j** $n = -7\frac{2}{5}$
- **k** $x = -\frac{1}{12}$
- **l** $m = 2\frac{3}{8}$

8 Teacher to check.

9 Various answers such as: $4(x + 2) = 20$, $-6(2x - 4) = -12$

10 Various answers such as: $5(x + 8) = 30$, $4(3x + 13) = 28$

Exercise 9-05

1 B   **2** C

3
- **a** T
- **b** F
- **c** F
- **d** T

4
- **a** $x = \pm 6$
- **b** $x = \pm 9$
- **c** $a = \pm 10$
- **d** $n = \pm 1$
- **e** $x = \pm\sqrt{7}$
- **f** $x = \pm\sqrt{11}$
- **g** $w = \pm\sqrt{14}$
- **h** $b = \pm\sqrt{22}$
- **i** $x = \pm 7$
- **j** $x = \pm\sqrt{38}$
- **k** $a = \pm\sqrt{106}$
- **l** $n = \pm\sqrt{19}$

m $a = \pm 12$ **n** $x = \pm\sqrt{68}$

o $m = \pm\sqrt{42}$ **p** No solution

q $x = \pm 16$ **r** No solution

s $a = \pm 100$ **t** $p = \pm\sqrt{111}$

5 **e** $x = \pm 2.6$ **f** $x = \pm 3.3$ **g** $w = \pm 3.7$

h $b = \pm 4.7$ **j** $x = \pm 6.2$ **k** $a = \pm 10.3$

l $n = \pm 4.4$ **n** $x = \pm 8.2$ **o** $m = \pm 6.5$

t $p = \pm 10.5$

6 **a** T   **b** F   **c** F   **d** T

   e F   **f** F

7 **a** $x = \pm 13.49$ **b** $n = \pm 20.76$

   c $w = \pm 4.32$ **d** $b = \pm 4.65$

   e $u = \pm 63.25$

8 **b** $n = \pm\sqrt{28}$ **c** $w = \pm\sqrt{54}$

   e $u = \pm\sqrt{100\,000}$ **g** $x = \pm\sqrt{212}$

Exercise 9-06

1 D

2 **a** $n + 6 = 8$ **b** $x - 4 = 18$

   c $8w = 24$ **d** $\dfrac{m}{6} = 14$

3 **a** $n = 2$ **b** $x = 22$ **c** $w = 3$ **d** $m = 84$

4 $n = 20$ **5** $p = 9$ **6** 60   **7** 5

8 7   **9** $s = 8$ **10** $n = 3$ **11** $n = 6$

12 Amy is 16 and Sophie is 48 years old.

13 Harry is 4 and Fatima is 8 years old.

14 2   **15** 28   **16** 14

Language activity

THEY BOTH HAVE TWO EQUAL SIDES

Practice test 9

Part A

1 $924   **2** $-x + 5y$ **3** 0.0905

4 32 cm^2 **5** 0.002   **6** 4.95×10^5

7 87.4%   **8** 29.3   **9** 10 000

10 $2l + 2w$ or $2(l + w)$

Part B

11 **a** $x = 5$ **b** $n = 10$ **c** $x = 6$

12 C   **13** B

14 **a** $x = -4$ **b** $c = -\dfrac{1}{2}$ **c** $x = -20$

15 **a** $x = 12$ **b** $a = -14$ **c** $n = -5$

16 **a** $x = 4$ **b** $u = -1$ **c** $x = \dfrac{1}{2}$

17 Teacher to check. Examples are: $2x - 12 = -16$ and $6(x + 4) = 12$

18 **a** $x = \pm 7$ **b** $x = \pm\sqrt{13}$

19 $n \approx \pm 7.81$

20 $3x + 12 = 33, x = 7$ **21** $x = 6$

22 Michael is 64, Sunny is 16.

CHAPTER 10

Exercise 10-01

1 C   **2** A

3 **a** $388.80   **b** $746.70

   c $591.81   **d** $913.31

4 **a** $777.60   **b** $1493.40

   c $1183.62   **d** $1826.62

5 **a** **i** $4160 **ii** $1920 **iii** $960

   b **i** $5980 **ii** $2760 **iii** $1380

   c **i** $9000 **ii** $4153.85 **iii** $2076.92

6 **a** $37 887.20   **b** $46 504.80

   c $59 441.46

7 **a** Georgia   **b** $36.92

8 **a** $899.08   **b** $1227.27

   c $984.23   **d** $1344.14

9 **a** $3120   **b** $3992.95

   c $3735.33   **d** $4439.85

10 Jack   **11** $57 890   **12** $2734.62

Exercise 10-02

1 C   **2** D

3 **a** $117 **b** $210.60

4 **a** $156 **b** $280.80

5 Normal wage = $23.60 \times 35 = 826

   Overtime = $23.60 \times 1.5 \times 4 + $23.60 \times 2 \times 1.5$

         = $212.40

   Total wage = $826 + $212.40

         = $1038.40

6 $565.90   **7** $644.88 **8** $2016

9 $1728.60   **10** $1654.43

Exercise 10-03

1 A   **2** B

3 **a** $2887.50   **b** $1360.80

   c $1306   **d** $1387.50

4 **a** $990   **b** $877.50

   c $923.52   **d** $778.72

5 **a** $7020   **b** $6570

   c $4454.13   **d** $6825

6 $395.50   **7** $14 750

8 $586.35   **9** $3720

10 **a** 560   **b** $840

Exercise 10-04

1 B   **2** B

3 **a** T   **b** F   **c** T

4 4 weeks wage = $725.50 \times 4 = 2902

   Annual leave loading = $17.5\% \times 2902 = 507.85

   Total holiday pay = $2902 + $507.85 = 3409.85

5 **a** $297.50   **b** $1997.50

6 **a** $882   **b** $3528

   c $617.40   **d** $4145.40

7   a $5538.46   b $969.23
    c $6507.69

8   $36 170.55

9   a $1110.58   b $7456.73
    c $4759.62

10   a $2486.25   b $446.25
    c $5482.50

Exercise 10-05

1   B   2   C

3   a $0   b $7400.50
    c $2766.40   d $18 287
    e $66 229

4   a $109 420, $28 432.40   b $22.6%

5   a $6149   b $12 516
    c $10 980   d $9289.60

6   a $36 351, $3448.69   b $68 984, $13 966.80
    c $144 020, $41 234.40   d $68 960, $13959

Exercise 10-06

1   C

2   B

3   a $450   b $304.50   c $89   d $88
    e $672   f $1700   g $340   h $338

4   $279.80

5   a $574   b $390.32

6   a $41 704   b $413.50

7

Name: Ali Mc Phee	Date: 20/07/15 to 24/7/15
Hourly rate: $21.60	Gross Pay: $680.40
Hours worked: 31.5	Tax: $166.70
Tax rate: 24.5%	Net Pay: $513.70

8   a $1925   b $435   c $1390

9   a $760.20   b   39.0%

Practice test 10

1   $4.60   2 $-6a-18$ 3   1.5%   4   30 cm²

5   $22   6   6   7   0.175   8   26

9   11, 13, 17 or 19   10   8h 15min

11   C   12   B   13   A

14   a $890.40   b $861.10

15   $15875   16 $987.84   17 $124 775

18   a $370.02   b $2484.42

19   a $1017.69
    b $6833.07 Note: a bonus is not included in holiday pay or loading.

20   a $10 683   b $3742.04

21   a $352   b $1690

22   a $272/week   b $843/fortnight

CHAPTER 11

Exercise 11-01

1   D   2   D

3   a   quantitative discrete
    b   quantitative continuous
    c   categorical
    d   quantitative discrete
    e   categorical
    f   quantitative continuous
    g   quantitative discrete
    h   categorical
    i   quantitative continuous
    j   categorical

4   a   D   b   D   c   C   d   C
    e   D   f   C   g   D   h   D
    i   D   j   C

5   a   T   b   T   c   F   d   F

6   Teacher to check.   7   Teacher to check.

Exercise 11-02

1   C   2   B

3   a   sample   b   census   c   sample   d   sample
    e   census   f   sample   g   census   h   sample

4   a   random   b   biased   c   biased   d   biased

5   a   807   b   3 : 2   c   30

6   Teacher to check.

7   Teacher to check.

Exercise 11-03

1   D   2   C

3   adding   dividing   number   often   popular   no   higher   lower

4   a   3.6   b   9   c   4   d   11.3
    e   15.5   f   34.5   g   4.3   h   12.4
    i   336.1   j   66.5

5   a   3   b   no mode
    c   2   d   12
    e   15 and 17   f   no mode
    g   2.6   h   21.5, 5.7, 11.1
    i   131   j   no mode

6   a $27.57   b $50   c $23.83
    d   The outlier made the mean higher than most amounts of pocket money.

7   a   40.27   b   18   c   higher
    d   42.5   e   Yes, 42

8   a   No mode   b $793 906.25
    c   Yes, $1 220 500   d   Increase

Exercise 11-04

1   A   2   B

3   highest   lowest   lowest   highest   middle   scores   even   average   middle

4 **a** 6   **b** 5.5   **c** 13
    d 24   **e** 55.5

5 **a** 7   **b** 7   **c** 8
    d 8   **e** 38

6 Red: median = 49.05, range = 8.9
    Yellow: median = 47.35, range = 15.8
    Green: median = 52.15, range = 13.9

7 **a** Yellow   **b** Green

8 **a** range = 4, median = 2
    b range = 12, median = 14
    c range = 6, median = 1
    d range = 4, median = 3
    e range = 4, median = 100
    f range = 4, median = 10.5
    g range = 8, median = 6.5
    h range = 5, median = 31
    i range = 5, median = 3.5
    j range = 5, median = 3

9 range = 3, median = 7

10 mode = 71, range = 8, median = 73.5, mean = 73.75

11 The mode, as it is the lowest statistic, compared to the mean and median.

Exercise 11-05

1 D   **2** B

3 horizontal   dots   score   hundreds   leaf

4 **a**

    b

    c

    d
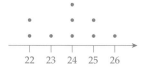

    e
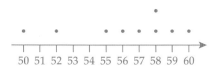

5 **a** **i** 3   **ii** 8   **iii** 7
    b **i** 4   **ii** 9   **iii** 8
    c **i** 9   **ii** 14   **iii** 11
    d **i** 22   **ii** 26   **iii** 24
    e **i** 50   **ii** 60   **iii** 58

6

(dot plot with horizontal axis labelled 180, 200, 220, 240 250 260, 280)

    a $280   **b** $250   **c** $720   **d** $240
    e A cluster from $220 to $250, no outliers.

7 **a**

Stem	Leaf
1	2 3 5
2	6 7 8
3	2 5
4	3 9

    b

Stem	Leaf
5	2 9
6	7 9
7	3 4 5
8	2 6 6

    c

Stem	Leaf
3	6 8
4	2 4 5
5	6 8 9
6	4
7	2

    d

Stem	Leaf
6	0 5 6
7	2 2 9
8	2 3
9	1 5

    e

Stem	Leaf
12	3 9
13	4 7 8
14	5 5
15	0 6
16	0

8 Range = 35, median = 75.5, mean = 76.5

9 **a**

Stem	Leaf
3	4 6 7 9
4	5 5 5
5	2 5 6
6	2 2 4 6
7	2 2 2 2 3 4
8	
9	8

    b Range = 64 and mode = 72
    c 58.6190 ... × $18 ≈ $1055
    d 62
    e Yes, from 60 to 70s.
    f Yes, 98

Exercise 11-06

1 D
2 C

ANSWERS

3 a

Score	Tally	Frequency
47	IIII	4
48	IIII II	7
49	IIII III	8
50	IIII IIII IIII	14
51	IIII II	7
52	IIII II	7
53	III	3

b Range = 6, mode = 50

4 a

Score (000s)	Tally	Frequency	fx
21	III	3	63
22	II	2	44
23	IIII IIII	10	230
24	II	2	48
25	II	2	50
26	IIII	5	130
27	II	2	54
28	III	3	84
29	I	1	29
	Totals	30	732

b range = 8000 km **c** mean = 24 400

5 a

Score	Tally	Frequency	fx
2	I	1	2
3	III	3	9
4	II	2	8
5	II	2	10
6	IIII	5	30
7	IIII IIII II	12	84
8	II	2	16
9	III	3	27
	Totals	30	186

b range = 7
c mode, mode = 7 **d** mean = 6.2

6 a

Score	Tally	Frequency	Cumulative frequency	fx
0	IIII	5	5	0
1	IIII	5	10	5
2	IIII IIII IIII IIII	20	30	40
3	IIII IIII III	13	43	39
4	IIII	5	48	20
5	II	2	50	10
	Totals	50		114

b mode = 2 **c** mean = 2.28 **d** median = 2

7 a Cumulative frequency: 3, 9, 18, 22, 25
   fx: 0, 6, 18, 12, 12, 48
   i 4 **ii** 2 **iii** 2 **iv** 1.92

b Cumulative frequency: 4, 13, 25, 32, 37, 40
   fx: 48, 117, 168, 105, 80, 51, 569
   i 5 **ii** 14 **iii** 14 **iv** 14.23

8

Score	Frequency	Cumulative frequency	fx
0	2	2	0
1	3	5	3
2	5	10	10
3	5	15	15
4	9	24	36
5	3	27	15
6	1	28	6
7	1	29	7
8	1	30	8
Totals	30		100

a 8 **b** 4 **c** 3.5
d mean = 3.333.. × $45 = $150

Exercise **11-07**

1 column joined line column
2 a

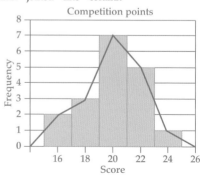

Competition points

b 20, mode **c** 24

3 a

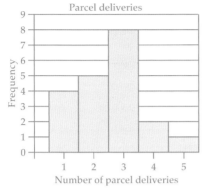

Parcel deliveries

b range = 4   mode = 3

c The mode is the most common number of parcels delivered.

d mean = 2.55

4 a

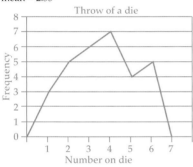

Throw of a die

b 4   **c** 4   **d** 1

e 5, which is the difference between the highest and lowest number on the die

f 53.3%

5 a

Mark	Frequency
72	3
73	3
74	6
75	2
76	3
77	0
78	3

b

Science test marks

c Range = 6, mode = 74

d mean = 74.55   **e** 40%

Exercise 11–08

1 B   **2** Teacher to check   **3** C

4 centre left   right   right   left   two

5 a i positively skewed

  ii clustered at 4-5, no outliers

b i negatively skewed

  ii clustered at 13-15, 1 outlier to the left

c i positively skewed

  ii clustered at 120-130 stems, 1 outlier 168

d i bimodal

  ii clusters at peaks 7 and 10, no outliers

6 a negatively skewed

b mode = 38, range = 29

c mean = 28.5

d median = 29

e In the 30s

7 mode = 38, range = 21, mean = 29.8 and median = 30 so it affects the range the most.

Language activity

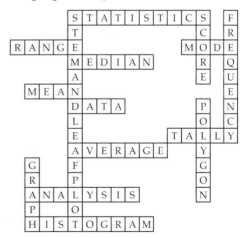

Practice test 11

Part A

1 $25.50 2 $\dfrac{25b^2}{a}$ 3 $1\dfrac{1}{12}$ 4 $x = 60$

5 –7 6 $x = 22\dfrac{1}{2}$

7 Several answers such as 2, 11, 7, 8. Teacher to check.

8 isosceles triangle 9 29 10 $\dfrac{7}{10}$

Part B

11 B 12 C 13 A

14 a sample b census

15 mode = 37, outlier = 78

16 a mean – 41.8

    b Yes, it makes it higher in this case.

17 range = 4.7, outlier = 2.7

18 a 5.85

    b It will have little effect on the median.

19 a

```
        •   •
      •   •   •
  •   •   •   •   •   •
  10  15  20  25  30  35
```

    b

Stem	Leaf
1	0 5 5 5
2	0 0 0 5 5
3	0 5

20 a 28 b 27.45 c 27.5

21

Score	Frequency
3	8
4	12
5	5
6	5

22 a negatively skewed b positively skewed

CHAPTER 12

Exercise 12–01

1 C

2 B

3 a mm b km c m d cm

    e cm f cm g kL h g

    i h j mL k t l s

4 a 3 mm b 850 km c 3 m d 15 cm

    e 12 cm f 170 cm g 25 kL h 50 g

    i 8 hours j 250 mL k 5 t l 50 sec

5

Millimetres	Centimetres	Metres	Kilometres
4000	400	4	0.004
65 000	6500	65	0.065
786 000	78 600	786	0.786
95 000 000	9 500 000	95 000	95
85 400	8540	85.4	0.0854
7900	790	7.9	0.0079
654 000	65 400	654	0.654

6 a 4 b 2.5 c 2.5 d 6500

    e 210 f 7850 g 7000 h 114

    i 9600 j 0.025 k 3.5 l 6800

7 a $\dfrac{1}{1000}$ b 1 000 000 c 1000

    d $\dfrac{1}{100}$ e $\dfrac{1}{1000000}$

    f 1 000 000 000 (one billion)

8 a 1000 b 1 000 c 1000

    d 2.75 MB e 1600 MB f terabyte (TB)

Exercise 12–02

1 B

2 D

3 a 1 mm b 1°C c 1 min d 100 g

    e 1 cm f 1° g 5 km/h h 250 g

    i 10 mL j 1 kL k $\dfrac{1}{4}$ tank

4 a ±0.5 mm b ±0.5 °C

    c ±0.5 min or ±30 s d ±50 g

    e ±0.5 cm or ±5 mm f ±0.5°

    g ±2.5 km/h h ±125 g

    i ±5 mL j ±0.5 kL or 500 L

    k $\pm\dfrac{1}{8}$ tank

5 a i ±0.5 kg ii 74.5 to 75.5 kg

    b i ±5 mL ii 275 to 285 mL

    c i ±10 m ii 210 to 230 m

    d i ±25 g ii 425 to 475 g

    e i ±0.05 cm ii 2.55 to 2.65 cm

    f i ±2.5 mm ii 52.5 to 57.5 mm

    g i ±25 mg ii 2725 to 2775 mg

    h i ±0.005 s ii 11.235 to 11.245 s

Exercise 12–03

1 B 2 A

3 a 34 cm b 25.4 m c 33.7 cm

4 Rectangle: P = 26 cm Square: P = 48 mm

    Equilateral triangle: P = 12.6 cm

    Rhombus: P = 33.6 m

    Parallelogram: P = 26 cm Kite: P = 18.6 cm

5 **a** 58.92 cm   **b** 42.26 cm   **c** 44 cm
   d 108 cm   **e** 44 cm   **f** 236 cm
6 **a** 300 m   **b** 30   **c** $315

Exercise **12-04**

1 A   **2** B
3 **a** 18.8 cm   **b** 25.1 mm   **c** 17.6 m
4 **a** 21.1 cm   **b** 123.4 mm   **c** 27.8 m
5 **a** **i** 20.57 m **ii** $(4\pi + 8)$ m
    b **i** 14.28 cm **ii** $(2\pi + 8)$ cm
6 **a** 41.1 cm   **b** 7.1 m   **c** 13.4 m
   d 12.1 cm   **e** 57.6 m   **f** 24.3 m
   g 31.7 m   **h** 21.4 cm   **i** 27.7 m
7 **a** 11.8 m   **b** $354

Exercise **12-05**

1 C   **2** B
3 **a** 0210   **b** 1748   **c** 0439   **d** 1852
   e 1200   **f** 0755   **g** 0046   **h** 2125
4 **a** 5 a.m.   **b** 9 p.m.   **c** 6:55 a.m.
   d 6:48 p.m.   **e** 8:50 a.m.   **f** 11:39 a.m.
   g 10:56 p.m.   **h** 12:54 a.m.
5 **a** 3 h   **b** 9 h   **c** 6 h   **d** 7 h
   e 12 min **f** 6 min **g** 7 min **h** 19 min
6 **a** 5 h 48 min   **b** 3 h 6 min   **c** 20 h 58 min
7 **a** 17 min 30 s   **b** 3 min 40 s   **c** 8 min 41 s
8 **a** 17 h 25 min   **b** 9 h 9 min   **c** 13 h 35 min
   d 13 h 19 min   **e** 10 h 21 min   **f** 18 h 21 min
9 5 h 55 min   **10** 6:51 p.m.   **11** 16 h 31 min

Exercise **12-06**

1 D   **2** B
3 **a** 6:03, 2 h 1 min   **b** 5:28, 2 h 37 min
   c 7:00, 2 h 7 min   **d** 6:05, 1 h 36 min
   e 7:18, 1 h 7 min
4 7:30 from Gosford
5 **a** 7:36   **b** 1 h 5 min   **c** 19 min
6 **a** 8:40 a.m.   **b** 7 h 40 min   **c** 7:30 p.m.

Exercise **12-07**

1 D   **2** B
3 **a** 12 midday   **b** 4:00 a.m.   **c** 2:30 p.m.
   d 6:00 p.m.   **e** 3:00 a.m.   **f** 7:00 p.m.
4 **a** 10:00 a.m.   **b** 11:00 a.m.   **c** 5:00 a.m.
   d 6:00 p.m.   **e** 6:00 p.m.   **f** 8:00 p.m.
5 6:20 a.m. Friday   **6** 2:30 p.m.
7 **a** 11:30 a.m.   **b** 11:30 a.m.
   c 9:30 a.m.   **d** 11:00 a.m.
8 **a** 10 p.m.   **b** 9:30 p.m.
   c 10 p.m.   **d** 10 p.m.
9 4:30 p.m.   **10** 4:00 a.m. to 8:15 a.m.
11 3 p.m.   **12** 12:30 a.m.

Practice test **12**

1 7:35 p.m.   **2** 7 : 4   **3** $\dfrac{1}{81}$

4 $x^2 - d^2 + r^2$ **5** $5xy(2x - 3y)$ **6** $7w$

7 $x = 5\dfrac{2}{5}$ **8** $12x - 60$

9 68%, $\dfrac{2}{3}$, 0.65, 0.6   **10** $\dfrac{2}{3}$

11 C   **12** D   **13** ±125 g
14 **a** 46.2 cm   **b** 25.5 cm
15 56.2 m
16 **a** 28.3 cm   **b** 20.1 m
17 50.6 cm
18 **a** 12 h 40 min   **b** 6h 28 min
19 **a** 22 h   **b** 9 h
20 **a** 6:40 a.m.   **b** 1 h 33 min
21 **a** 5:50 p.m. Saturday   **b** 3:15 p.m.

CHAPTER 13

Exercise **13-01**

1 C   **2** B
3 **a** 25 m²   **b** 10.08 cm²   **c** 9.01 cm²
   d 10.24 m²   **e** 4 m²   **f** 18 cm²
   g 48 m²   **h** 31.5 cm²   **i** 20 m²
   j 24 cm²   **k** 24 cm²   **l** 36 m²
   m 16.56 cm² **n** 65.92 cm²
4 **a** 80 m²   **b** 48 cm²
   c 103.32 m²   **d** 57.12 cm²
   e 22 m²   **f** 74.39 mm²

Exercise **13-02**

1 B   **2** A
3 **a** 254.47 cm²   **b** 50.27 m²   **c** 113.10 mm²
4 **a** 314 m²
5 **a** 25.1 m²   **b** 7.1 m²   **c** 56.5 cm²
   d 28.3 m²   **e** 190.1 m²   **f** 254.5 m²
6 10.18
7 **a** 76.97 m²   **b** 33.18 m²   **c** 63.2 cm²
   d 31.49 m²   **e** 50.3 cm²   **f** 36.9 m²
8 $445.91
9 **a** 628.3 mm²
   b **i** 25132 mm² **ii** 32000 mm² **iii** 6868 mm²

Exercise **13-03**

1 C   **2** A
3 **a** 160 cm²   **b** 2.16 mm²   **c** 2.5 m²
   d 0.05 cm²   **e** 96.76 cm²   **f** 70.29 m²
4 **a** 76 cm²   **b** 187 cm²   **c** 52.08 cm²
   d 76.1 cm²
5 **a** 16.16 m²   **b** 32   **c** $147.20

Exercise **13-04**

1 A 2 C
3 **a** 96 cm² **b** 190 cm² **c** 136 cm²
4 **a**

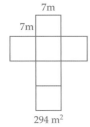

294 m²

b

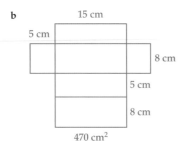

470 cm²

5 **a** 278 cm² **b** 460 cm² **c** 224 m²
   d 292 m² **e** 200.88 cm² **f** 121.5 m²
6 **a** 86.64 mm² **b** No **c** 101.08 mm²

Exercise **13-05**

1 D 2 D
3 **a**

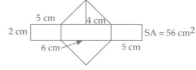

SA = 56 cm²

b

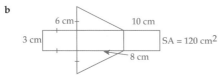

SA = 120 cm²

4 **a** 240 cm² **b** 108 m² **c** 576 cm²
   d 1000 m² **e** 94.68 cm² **f** 220.08 m²
5 240 cm²

Exercise **13-06**

1 B 2 B
3 **a**

b 376.99 cm²
5 **a** 327.2 m² **b** 414.3 cm² **c** 263.9 m²
   d 39.6 m² **e** 174.9 cm² **f** 329.9 cm ²
   g 485.2 m²
6 **a** 332.51 m² **b** 14 cans **c** $896

Exercise **13-07**

1 C 2 C
3 **a** 20 **b** 450 000 **c** 72.5
   d 655 **e** 2.56 **f** 5.7
4 **a** **b**

c

5 **a** 480 cm³ **b** 108 m³ **c** 123.2 m³
   d 61.1 m³ **e** 960 cm³ **f** 1200 cm³
   g 1632 cm³ **h** 2033.6 cm³
6 **a** 640 cm³ **b** 640 mL **c** 0.64 L
7 **a** 99 m³ **b** 99 000 L

Exercise **13-08**

1 A 2 B
3 **a** 785.4 cm³ **b** 274.9 mm³ **c** 1017.9 m³
   d 196 349.5 cm³ **e** 2481.9 cm³ **f** 247.9 cm³
4 **a** 327.1 cm³ **b** 327.1 mL
5 1357 mm³
6 **a** 117.81 m³ **b** 117 810 L
7 4241 cm³

Language activity

THEY ARE BOTH A SUM OF FACES.

Practice test **13**

1 26 **2** $-6ab - 6ab^2$ **3** 24
4 $y = 122$ **5** $6\frac{2}{3}$ **6** $x = 1$ **7** y-axis
8 −65 **9** $\frac{1}{42}$ **10** $800
11 **a** 16.38 cm² **b** 91.98 m²
12 B **13** A
14 **a** 29.04 cm² **b** 33.01 cm²
15 **a** 27.5 m² **b** 6.825 m²
16 58.88 cm²
17 **a** 73.5 cm² **b** 108.08 m²
18

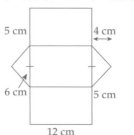

SA = 216 cm²
19 **a** 640.9 cm² **b** 134.2 m²
20 **a** 183.75 cm³ **b** 48.5 m³
21 **a** 108.875 m³ **b** 108 875 L

CHAPTER 14

Exercise **14-01**

1 C   **2** B

3 **a** $y = 2x$

x	0	1	2	3
y	0	2	4	6

b $y = x - 1$

x	10	8	6	4
y	5	4	3	2

c $y = x \div 2$

x	4	3	2	1
y	3	2	1	0

d $y = x + 3$

x	0	1	2	3	4
y	3	4	5	6	7

e $y = 3x$

x	1	2	3	4
y	3	6	9	12

f $y = x - 3$

x	7	6	5	4	0	−1
y	4	3	2	1	−3	−4

g $y = 2 + x$

x	0	1	2	3	4
y	2	3	4	5	6

h $y = x \div 3$

x	12	9	6	3	0	−3	−6
y	4	3	2	1	0	−1	−2

i $y = 2x + 1$

x	1	2	3	4
y	3	5	7	9

j $y = 3x - 1$

x	4	3	2	1	0	−1	−2	−3
y	11	8	5	2	−1	−4	−7	−10

4 **a** $d = 5c + 1$

c	−1	0	1
d	−4	1	6

b $h = 2g - 3$

g	1	2	3
h	−1	1	3

c $q = 6 + 2p$

p	−1	0	1
q	4	6	8

d $t = 12 - 4s$

s	1	2	3
t	8	4	0

e $n = 3 - 2m$

m	−2	0	2
n	7	3	−1

f $y = 5x - 6$

x	1	3	5
y	−1	9	19

Exercise **14-02**

1 3rd   **2** C

3 **a** $A\,(3, -2)$ $B\,(-2, -1)$ $C\,(-4, 0)$ $D\,(-3, 3)$
    $E\,(4, 1)$ $F\,(0, -2)$

b A 4th quadrant B 3rd quadrant
    C x-axis D 2nd quadrant
    E 1st quadrant F y-axis

4 **a**

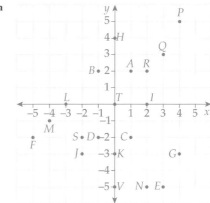

b **i** a rectangle   **ii** a triangle
    iii a parallelogram

5

The Sydney Harbour Bridge

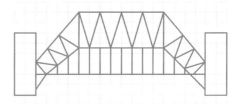

Exercise **14-03**

1 B

2 **a**

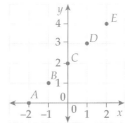

b

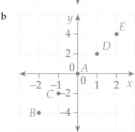

c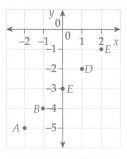

d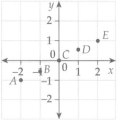

e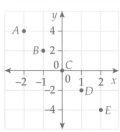

f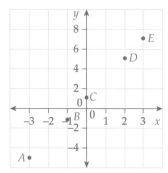

g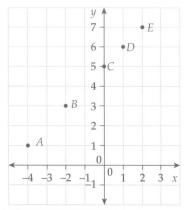

h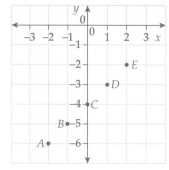

i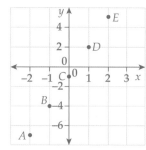

j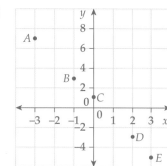

3 They form a straight line.

4 **a** $y = x + 3$

x	−1	0	1	2
y	2	3	4	5

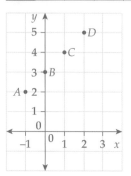

b $y = 6 - x$

x	−1	0	1	2
y	7	6	5	4

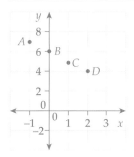

c $y = 2x + 1$

x	−2	0	1	2
y	−3	1	3	5

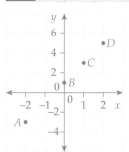

d $y = 4x - 3$

x	−1	1	0	2
y	−7	1	−3	5

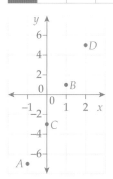

e $y = -3 + x$

x	−1	0	1	2
y	−4	−3	−2	−1

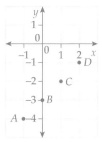

f $y = 12 - 5x$

x	1	0	3	2
y	7	12	−3	2

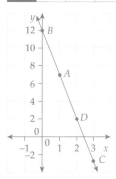

Exercise 14-04

1   C
2   values   table   points   line
3   **a** x-intercept = 2, y-intercept = −2
    b x-intercept = 4, y-intercept = −2
    c x-intercept = 1, y-intercept = 2
    d x-intercept = 4, y-intercept = 1
4   **a** $y = 2x$

x	0	1	2
y	0	2	4

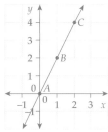

x-intercept = 0, y-intercept = 0

b $y = x - 3$

x	0	1	2
y	−3	−2	−1

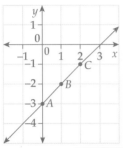

x-intercept = 3, y-intercept = −3

c $y = 5 - x$

x	0	1	2
y	5	4	3

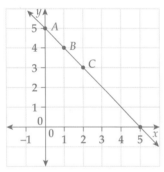

x-intercept = 5, y-intercept = 5

d $y = 3x - 1$

x	0	1	2
y	−1	2	5

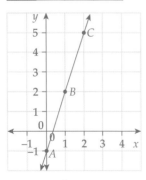

x-intercept = $\dfrac{1}{3}$, y-intercept = −1

e $y = 6 - 2x$

x	0	1	2
y	6	4	2

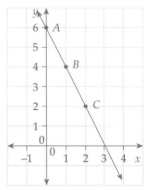

x-intercept = 3, y-intercept = 6

f $y = 4 - 2x$

x	0	1	2
y	4	2	0

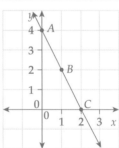

x-intercept = 2, y-intercept = 4

g $y = -3x + 6$

x	0	1	2
y	6	3	0

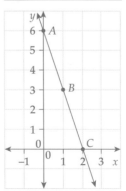

x-intercept = 2, y-intercept = 6

ISBN 9780170351027

h $y = -4x - 1$

x	0	1	2
y	–1	–5	–9

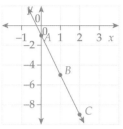

x-intercept $= -\dfrac{1}{4}$, y-intercept $= -1$

Exercise 14-05

1 C   **2** B

3 y-value   lies   false   lie

4 a, d and e.

5

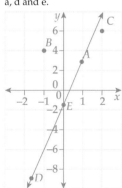

6 a and b.

7 **a** No Yes **b** No Yes **c** No Yes **d** No No
    e No Yes **f** Yes No **g** Yes No **h** No No

Exercise 14-06

1 B   **2** A

3 **a** F   **b** T   **c** T   **d** F

4 **a** $x = -4$, $x = -1$, $x = 3$ **b** $y = 4$, $y = 1$, $y = -2$

5

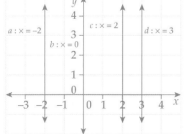

6

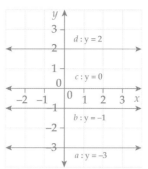

7 **a** $y = 3$ **b** $x = -2$ **c** $y = 2$ **d** $x = 2$

8 **a**

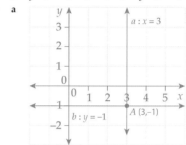

Exercise 14-07

1 C   **2** B

3 $y = 4x - 2$ graphs x-value

4 **a** **i** $(2, 4)$ **ii** $3x - 2 = 4$ **iii** $x = 2$
    b **i** $(-3, -2)$ **ii** $2x + 4 = -2$ **iii** $x = -3$

5 **a** $x = 3$ **b** $x = 5$ **c** $x = -1$

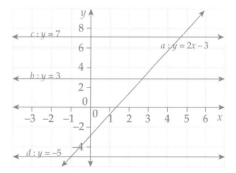

6   **a** $x = 1$ **b** $x = 2$ **c** $x = -2$

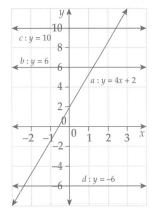

7   **a** $x = -1$ **b** $x = 2$ **c** $x = 5$

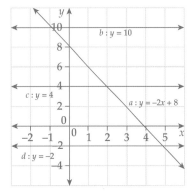

Language activity

1   PLANE BRAIN
2   NUMBER SLUMBER
3   POINT JOINT
4   DOT PLOT
5   RIGHT HEIGHT
6   DOWN TOWN

Practice test **14**

1 $22.30   **2**   7.05   **3** 9.51×10^{-3}

4   24 cm²   **5** $\dfrac{5}{12}$ **6** $153   **7**   420

8   8.625   **9**   192.460   **10**   30

11   **a**

x	6	4	2
y	1	−1	−3

     b

x	4	3	2
y	8	5	2

12   C   **13**   B

14

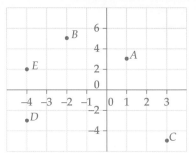

15

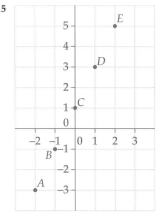

16   **a** x-intercept = 1, y-intercept = −1
     b x-intercept = 4, y-intercept = 2

17   **a** and **b**

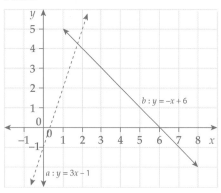

18   c lies on the line

19

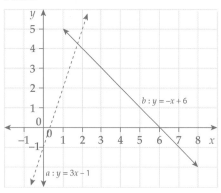

20 a $y = 2x - 5$ and $y = 3$

b

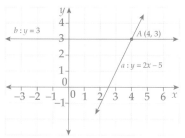

Solution is $x = 4$

CHAPTER 15

Exercise **15-01**

1 B   2 D

3 a even chance   b impossible
   c likely   d even chance
   e certain   f even chance
   g likely   h impossible
   i unlikely   j unlikely

4 a {head, tail}
   b {hearts, diamonds, clubs, spades}
   c {red, amber, green}   d {14, 16, 18, 20, 22}
   e {$5, $10, $20, $50, $100}
   f {HH, HT, TH, TT}

5 a Yes   b No   c Yes
   d No   e Yes   f yes

6 a $\frac{1}{6}$ b $\frac{1}{6}$ c $\frac{2}{3}$
   d $\frac{1}{2}$ e $\frac{1}{2}$ f $\frac{1}{2}$

7 a i $\frac{3}{7}$ ii $\frac{4}{7}$
   b 1

8 a $\frac{2}{9}$ b $\frac{1}{3}$ c $\frac{4}{9}$
   d $\frac{7}{9}$ e $\frac{2}{3}$ f $\frac{5}{9}$

9 a $\frac{1}{4}$ b $\frac{1}{4}$ c $\frac{1}{13}$
   d $\frac{1}{13}$ e $\frac{1}{52}$ f $\frac{1}{13}$

Exercise **15-02**

1 B   2 C

3 a Not getting a 3
   b Choosing a yellow sock
   c Getting a tail   d Not coming 2nd
   e Not choosing a heart   f Not picking a winner

4 event one

5 a $\frac{1}{6}$ b $\frac{5}{6}$ c $\frac{1}{3}$ d $\frac{2}{3}$

6 a $\frac{1}{3}$ b $\frac{2}{3}$ c $\frac{5}{18}$
   d $\frac{13}{18}$ e $\frac{7}{18}$ f $\frac{11}{18}$

7 0.2   8 35%

9 a $\frac{99}{100}$ b 200

Exercise **15-03**

1 $\frac{13}{30}$ 2 A

3 a probability event frequency
     frequencies 1

4 a

Number of heads	Frequency	Relative frequency
0	16	$\frac{16}{80} = \frac{1}{5}$
1	42	$\frac{42}{80} = \frac{21}{40}$
2	22	$\frac{22}{80} = \frac{11}{40}$
Total	80	1

   b $\frac{21}{40}$ c 168   d $\frac{1}{5}$ e 1

5 a

Score	Frequency	Relative frequency
1	18	$\frac{18}{120} = \frac{3}{20}$
2	23	$\frac{23}{120}$
3	16	$\frac{16}{120} = \frac{2}{15}$
4	24	$\frac{24}{120} = \frac{1}{5}$
5	22	$\frac{22}{120} = \frac{11}{60}$
6	17	$\frac{17}{120}$
Total	120	1

   b $\frac{1}{5}$ c 120   d $\frac{41}{120}$ e 1

Exercise **15-04**

1 D   2 B

3 a 70
   b i 23   ii 8   iii 43   iv 4   v 39
   c No

4 a 58
  b i 34   **ii** 18   **iii** 0   **iv** 6   **v** 52
  c Yes

5

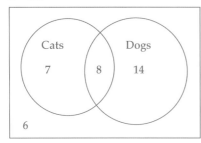

  a 7   **b** 14   **c** 6
6 a $\frac{16}{55}$ **b** $\frac{7}{55}$ **c** $\frac{44}{55} = \frac{4}{5}$ **d** $\frac{48}{55}$

7 a

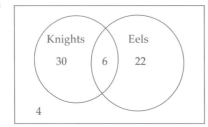

  b i $\frac{22}{62} = \frac{11}{31}$ **ii** $\frac{6}{62} = \frac{3}{31}$ **iii** $\frac{52}{62} = \frac{26}{31}$
  iv $\frac{30}{62} = \frac{15}{31}$

Exercise 15-05

1 B   **2** A

3

	Female	Male	Total
Blue Eyes	8	10	18
Not Blue Eyes	18	12	30
Total	26	22	48

  a 18   **b** 12   **c** $\frac{8}{48} = \frac{1}{6}$ **d** $\frac{22}{48} = \frac{11}{24}$

4

	Girls	Boys	Total
Football	12	18	30
Not Football	11	9	20
Total	23	27	50

  a 30   **b** 9   **c** $\frac{12}{50} = \frac{6}{25}$ **d** $\frac{18}{50} = \frac{9}{25}$

5 a

	Cake	Not Cake	Total
Lollies	9	22	31
Not Lollies	15	4	19
Total	24	26	50

  b i $\frac{9}{50}$ **ii** $\frac{37}{50}$

6 a

	Swimming	Not Swimming	Total
Running	32	28	60
Not Running	14	16	30
Total	46	44	90

  b i $\frac{14}{90} = \frac{7}{45}$ **ii** $\frac{32}{90} = \frac{16}{45}$

Exercise 15-06

1 C   **2** D

3 a $\frac{3}{10}$ **b** $\frac{7}{10}$ **c** $\frac{3}{10}$
  d $\frac{4}{10} = \frac{2}{5}$ **e** $\frac{6}{10} = \frac{3}{5}$

4 a $\frac{4}{30} = \frac{2}{15}$ **b** $\frac{6}{30} = \frac{1}{5}$ **c** $\frac{13}{30}$ **d** $\frac{23}{30}$

5 a i $\frac{36}{200} = \frac{9}{50}$ **ii** 0.18

  b i $\frac{1}{4}$ **ii** 0.25

  c No

  d It should become closer to the theoretical probability.

6 a Incorrect, as 3 takes up 2 sections out of 6 on the spinner.

  b $\frac{2}{6} = \frac{1}{3}$

7 a i 24   **ii** 62
  b i $\frac{38}{77}$ **ii** $\frac{6}{77}$

8

	Girls	Boys	Total
Dancing	23	12	35
Not Dancing	8	18	26
Total	31	30	61

  a $\frac{23}{61}$ **b** $\frac{12}{61}$

Language activity

1	CHANCE	2	CERTAIN
3	IMPOSSIBLE	4	UNLIKELY
5	EVENT	6	OUTCOME
7	PROBABILITY	8	THEORETICAL
9	EQUALLY LIKELY	10	SAMPLE SPACE

Practice test 15

1 $42.50 **2** 21 **3** −6 **4** 28.8 m²

5 $8y(2xy − k)$ **6** $2.14\dot{6}$ **7** a horizontal line

8 $\dfrac{p^2}{4y^2}$ **9** 12.4 **10** 5:55 a.m.

11 A **12** B **13** D

14 a $\dfrac{1}{6}$ **b** $\dfrac{5}{6}$

15 a

Number of heads	Frequency	Relative Frequency
0	8	$\dfrac{8}{40} = \dfrac{1}{5}$
1	23	$\dfrac{23}{40}$
2	9	$\dfrac{9}{40}$
Total	40	1

b $\dfrac{23}{40}$ **c** 57 times **d** $\dfrac{1}{5}$

16 a 58 **b** 11 **c** 42 **d** $\dfrac{31}{58}$

17

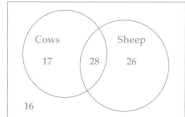

18

	Smartphone	No smartphone	Total
Tablet	12	11	23
No tablet	31	4	35
Total	43	15	58

19

	Male	Female	Total
Licence	24	18	42
No Licence	5	3	8
Total	29	21	50

a 42 **b** 5 **c** $\dfrac{18}{50} = \dfrac{9}{25}$

20 a 0.26 **b** 0.25 **c** Yes.

CHAPTER 16

Exercise **16-01**

1 B **2** D

3 a 15 **b** 12 **c** 12 **d** 72

e 20 **f** 6 **g** 2 **h** 8

i 4 **j** 9 **k** 20 : 32 **l** 9 : 15

4 a 1 : 2 **b** 1 : 3 **c** 1 : 3 **d** 5 : 1

e 1 : 4 **f** 1 : 3 **g** 1 : 4 **h** 6 : 1

i 4 : 1 **j** 1 : 8 **k** 1 : 6 **l** 1 : 3

m 3 : 8 **n** 2 : 3 **o** 8 : 9 **p** 3 : 4

q 9 : 20 **r** 1 : 125 **s** 20 : 3 **t** 15 : 1

5 a 1 : 2 **b** 1 : 100 **c** 1 : 25 **d** 1 : 4

e 6 : 7 **f** 1 : 4 **g** 1 : 6 **h** 3 : 100

i 10 : 1 **j** 49 : 60 **k** 1 : 6 **l** 1 : 2

m 1 : 2 **n** 7 : 24 **o** 2 : 15 **p** 1 : 4

q 1 : 12 **r** 1 : 10 **s** 3 : 2 **t** 25 : 2

Exercise **16-02**

1 A **2** B

3 a 1 : 3 **b** 1 : 6 **c** 3 : 1

4 a 12 : 25 **b** 6 : 25 **c** 25 : 6

5 $164, $369 **6** 45 **7** 35

8 $42 500 **9** 480 kg **10** 78

11 104 **12** $539 **13** 15 cups

14 a 28 **b** $\dfrac{1}{2}$

Exercise **16-03**

1 B **2** D

3 a 120 cm **b** 280 cm **c** 260 cm

4 a 9 cm **b** 120 cm **c** 5.7 cm

d 34 720 cm = 347.2 m

5 a 3 cm **b** 4.5 cm

6 a A rectangle 12 cm long and 6.5 cm wide.

b Singles court is 12 cm long and 6.3 cm wide.

c 120 m long and 63 m wide.

7 a 1 : 500 **b** 1 : 400 **c** 2 : 500 000

d 1 : 40

8 a 16.75 m **b** 5.5 m

9 6.8 m

10 a 1.2 km **b** 1.4 km **c** 1.1 km

d 1.4 km **e** 2.0 km

ANSWERS

Exercise 16-04

1 B   **2** D

3

Ratio	Total number of parts	Total amount	One part	Ratio
3 : 1	3 + 1 = 4	64	64 ÷ 4 = 16	48 : 16
2 : 3	5	75	15	30 : 45
1 : 5	6	600	100	100 : 500
3 : 5	8	720	90	270 : 450
4 : 5	9	1800	200	800 : 1000
7 : 6	13	3900	300	2100 : 1800

4 Asha $10, Juno $15
5 Nathan $2700, Tom $4500
6 a $480   b $160
7 $24 000 : $32 000 : $40 000
8 Tara $56 250, Gemma $93 750
9 3 matches   **10** $227.20
11 280 : 420 : 560   **12** True

Exercise 16-05

1 D   **2** A

3 a 30   b 50   c 1.84   d 10
   e 12.10   f 300   g 20   h 1400
   i 232   j 23

4 8 km/L
5 a $166/day   b $140/day
   c The motel in b.
6 $23/h   **7** $66/h   **8** 28 beats / min
9 55.7 km/h   **10** 680 words in 8 min.

Exercise 16-06

1 D   **2** A

3 a $\dfrac{\$}{h}$ b $550   c 13 h
4 a 96 words/min   b $\dfrac{words}{minutes}$
   c 42 mins   d 2880
5 a $5214   b $28 000
6 a $8.58   b 816 g
7 $1164.45   **8** 104 minutes
9 11.25 L/100km   **10** 12 L/100 km
11 a 550 minutes   b 22 mins/car
   c 7 hours 20 mins

Exercise 16-07

1 B   **2** A

3

Distance	Time	Speed
120 km	3 h	40 km/h
650 km	5 h	130 km/h
1250 km	25 h	50 km/h
680 m	10 s	68 m/s
4860 m	80 s	60.75 m/s

4 55 km/h   **5** 70 km/h
6 a 16 km   b 2 h   c 8 km/h

7

Distance	Time	Speed
450 km	9 h	50 km/h
1600 m	80 s	20 m/s
472.5 km	4.5 h	105 km/h

8 a 75 km/h   b 300 km
   c 5.5 h
9 a 165 km   b 165 km/h
   c 2750 m   d 45.8 m/s
10 a 1 km   b 1000 m   c 16.7 m/s
11 a 240 km   b 240 km/h
   c 4000 m   d 66.7 m/s

Language activity

1 ROUND ABOUT   Ratio is 1 : 1
2 TIME TABLE   Ratio is 4 : 5
3 PIECE WORK   Ratio is 5 : 4
4 BREAK AGE   Ratio is 5 : 3
5 OVER ALL   Ratio is 4 : 3
6 ANY WHERE   Ratio is 3 : 5
7 SUN SET   Ratio is 1 : 1
8 MOON LIGHT   Ratio is 4 : 5
9 RATTLE SNAKE   Ratio is 6 : 5
10 SEA SHORE   Ratio is 3 : 5
11 TEXT BOOK   Ratio is 1 : 1
12 HONEY SUCKLE   Ratio is 5 : 6
13 MATCH BOX   Ratio is 5 : 3
14 SUN GLASSES   Ratio is 3 : 7
15 TYPE WRITER   Ratio is 2 : 3
16 CHEESE BURGER   Ratio is 1 : 1
17 WAVE LENGTH   Ratio is 2 : 3

Practice test 16

1 2025   **2** 3 : 4   **3** $\dfrac{1}{64}$
4 $x^2 = 8^2 + 15^2$ **5** $5xy(4x + 3y)$
6 11, 13. 17, 19   **7** $22   **8** $-6a - 18$
9 $\dfrac{4}{5}$, 0.803, 0.85, 86%   **10** $\dfrac{1}{3}$
11 D   **12** B   **13** B
14 a 520 mL   b 1.56 L
15 a 900 cm   b 42 cm
16 Lachlan $5000, Courtney $20 000
17 Ruben $36 000   John $27 000   Tim $48 000
18 a $40/h   b 60 c/apple
   c 62 words/min
19 a $6.51   b 849 g
20 a 198 km   b 13 h 20 min

Developmental Mathematics Book 3

ISBN 9780170351027